AF352975

Advances of Machine Design in Italy 2022

Advances of Machine Design in Italy 2022

Editors

Marco Ceccarelli
Giuseppe Carbone
Alessandro Gasparetto

MDPI • Basel • Beijing • Wuhan • Barcelona • Belgrade • Manchester • Tokyo • Cluj • Tianjin

Editors
Marco Ceccarelli
University of Rome Tor
Vergata
Italy

Giuseppe Carbone
University of Calabria
Italy

Alessandro Gasparetto
University of Udine
Italy

Editorial Office
MDPI
St. Alban-Anlage 66
4052 Basel, Switzerland

This is a reprint of articles from the Special Issue published online in the open access journal *Machines* (ISSN 2075-1702) (available at: https://www.mdpi.com/journal/machines/special_issues/Machine_Design_Italy2022).

For citation purposes, cite each article independently as indicated on the article page online and as indicated below:

LastName, A.A.; LastName, B.B.; LastName, C.C. Article Title. *Journal Name* **Year**, *Volume Number*, Page Range.

ISBN 978-3-0365-6754-9 (Hbk)
ISBN 978-3-0365-6755-6 (PDF)

Contents

About the Editors

Marco Ceccarelli

Marco Ceccarelli , born in Rome in 1958, is a Professor of the Mechanics of Machines at the University of Rome Tor Vergata, Italy, where he chairs LARM2: Laboratory of Robot Mechatronics. His research interests cover subjects of robotics, mechanism design, experimental mechanics, and the history of mechanical engineering, with several published papers in the fields of robotics and mechanical engineering. He has been a Visiting Professor at several universities in the world. He is also an ASME fellow. Professor Ceccarelli serves on several journals' editorial boards and scientific conference committees. He is the Editor-in-Chief of the MDPI journal *Robotics* and of the SAGE *International Journal of Advanced Robotic Systems* for the area of Service Robotics. He is an editor of the Springer book series on mechanism and machine science (MMS) and the history of MMS. He has been the President of the IFToMM, the International Federation for the Promotion of MMS, in 2008-11 and 2016-19. More information can be found at the following LARM2 webpage: https://larm2.ing.uniroma2.it/marco-ceccarelli/.

Giuseppe Carbone

Giuseppe Carbone is an Associate Professor at the Department of Mechanics, Energy, and Management, University of Calabria, Italy, since 2018. His research interests cover aspects of robotics, mechatronics, engineering design, and the mechanics of manipulation and grasp, with over 400 published research papers and 20 patents. He serves on the Editorial Board of several international journals, including being the Editor-in-Chief of *ROBOTICA* (Cambridge Univ. Press), the Section Editor-in-Chief of the *Journal of Bionic Engineering* (Springer), MDPI's *Machines*, MDPI's *Robotics*, a Technical Editor of *IEEE/ASME Transactions on Mechatronics*, and an AE of the ASME's *Journal of Autonomous Vehicles and Systems* in addition to *Advanced Robotic Systems*. He has received several honors and awards, including 30 best paper awards and 10 international patent awards. He has recently been awarded a Honoris Causa Doctoral Degree from the Technical University of Cluj-Napoca, Romania. More information is available at https://www.researchgate.net/profile/Giuseppe-Carbone-3.

Alessandro Gasparetto

Alessandro Gasparetto (Rovigo, Italy, 26 October 1968) received his MSc in electronic engineering from the University of Padova, Italy, in 1992; his MSc in mathematics from the University of Padova, Italy, in 2003; and his PhD in the mechanics of machines from the University of Brescia, Italy, in 1996. He is a Full Professor of the Mechanics of Machines at the Polytechnic Department of Engineering and Architecture, University of Udine (Udine, Italy), where he is the head of the research group on mechatronics and robotics, as well as the Head of the Department (since 2021). He has been included in the ranking of the top 2% most quoted and authoritative scientists in the world, published by researchers at Stanford University (2019 and 2021). Since 2017 he has been the Chair of IFToMM Italy, the Italian branch of the IFToMM (the International Federation for the Promotion of Mechanism and Machine Science). Since 2018 he has been the Chair of the IFToMM Permanent Commission for the History of Mechanism and Machine Science. His research interests are in the fields of the modeling and control of mechatronic systems, robotics, mechanical design, industrial automation, and mechanical vibrations. He is the author of more than 200 international publications, and has been involved in the scientific and organizing committees of several conferences, as well as in many research projects, at the regional, national, and European levels.

Preface to "Advances of Machine Design in Italy 2022"

This book contains a Special Issue of the MDPI journal Machines on Italian advances in mechanism and machine science through a collection of selected papers from the Fourth International Conference of IFToMM Italy, IFIT 2022, that was held in Naples on 7–9 September 2022.

The peer-reviewed papers are extended, revised versions of those presented at IFIT 2022, with the aim of giving an overview of the variety of activities and achievements from the Italian IFToMM community, even with international collaborations. The authors are mainly from universities from all around Italy.

IFToMM Italy is one of the founding member organizations of the IFToMM, the International Federation for the Promotion of Mechanism and Machine Science, which was founded in 1969. Since then, the community of IFToMM Italy has been active with contributions at the national and international levels. In 2014, IFToMM Italy was legally established as the Italian IFToMM society, with research and development activity in mechanism and machine science (MMS) in addition to a biennial conference as the core conference event.

This book includes papers belonging to a broad range of disciplines in MMS, with research and design results that can be of interest not only to scholars in the field of MMS and mechanical engineering but also to professionals and even students, broadening their understanding of the problems and solutions under development, mainly, but not only, from the Italian community.

The Guest Editors wish to thank the authors for their efforts and time spent on their valuable scientific contributions in addition to the MDPI editorial staff for their support and assistance in preparing the Special Issue and then this book.

Marco Ceccarelli, Giuseppe Carbone, and Alessandro Gasparetto
Editors

Editorial

Advances of Machine Design in Italy 2022

Marco Ceccarelli [1,*], Giuseppe Carbone [2,*] and Alessandro Gasparetto [3,*]

[1] Department of Industrial Engineering, University of Rome Tor Vergata, 00133 Rome, Italy
[2] Department of Mechanical Engineering, Energy Engineering and Management, University of Calabria, 87036 Rende, Italy
[3] Polytechnic Department of Engineering and Architecture, University of Udine, 33100 Udine, Italy
* Correspondence: marco.ceccarelli@uniroma2.it (M.C.); giuseppe.carbone@unical.it (G.C.); alessandro.gasparetto@uniud.it (A.G.)

Citation: Ceccarelli, M.; Carbone, G.; Gasparetto, A. Advances of Machine Design in Italy 2022. *Machines* **2023**, *11*, 64. https://doi.org/10.3390/machines11010064

Received: 23 December 2022
Accepted: 3 January 2023
Published: 4 January 2023

This Special Issue is aimed to promote and circulate recent developments and achievements in the field of Mechanism and Machine Science coming from the Italian community with international collaborations and ranging from theoretical contributions to experimental and practical applications. It contains selected contributions that were accepted for presentation at the Fourth International Conference of IFToMM Italy, IFIT 2022, that has been held in Naples on 7–9 September 2022 [1]. This IFIT2022 conference is the fourth event of a series that was started in 2016 by IFToMM Italy in Vicenza, and continued with IFIT 2018 in Cassino, and with IFIT 2022 in Naples but held online. The IFIT conferences were established to bring together researchers, industry professionals and students, from the Italian community with international participation in an intimate, collegial, and stimulating environment.

IFToMM Italy is one of the founding member organizations of IFToMM, the International Federation for the Promotion of Mechanism and Machine Science, which was founded in 1969. Since then, the member organization IFToMM Italy has been active with contributions at national and international levels. In 2014, IFToMM Italy was legally established as the Italian IFToMM society with research and development activity in Mechanism and Machine Science (MMS).

This Special Issue includes papers belonging to a broad range of disciplines in MMS, such as collaborative robotics, vibration analysis, rotor dynamics, gear design, control of vehicles, design of parallel mechanisms, tribology, lubrication, UAVs, mobile manipulators for agricultural applications, kinematics and dynamics of mechanisms, history of mechanisms.

These contributions have been selected from among the 105 papers that were presented at IFIT 2022 conference [1]. Authors have been invited to contribute extended revised versions of the presented conference works. Contributions have been mostly selected among those receiving award recognition in one of the three IFToMM categories of research, applications, and student. These papers were evaluated again with a blind peer-review process to confirm the high quality of the works. In particular, paper [2] presents an historical analysis of developments for the creation and usage of models of mechanisms in academic teaching fields. Manuscript [3] is a review of methodologies for the analysis and synthesis of planar mechanisms developed at the Laboratory of Mechatronics and Virtual Prototyping of the University of Ferrara. In [4], the kinematic model and a motion planning pipeline for a mobile manipulator specifically designed for precision agriculture applications, such as crop sampling and monitoring, formed by a novel articulated mobile base and a commercial collaborative manipulator with seven degrees of freedom, is presented. Paper [5] investigates payload solutions for medium and small package delivery (up to 5 kg) with a medium-sized UAV. From this analysis, a prototype for an industrialized package is obtained. A relevant tribology problem is investigated in [6], namely, a methodology for estimating the coefficient of friction with a semi-empirical formulation is presented so that its results are employed to analyze mechanical efficiency losses in a hypoid gearset. A design of a planar parallel mechanism installed on a fast-operating automatic machine is discussed in [7], where the mechanism design is optimized in order to reduce the observed vibrations. An advanced methodology, such as multi-objective design optimization, is employed in [8] to efficiently and reliably achieve an optimal lightweight gear design. Paper [9] looks at the squeeze film dampers (SFDs), i.e., components used in many industrial applications, ranging from turbochargers to jet engines. A numerical model based on the Reynolds equation, discretized with the finite difference method, is proposed, and validated by means of experimental tests. The conceptualization and implementation of a versatile and modular unmanned ground vehicle prototype, aimed at testing and assessing new motion planning and control algorithms for different Precision Agriculture applications, is presented in [10]. Manuscript [11] describes a model-based formulation to analytically evaluate the load transfer dynamics of a vehicle and its variation due to the presence of road perturbations. Paper [12] deals with combined experimental and statistical approaches to evaluate the vibrational behavior of a gas micro-turbine supplied with different pure fuels and admixed with rapeseed oils. In [13], some case studies are presented and discussed concerning tools and methodologies for Human–robot collaboration that are developed at the I-Lab laboratory of Università Politecnica delle Marche (Ancona, Italy). In [14], the dynamic sub structuring technique is employed to predict the vibrational behavior of a three-point linkage, in order to determine the frequency response functions, the natural frequencies and the mode shapes of the mechanism in a wide range of configurations. Paper [15] presents the design of a novel feeding device for industrial applications that is composed of a rotary distributor and a four-bar linkage mechanism. The mechanism is designed as result of a specific functional synthesis and the movement of the conveyor blades is driven by the kinematics of the four-bar linkage with a fairly simple control system. In the paper [16], a survey is presented on strategies that were proposed in the last few years to tackle contact mechanics problems involving rough interfaces. Attention is focused on Boundary Element Methods capable of solving the contact with great accuracy and with a proper computational efficiency. Particular attention is addressed to non-linearly elastic constitutive relations and linearly viscoelastic rheology with important practical implications in all the systems, such as, for example, in vibration isolators, dynamic seals, pick and place devices.

We would like to thank the members of the Scientific Committees for strong support for the success of IFIT 2022:

Alessandro Gasparetto (University of Udine) Chair
Nicola Pio Belfiore (University of Roma)
Massimo Callegari (Polytechnical University of Marche)
Roberto Caracciolo (University of Padova)
Giuseppe Carbone (University of Calabria)
Marco Ceccarelli (University of Rome Tor Vergata)
Enrico Ciulli (University of Pisa)
Raffaele di Gregorio (University of Ferrara)
Pietro Fanghella (University of Genova)
Andrea Manuello Bertetto (Politecnico di Torino)
Arcangelo Messina (University of Salento)
Domenico Mundo (University of Calabria)
Vincenzo Niola (University of Napoli)
Paolo Pennacchi (Politecnico di Milano)
Giuseppe Quaglia (Politecnico di Torino)
Rosario Sinatra (University of Catania)
Alberto Trevisani (University of Padova)

The guest editors of this Special Issue thank the authors and reviewers for their efforts and time spent in the valuable scientific contributions and useful feedback that have confirmed the high scientific quality of the IFIT 2022 papers. We also take this opportunity to invite contributions for the next IFIT 2024 event that is scheduled to be held at Politecnico di Torino, Turin, Italy.

Conflicts of Interest: The authors declare no conflict of interest.

References

1. Niola, E.; Gasparetto, A.; Quaglia, G.; Carbone, G. (Eds.) Advances in Italian mechanism science. In Proceedings of the Fourth International Conference of IFToMM Italy, Napoli, Italy, 7–9 September 2022; Springer: Cham, Switzerland, 2022; Volume 122.
2. Ceccarelli, M.; Cocconcelli, M. Italian historical developments of teaching and museum valorization of mechanism models. *Machines* **2022**, *10*, 628. [CrossRef]
3. Di Gregorio, R. The role of instant centers in kinematics and dynamics of planar mechanisms: Review of LaMaVIP's contributions. *Machines* **2022**, *10*, 732. [CrossRef]
4. Colucci, G.; Botta, A.; Tagliavini, L.; Cavallone, P.; Baglieri, L.; Quaglia, G. Kinematic modeling and motion planning of the mobile manipulator Agri.Q for precision agriculture. *Machines* **2022**, *10*, 321. [CrossRef]
5. Saponi, M.; Borboni, A.; Adamini, R.; Faglia, R.; Amici, C. Embedded payload solutions in UAVs for medium and small package delivery. *Machines* **2022**, *10*, 737. [CrossRef]
6. Grabovic, E.; Artoni, A.; Gabiccini, M.; Guiggiani, M.; Mattei, L.; Di Puccio, F.; Ciulli, E. Friction-induced efficiency losses and wear evolution in hypoid gears. *Machines* **2022**, *10*, 748. [CrossRef]
7. Zaccaria, F.; Quarta, E.; Badini, S.; Carricato, M. Optimal design for vibration mitigation of a planar parallel mechanism for a fast automatic machine. *Machines* **2022**, *10*, 770. [CrossRef]
8. Cosco, F.; Adduci, R.; Muzzi, L.; Rezayat, A.; Mundo, D. Multiobjective design optimization of lightweight gears. *Machines* **2022**, *10*, 779. [CrossRef]
9. Gheller, E.; Chatterton, S.; Vania, A.; Pennacchi, P. Squeeze film damper modeling: A comprehensive approach. *Machines* **2022**, *10*, 781. [CrossRef]
10. Saeed, R.A.; Tomasi, G.; Carabin, G.; Vidoni, R.; Von Ellenrieder, K.D. Conceptualization and implementation of a reconfigurable unmanned ground vehicle for emulated agricultural tasks. *Machines* **2022**, *10*, 817. [CrossRef]
11. Tota, A.; Dimauro, L.; Velardocchia, F.; Paciullo, G.; Velardocchia, M. An intelligent predictive algorithm for the anti-rollover prevention of heavy vehicles for off-road applications. *Machines* **2022**, *10*, 835. [CrossRef]
12. Niola, V.; Savino, S.; Quaremba, G.; Cosenza, C.; Nicolella, A.; Spirto, M. Discriminant analysis of the vibrational behavior of a gas micro-turbine as a function of fuel. *Machines* **2022**, *10*, 925. [CrossRef]
13. Callegari, M.; Carbonari, L.; Costa, D.; Palmieri, G.; Palpacelli, M.-C.; Papetti, A.; Scoccia, C. Tools and methods for human robot collaboration: Case studies at i-LABS. *Machines* **2022**, *10*, 997. [CrossRef]
14. Brunetti, J.; D'Ambrogio, W.; Fregolent, A. Configuration-dependent substructuring as a tool to predict the vibrational response of mechanisms. *Machines* **2022**, *10*, 1146. [CrossRef]
15. Bottin, M.; Minto, R.; Rosati, G. Design of the drive mechanism of a rotating feeding device. *Machines* **2022**, *10*, 1160. [CrossRef]
16. Menga, N.; Putignano, C.; Carbone, G. Recent advancements in the tribological modelling of rough interfaces. *Machines* **2022**, *10*, 1205. [CrossRef]

Review

Italian Historical Developments of Teaching and Museum Valorization of Mechanism Models

Marco Ceccarelli [1],* and Marco Cocconcelli [2]

[1] Laboratory of Robot Mechatronics, Department of Industrial Engineering, University of Rome Tor Vergata, Via del Politecnico 1, 00133 Roma, Italy
[2] Department of Sciences and Methods for Engineering, University of Modena and Reggio Emilia, Via G. Amendola 2-Pad. Morselli, 42123 Reggio Emilia, Italy; marco.cocconcelli@unimore.it
* Correspondence: marco.ceccarelli@uniroma2.it

Abstract: This paper presents an historical analysis of developments for the creation and usage of models of mechanisms in academic teaching fields, with the aim of re-evaluating the interest and usefulness of models in teaching and research, and of promoting their merits as a cultural heritage worthy of being preserved. The historical analysis is focused on developments in Italy, with specific attention given to physical models created and used for training young engineers in Italian engineering schools, using commercial products, but also original Italian creations. Examples are reported from the main Italian academic sites, where examples of such models of mechanisms have been preserved or have survived, also, as first attempts at museum valorization in terms of historical memorabilia of educational developments on mechanism design issues.

Keywords: history of mms; italian history of mechanism design; models of mechanisms; history of teaching; italian mechanism collections

Citation: Ceccarelli, M.; Cocconcelli, M. Italian Historical Developments of Teaching and Museum Valorization of Mechanism Models. *Machines* **2022**, *10*, 628. https://doi.org/10.3390/machines10080628

Academic Editor: Antonio J. Marques Cardoso

Received: 28 June 2022
Accepted: 26 July 2022
Published: 29 July 2022

1. Introduction

Modeling, which can be understood as the conceptualization of structural and functional characteristics, was and still is the basis of any analysis and design procedure, especially in mechanical engineering. In particular, the development of mechanical machinery, both in theory and in practice, has been and continues to be based on the development of appropriate models of various forms and insights. Usually, in the history of mechanical engineering and machines, no reference is made to the history of models, even if they are used mainly in the form of graphical solutions to illustrate the evolution of knowledge and design solutions, that have had an impact on both science and technology in a broad spectrum, such as, for example, in encyclopedic treatises [1,2], and bibliographic overviews [3]. In the history of machine science and technology, models have had a rapid development, especially during the Industrial Revolution, as demonstrated by the numerous cabinets and exhibitions that were created in various locations, and especially in Europe [4]. Furthermore, specific attention given to the models of machines and mechanisms reached widespread and professional credibility with the work of Franz Releaux [5], who, in the second half of the 19th century, created a set of models of mechanisms for didactic use [6] through the Voigt company, obtaining global success, which is demonstrated by the fact that examples of his commercial collection are still present in many academic sites and museums across the world. In the same period, there were also local initiatives for the development of models of mechanisms, such as the collection of mechanisms at the University of Dresden, Germany [7], and at the Bauman Polytechnic, Moscow [8].

With the aim of refreshing attention to this technical and cultural heritage, this work presents models of mechanisms collections from Italian academic fields, which have had, and still have, not only technical but also historical interest. In general, examination of the historical development of Italian models of mechanisms has not been specifically addressed as far as can be understood from the few publications available, especially with regard to recent papers, that have, for example [9], reported technical-historic re-evaluation studies of these collections. In this work, the focus is on Italian initiatives and experiences, referring to the main historical sites of engineering schools in Italy in order to outline an historical interest, and to characterize the development of models of mechanisms as a cultural, and not only technological heritage, that can be considered of public worth and therefore deserves to be preserved and made usable, as highlighted in [10–13].

2. Consideration on Models

A mechanism is defined in the IFToMM terminology [14], as "Constrained system of bodies designed to convert motions of, and forces on, one or several bodies into motions of, and forces on, the remaining bodies". From a teaching viewpoint, as stated in several textbooks worldwide, for example [15,16], a mechanism is defined as a set of rigid bodies in the form of links, that are connected to each other with the purpose of converting the mechanical energy given to an input link into mechanical energy with a different value in the output link. Thus, a mechanism is used for its capacity in motion and force transmission, as a motion generator or action generator, respectively. In this way, a mechanism can be a fundamental component of machine design for its function in elaborating mechanical energy at proper levels, according to a machine task.

Consequently, a mechanism can be modelled by a scheme of its structure and operation parameters, with graphical drawings and/or mathematical formulation that can be used for performance analysis and operation simulation of the tasks, for which the mechanism is designed and operated. In general, a model can be a simplified representation of a system, and in the case of mechanical systems it is an abstract representation of the mechanism structure with its main mechanical features and parameters, so that it can be useful in teaching, design, and advertising through different modes, as outlined in [17].

Because of the above considerations, the modelling of a mechanism can be useful not only for design purposes but also for analyzing the mechanism working with its peculiar performance. Thus, a mechanism model can be created for design, simulation, and testing purposes with solutions that today range from virtual representations in computer-oriented software packages, to scaled simplified mechanical constructions, including prototypes or demonstrators, when referring to the same model as an evolution of a design solution.

As mentioned above, in general, mechanism models are useful and indeed used in several kinds of activity in teaching, study, investigation, design, demonstrative exhibitions, simulation, testing of operation characteristics, experimental validations, and even promotion and marketing publicity. They are also used as a means for design and construction of mechanical systems, and as tools or system components in activities for other engineering subjects with similar goals or complementary actions, as outlined in [18], referring to service operations.

Today, mechanism models are created as sketches or drawings (by hand or using computer-graphics apps); by mechanical constructions in either reduced sizes or prototypes; and in virtual computer-based solutions using different software with visualization for human interaction. Figure 1 summarizes the historical evolution of mechanism, indicating solutions that are linked to the progress of technological tools [17].

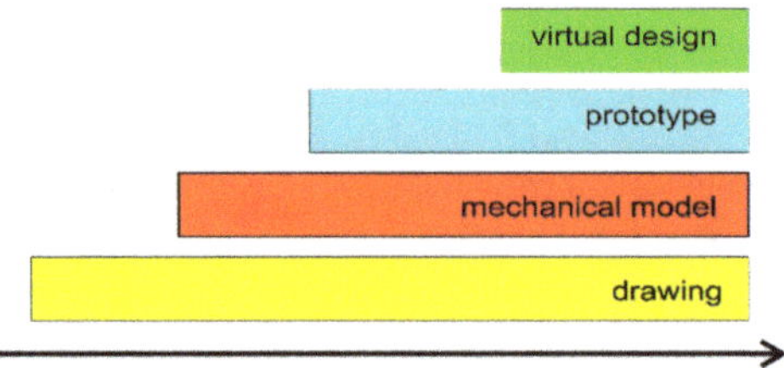

Figure 1. A timeline of conceptual evolution of mechanism models [17].

In Figure 1, the 'drawing' type of model refers to a graphical scheme that, since Antiquity, was produced as a pictorial representation with several other details. Even today, first design concepts are outlined with sketches (by hand or by graphics software), which are considered the first results in formulating and designing mechanical solutions, or in expressing concepts and solutions. A graphical drawing model can be produced in several different formats, with several different levels of accuracy in terms of graphical expression, as used in several disciplines other than mechanical engineering. Over time, drawing models have evolved from base-line abstraction, to naturalistic scenarios, and up to technical drawings with standardized rules (see, for example, isometric views and cross-sections with standardized colors for material and lines), such as, for example, in the representation of gear designs [19]. Indeed, the technical drawing technique is established with standards such as drawing rules, so that teaching activity can be planned for specific courses. In practically all engineering curricula, and mainly in the first year of study, the significance of the drawing tool in learning and explaining technical machine designs is a fundamental topic.

The second type, 'mechanical model', as seen in Figure 1, indicates mechanical constructions that can be a prototype of machine design or construction for experiments of phenomena still under investigation and in need of explanation. During the historical evolution of science and technology, mechanical models were extensively used before other more significant activities. Since today's prominence of informatics development, they have again acquired significant attention in teaching and research activity, mainly when prototypes are used in developing machine products. Thus, a mechanical model is again expressed with a mathematical formulation emphasizing mechanical features, in order to be used in reiterated design solutions and in teaching characterization of design parameters and operation performance.

A 'prototype' is the third type of model as seen in Figure 1, and is intended to be a mechanical-built solution that can be realized in different size scales. A prototype is produced as a first design, before a final solution of a product, and it is often considered as a tool within the design process to validate operation and performance, before having the final product ready for manufacture and production, and/or before proposing it to the market, or to the invention intellectual-protection process. Prototype models are extensively used in the design process for determining machine functionality; this may require analysis of a machine divided into subsystems devoted to specific aspects. For example, the design of a car engine can be organized with mechanical, electrical, and thermal parts, and for each of them a specific prototype can be considered within the design process.

Mechanical models and prototypes require efforts in manufacturing and processing, however, an understanding of computer science is useful for 'virtual design' models, that are designed, operated, and experienced virtuality with systems ranging from simple CAD designs up to sophisticated, haptic 3D simulations. Today, virtual models are widely implemented in the design process and teaching, and are produced following previous solutions based on modelling or drawing. Although not explicit, drawing modelling was a necessary development. The virtual model approach in Figure 1 is useful for functionality analysis and for finalizing product design, with considerations of durability and economic convenience by evaluating the effects of multidisciplinary and component integration in several applications, such as advertising, study, design, and teaching. Thus, virtual models can also be considered to be a representation of other types of models, such as modern expressions of significant content within science and technology, and also in the history of mechanical engineering. As a value of cultural heritage, it is recognized as being worthy of being preserved, thanks to its design concepts (in graphical or mathematical formats) and product features. A virtual model is a fruitful means of preserving intangible aspects of cultural heritage when original solutions cannot be properly expressed, so that they can be used to show the historical evolution of knowledge and be a source for further developments.

As summarized in Figure 1, mechanism models were, and still are, created in different formats for different aims, within activities such as teaching, analysis, design, experimental activity, and historical investigations, and they may still pay a key role not only in mechanical systems, but in other technical and scientific activities.

3. A Short Account of Italian History of Mechanism Design

In Italy, the Industrial Revolution occurred during a period of national unification in political, social, and cultural aspects, with events that also affected academic research and training programs, as summarized in [20].

During the first half of the 19th century, Italy still suffered significant political fragmentation, which also affected curriculum teaching, with different plans in each state. After political unification was achieved, the standardization of university programs was planned as a priority, in order to achieve a cultural unification. This required robust plans and significant adjustments in most Italian universities, especially with regards to the programming curriculum content given by teachers, who often worked in more than one university. In addition to restructuring activity, the Italian academic communities, and particularly those working in mechanical engineering, made significant contributions to Italian industrialization with regards to technical-scientific achievements, and in the organization and support of entrepreneurial initiatives.

During the first half of the 19th century, following advances of the metallurgical industry, the first significant industrialization initiative was developed in Southern Italy, within the organization of the Kingdom of the Two Sicilies, for the production of civil constructions and infrastructures, steam engines, and several types of industrial machines. While this first excellent achievement is recognized today, as briefly outlined in [21], the academic support of those industrial initiatives is still not well known, particularly when referring to specific figures who were not only devoted to theoretical investigations and fundamental mechanical designs. At the same time, in Northern Italy, universities activated specific programs on the mechanics of machines, recognizing the significant value of scientific formation for the nascent industrial frame, as a result of more contact with neighboring industrializing European countries. Examples of this can be considered the milestone work of Giuseppe Antonio Borgnis [22], who was formed at the Ecole Polytechnique, Paris, while also being professor at Pavia University from 1818 to 1821, and who published his famous 9-volume handbook on machines, that included the first technical dictionary on machines. Another example is the milestone work by Carlo Ignazio Giulio who, at the same time as Willis's work, published the first Italian textbook Theory of Mechanisms [23]. One more example of northern prolific theoretical activities is the work by Gaetano Giorgini who formed at Ecole Polytechnique, Paris, while working in Modena elaborated the first modern investigation [24] on the theory of screw (general helicalicoidal) motion of rigid bodies, with a postscript note on the 1763 book by Giulio Mozzi.

Even more significant are works and achievements in the second half of the 19th century, because of the cultural unity of academic frames that was organized and consolidated through the Royal Application Schools for Engineers in the main universities, which further supported the maturity and growth of industrialization initiatives. In the formation curriculum of industrial engineers, Mechanics Applied to Machines was planned with a central role, as it was in all the European universities; later, the augmented discipline of Kinematics Applied to Machines was recognized as mandatory for all curricula of industrial engineering at ministry level, with the Decree of 3 July 1879.

The university sites were in Turin, Milan, Bologna, Rome, and Naples. Domenico Tessari (1837–1909) and Scipione Cappa (1857–1910) were active in Turin. It is also worthy to note that Elia Ovazza (1852–1928), who moved to Palermo in 1890, successfully started a significant center there, including a laboratory which is still of historical interest today [25]. In Milan, Giuseppe Colombo (1836–1921) and Ugo Ancona ((1867–1936) were significant figures in mechanical engineering at the time. In Bologna, Giuseppe Barilli (1812–1894) and Francesco Masi (1852–1944) were similarly very active and well reputed. In Rome, the field was developed by Valentino Francesco Cerruti (1850–1909) and Carlo Saviotti (1845–1928). Fortunato Padula (1815–1881) and Ernesto Cavalli (1852–1911) worked in mechanical engineering in Naples, where Giovanni Battaglini (1826–1894) and Dino Padelletti (1852–1892) paid particular attention to more theoretical subjects. Emblematic is Ernesto Cavalli (1852–1911), who simultaneously worked as a teacher in Milan, Livorno, Pisa, and Naples, with the clear purpose of promoting common plans for teaching the subject of Mechanism Design.

Those academic works also stimulated initiatives for new companies since several graduates also initiated industrial activities with the help of their teachers, for example, in Milan, as discussed in [26]. Also prominent were Giuseppe Colombo, Francesco Masi and Lorenzo Allievi (1856–1941), as emblematic examples of the cultural prolificacy and success of the formation of Mechanics Applied to Machines, with a large range of applications: Colombo was a strong promoter of the industry, as documented in the Italian technical handbook (still in use today, with modern editions), with direct entrepreneur action even at governmental levels when he was a politician with important ministerial positions; Masi was a rigorous teacher who classified the variety of mechanisms, as well as designing the innovative machines for practical implementation [27]; Allievi, who was based in Rome, developed the milestone work [28] on the kinematics of machines, which is still of interest today [29]. Later, Allievi applied his skill in the management of industrial plants while considering the technical problems of the water hammer in hydraulic engineering, which he solved practically and theoretically, with a theorem ascribed to him [30].

After the First World War, Mechanics Applied to Machines was confirmed as being of primary significance as a driving technological development, and was given a central role in the formation programs of technicians and of all engineering professions. Between world wars and during wartime, activities were carried out intensively but with very limited dissemination. Nevertheless, achievements were fundamental for advances during wartime, and at the start of the post-war booms. The major university sites before 1940 were still located in Turin, Milan, Bologna, Rome, and Naples. Modesto Panetti (1875–1957) was in Turin. Ugo Ancona and Igino Saraceni were in Milan. Aristide Prosciutto (1895–1954) was in Bologna. In Rome, Anastasio Anastasi (1877–1969) was the main teacher. In Naples, activities were coordinated by Giovanni Domenico Mayer (1868–1925). Other significant activities were in Pisa, carried out by Enrico Pistolesi (1889–1968), and in Genoa, by Agostino Antonio Capocaccia (1901–1978).

Teaching teams, which were growing in member numbers and works, were active in the most university sites; in 1940 they were present in all universities with engineering schools, such as in Turin, Milan, Genoa, Padua, Bologna, Pisa, Rome, Naples, and Palermo. In the 1950s, new schools of engineering were started in Cagliari, Trieste, and Bari, with teachers in Mechanics of Machinery. Interests spanned from today's traditional subjects of the Theory of Mechanisms, linked to analytical and rational mechanics, to more challenging fields referring to the design and development of mechanical systems for industry and services. Specific attention was given to new subjects related to aeronautic systems, aerodynamics, biomechanics and medical devices, fluid dynamics, industrial automation, machine regulation, and finally robotics, with achievements and designs that were among the most advanced in the world. Industrial mechanical engineering was also improved by this vivacity and cultural prolificacy, with achievements in applied research and even with the formation of creative professionals and industrial managers. A brilliant example is Corradino D'Ascanio (1891–1981), who spanned design skills from mechanical industry to aeronautics and transport vehicles, referring to his pioneering helicopter designs and Vespa scooter by Piaggio [31]. All of these Italian activities produced significant literature in national and international frames, both for teaching and research, with textbooks, such as those indicated in [32], edited books, journals and conference papers. Even today, there are major Italian university teams working in Mechanics of Machinery in Turin, Milan, Padua, Genoa, Bologna, Rome, Naples, and Bari.

4. Modern Italian Mechanism Models in MMS Teaching

The modern teaching of the Mechanisms and Machine Science (MMS) in Italy dates back to post World War II. In this period, we witnessed the establishment of modern universities, a subdivision of knowledge into organic courses, and a formalization of the proposed contents. In Italy, the courses dedicated to the theory of mechanisms and machines take the name of "Mechanics Applied to Machines" (sometimes also known as "Machine Mechanics", "Applied Mechanics", "Mechanics Applied to the Arts"), with content that has remained substantially unchanged to the current day.

The structure of the applied mechanics course can normally be considered to be divided into two main parts: the Theory of Mechanisms and the Kinetics of Machines. In the first part, the machine elements and mechanisms are considered from a purely geometric aspect, that is, their kinematic composition and the transformation of movements with procedure for Kinematics and Dynamics Analysis of the mechanism functioning in theoretical and practical aspects. In the second part, aspects of the transformation of energy are considered, referring to the specific design of machine elements with the aim of determining optimal operating conditions. Often, the kinetics of machines includes a relevant part of metrology, for example, for the measurement of friction or mechanical work, which nowadays has taken on its own identity in separate specific courses.

As an example, we can consider the textbook written by Modesto Panetti, which is organized in three volumes. The index of the textbook probably covers most of the contents, which is addressed in the course of Mechanics Applied to Machines in most Italian universities in those years (few differences can be recognized in the theory dealt with, and in the application examples that are characterized as functions of the individual universities [32]).

- Volume 1: Kinematics of machines. Friction resistance and resistance of the medium. Elementary friction and lubricated pairs. Rolling pairs. Vibrations.
- Volume 2: Gears and wheels.
- Volume 3: Flexible organs. Funicular machines. Cableways. Transmissions with belts.

From the above list, it follows what are the main mechanisms, whose models were made for teaching purposes.

The development of Models of Mechanisms for teaching is aimed at providing a practical example of the theoretical models that are introduced during the course, so that a student can better understand the operating principle of a mechanism, mainly referring to the kinematics of the motion and force transmission. From this point of view, the models are an integration only to the technical drawings in textbooks, which, due to their two-dimensional nature, can generate difficulties in student understanding. Finally, note should be given to the application-professional nature of the teaching that was expected for a degree in engineering, and in particular for a course in machine mechanics. The first models used in universities in Italy were made of wood, as it was cheaper than metal and lighter to transport to the classroom. The models followed the well-established classification of the mechanisms as proposed by Franz Reuleaux [33], and the subsequent elaborations of Carlo Giulio [23] and Francesco Masi [6,34,35], to cite two Italian significant contributors, as outlined in [36]. The main models refer mainly to:

- Linkage mechanisms
- Cam mechanisms
- Gear train mechanisms
- Friction based mechanisms
- Wedge mechanisms
- Screw mechanisms
- Belt and chain drives
- Mechanism, containing pressurizing medium
- Step mechanisms

As an example, Figure 2 shows the comparison between a drawing in the treatise of Masi [34], and the corresponding wooden model of a cylindrical cam disk, still existing in Bologna. According to the description by Masi: "The figure represents an eccentric in which the line is double curved and is placed on a cylindrical surface; hence the eccentric is also called cylindrical cam. By rotating the cam around its axis, the temple moves alternately in a direction parallel to the axis of rotation. The chain is closed by force by the action of a spring" [5,34]. By means of the knob on the cam axle, a student or a teacher can rotate the cylindrical cam with a continuous motion and observe the alternate motion of the output rod.

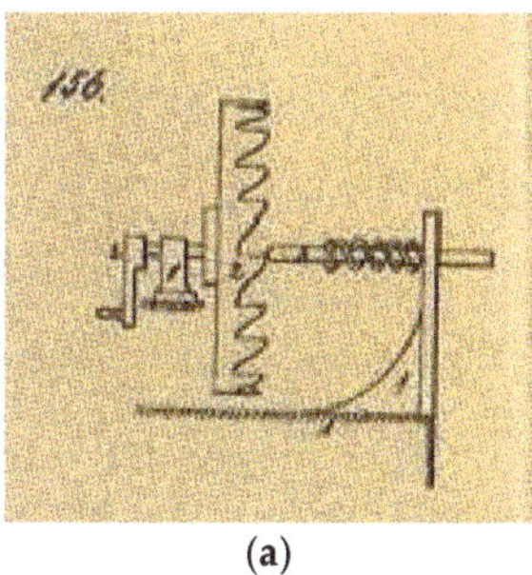

(a)

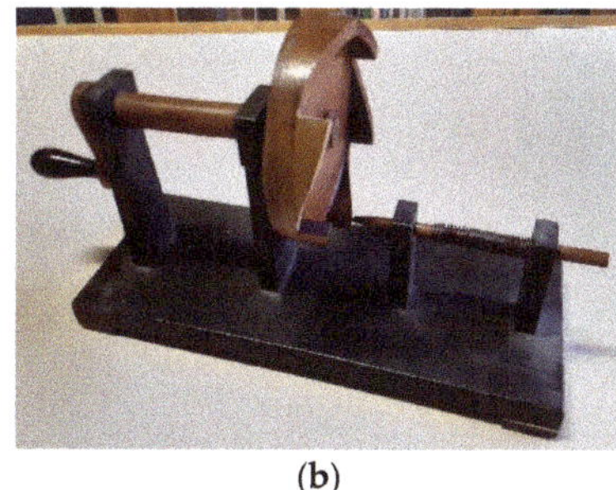

(b)

Figure 2. A model of lylindrical cam disk: (**a**) in the treatise of Masi [6]; (**b**) the corresponding wooden model at the University of Bologna.

Sometimes, the models have more than one input as in the double Hooke's joint, shown in Figure 3. The input and output shafts can be rotated with respect to the middle one, to show students the relationship between speed ratio and the input/output angular misalignment. Indeed, the model of the mechanism allows a quick and easy demonstration of the theory behind the simple drawing in Figure 3a.

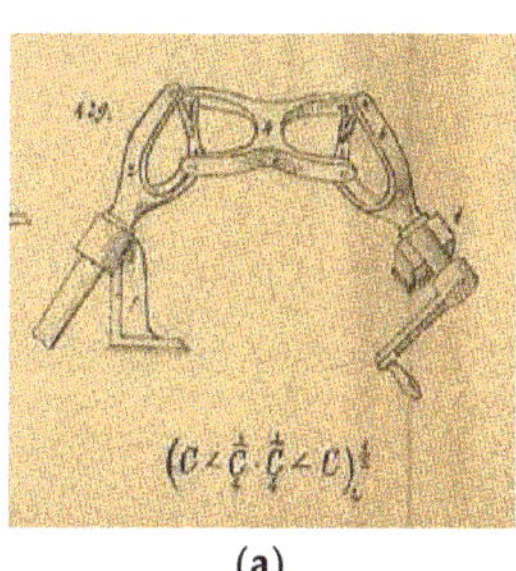

(a)

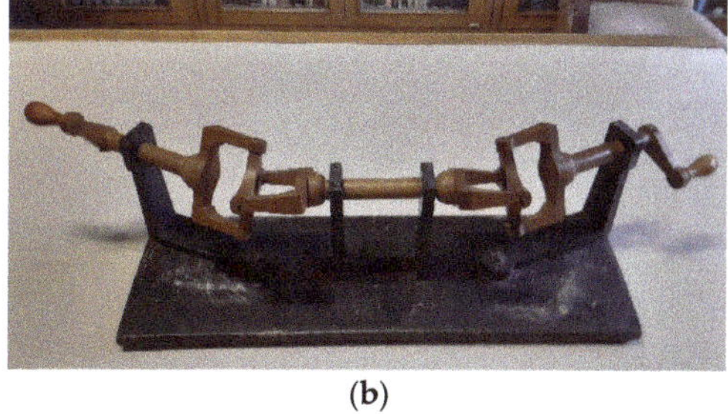

(b)

Figure 3. A model of the double Hooke's joint: (**a**) in the treatise of Masi [34]; (**b**) the corresponding wooden model at the University of Bologna.

More recently, plastic materials have been used for models, such as bakelite or plexiglass. The advantages are for lighter and cheaper models than wooden ones, especially for planar mechanisms that can be obtained from the processing of plastic sheets. Figure 4 shows the model of a Geneva drive at the University of Bologna: an intermittent mechanism made by five bodies, namely the drive wheel (red), the crank (yellow), the rocker arm (green), and the driven wheel (white). The left extreme of the rocker arm is connected to the drive wheel, while there is a pin on the right one that engages in the slots of the driven wheel.

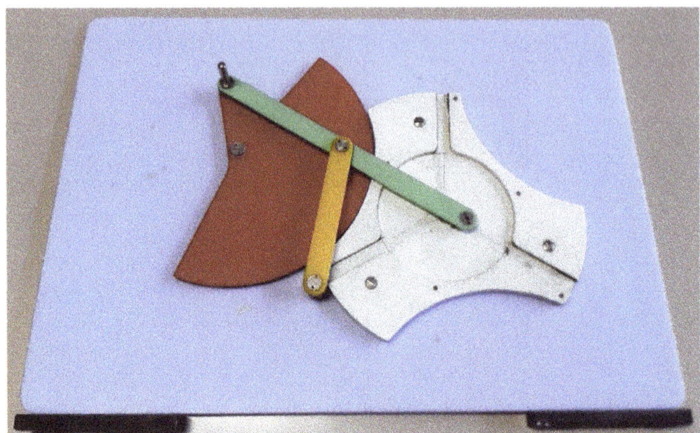

Figure 4. Intermittent mechanism model made of colored bakelite at the University of Bologna.

Iron models, more expensive than the previous ones, are often used to show the complexity of a real machine rather than a specific mechanism. They are models with cutaways to reveal the internal components. More realistic than previous types, these models are intended to show the real components that are the result of different types of analysis, such as the cooling fins of an engine, which have implications in the thermal analysis but not in mechanical ones. Figure 5 shows a model of the iconic motorcycle, Lambretta, a 2-stroke single-cylinder engine, piston, head and crankcase made of aluminum alloy with a cast iron cylinder. The engine is part of the collection at the Museum of Engines and Mechanisms of the University of Palermo [37].

Figure 5. Engine of the Innocenti motorcycle, Lambretta 125 C (1950–1951) at the Museum of Engines and Mechanisms, University of Palermo.

With the birth and diffusion of computers for personal use, the continuous increase of computational performance and decrease of the costs of the processors, 3D graphics software codes can be made available to universities and students. These 3D CAD software packages allow the modeling of rigid bodies, to assembly several bodies to model a complex mechanism, and to perform multibody analysis, combining Kinematics and Dynamics.

This is the birth of the first virtual models of mechanisms: for the first time it is not possible to take in hand or weigh a mechanism model, but its details can be observed through a graphic model on a screen The advantages are the elimination of production costs (with the exception of costs related to the software license), the reduction times of model realization, the possibility of sectioning views to show the contents of an assembly, or to create an exploded view of the component. From the didactic point of view, a student can easily modify or scale the parametric CAD model to finally appreciate the different resulting kinematic behaviors. To date, the main software manufacturers offer integrated solutions, to develop the design phase of the mechanism, the kinematic analysis, the dynamic analysis, and resistance tests in the same pc environment. On the other hand, with virtual models, more practical aspects such as inertial effects or frictions are lost, together with the reality of the mechanism structure and operation with all its aspects. Figure 6 shows an example of a three-speed gearbox model that was developed in Solidworks environment by two students at the University of Ferrara [38].

Figure 6. CAD model of a three-speed gearbox at University of Ferrara [38].

Today, the possibility of building educational models is within everyone's reach, thanks to 3D printing and the concept of additive manufacturing. Fused deposition modeling (FDM) printers of materials such as PLA and ABS are used by makers and hobbyists, and could easily be adopted by universities to model complex mechanisms and machines. Figure 7 shows the model of a V8 engine that was developed as a final project by a Bachelor student at the University of Ferrara. The model has been designed in a CAD environment (Solidworks) and printed in ABS (Makerbot Replicator 2X) [39].

Examples are discussed in the following sections to better explain the above concepts, and to show cases that can be of inspiration for further research work on landmarks of MMS heritage, referring specifically to Italian experiences.

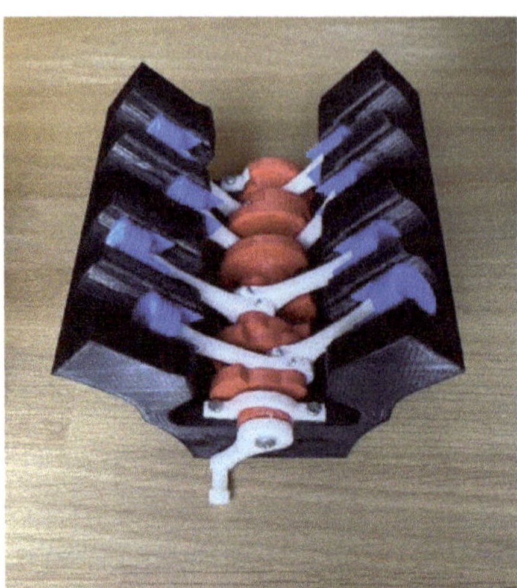

Figure 7. Construction of a V8 engine model by 3D printing at University of Ferrara [39].

5. Mechanism Models for Museum Valorization

Recording past developments and achievements is recognized as having archival value for clear identification of subjects, communities, and people, and is today evaluated as being heritage and worthy of being preserved for understanding the past by future generations. This is what is identified as cultural heritage, which is useful not only for museum preservation and exhibition, but for tracking and maintaining the memory of human evolutions [40].

In technical areas, this attention to past achievement is identified in the discipline of History of Science and Technology, which also includes the History of Engineering, with a large variety of topics and interests. The specific History of Mechanism and Machine Science (MMS) is focused on historical developments in MMS; only recently was it considered as part of historical technical-scientific subjects, though with strong technical contents for memory purposes when MMS inventors and scientists are recognized, and engineering procedures and theories are recorded. Nevertheless, MMS historical value seems not to be yet fully recognized within technical cultural heritage, since most MMS achievements are not fully available and understood by the wider public; therefore, historical awareness is still limited within the specific technical community, as stressed in [13,41].

Indeed, cultural heritage is usually indicated as "the legacy of products of human ingenuity in the form of physical artifacts and intangible attributes of a group or community that are inherited from past generations, are maintained in the present and are preserved for the benefit of future generations", as stated in UNESCO documents [40]. Thus, physical, or tangible products, with cultural heritage value are buildings and historic constructions like monuments and artifacts, so that they are suitable for future preservation. These products can be objects in their full state or only remains that contain significant aspects related to the archaeology, architecture, science or technology of a specific culture, with socioeconomic, political, ethnic, religious, and philosophical impacts on people with a community. Intangible products of cultural heritage value are those achievements that are not expressed with physical objects, but refer to behaviors, attitudes, knowledge, and other intellectual activities and are therefore difficult to be preserved with physical products.

Tangible MMS artifacts can be machines that were built and operated successfully, or even unsuccessfully, but made a significant contribution to the development of technology and society at a local and/or worldwide level. Even project documents, drawings, and patents can be considered as MMS products of cultural heritage value and, in fact, they are often preserved and exhibited in museums and in explanations of the history of science and technology. On the other hand, intangible MMS products can be considered as the achievement in acquiring knowledge, expertise, and skills that are expressed in theories, algorithms, and procedures for designing and operating mechanical systems. These intangible MMS products are difficult to be indicated, and even more difficult to be preserved, when they were not published in written reports or books, or not referring to tangible MMS products. Even if they were documented in written publications, most of the intangible MMS products are worth full of consideration with regards to cultural heritage value since their technical content is very specialistic, and not easily understood by a non-expert public [13]. Therefore, a valuable contribution remains hidden, or even forgotten, in books or manuscripts in old libraries or personal archives.

A cultural heritage value of products of MMS activities can be expressed today with the concept of landmark referring to a technical content that can be understood even by the larger public, and can be useful to current professionals. This paper is specifically addressed to Italian aspects in the history of MMS, by referring to mechanism design and its modelling.

An MMS milestone of cultural heritage merit can be an achievement, or a figure, with significant impact on machine technology and/or in MMS development. For example, a technical-scientific cultural heritage value in the area of mechanical engineering was recognized by the American Society of Mechanical Engineers (ASME) since the 1970s with a specific Landmark program, specifically related only to mechanical systems, as stated in [42]. Besides indicating criteria for landmark merits, ASME's History and Heritage Committee (HHC) has promoted those merits with publications and books as in [11,12], referring to the most significant prized landmarks with the aim of disseminating to the public those of mechanical achievement and impact.

The milestone value of MMS achievements can be indicated with technical peculiarities in the historical development of MMS, referring to theoretical and practical aspects. Figure 8 summarizes those aspects that can be recognized as fundamental in giving a value of MMS milestone to an MMS product [43].

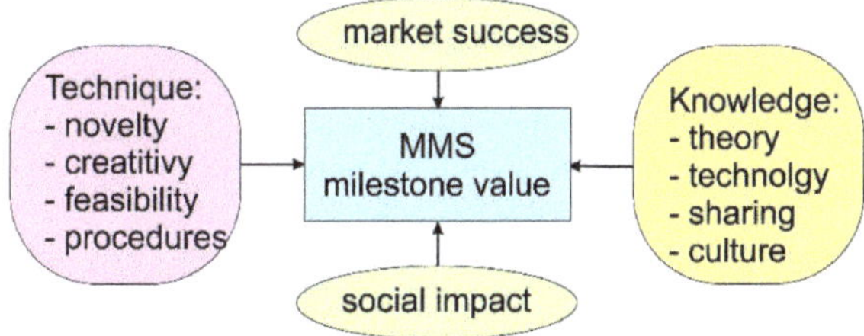

Figure 8. A scheme for milestone values of MMS achievements.

In addition to technique merits, milestone values in achievements and people can be identified by the significance and impact on the larger public. Public understanding and appreciation can be achieved with exhibitions of those achievements as tangible products of cultural heritage value. In the case of people, their contributions as intangible characters can be understood and appreciated by the influential legacy that they have left in the evolution of specific fields.

In the specific case of MMS milestones, specific technical content refers to machines and mechanisms as related to the topics that contributed to developments in mechanical engineering, through achievements by figures and community in those specific topics. The cultural value in MMS milestones can be evaluated referring to criteria as summarized in Figure 8, when considering evaluation of technical impacts in theoretical aspects, design issues, and applications. As stressed in Figure 8, technical aspects and knowledge values are considered together, with results on market valorization and social acceptance, in order to demonstrate practical success of an achievement or figure.

Technical merits can be summarized in aspects that are related to:

- Novelty, referring also to originality, when a milestone product was conceived and proposed, with new features referring to new concepts and solutions;
- Creativity, when a milestone product was obtained from creative activity;
- Feasibility, when a milestone product was applied efficiently with technical acceptance;
- Procedures with technical impacts, when a milestone product improved technical practices and applications.

Merits on knowledge acquisition can be summarized in aspects that are related to:

- Theory news, when a milestone product was a theoretical advance even producing further MMS developments;
- Technological transfer, when a milestone product produced a technical dissemination of achievements within designer/inventor communities;
- Dissemination, when a milestone product was disseminated producing further knowledge acquisitions and achievements;
- Cultural impact, when a milestone product affected cultural aspects at large with the innovation characters that have been fully accepted by users in the society.

The MMS products with milestone significance in technical achievements can be not only in machine designs and machinery constructions with physical remains, but even in knowledge and procedures used in conception, design, and production. As per Italian machinery milestone identification, attempts have been produced only recently, as reported in [43], as the example referring to the engine mechanism collection at the University of Palermo [44], and with an international recognition of the ASME landmark program.

All the above-mentioned considerations can be addressed to the models of mechanism with cultural heritage value, referring to their tangible and intangible contents. Therefore, mechanism models, in any of their formats, can be suitable and valuable for museum valorization, far more than is offered today, with only a few examples exhibited as an attraction of past technology.

6. Mechanism Models in Italian Universities

In this section, examples of mechanism models with their values are discussed to illustrate the Italian collections still existing with those values of cultural heritage products, but also as a useful means for teaching and design purposes. The reported examples are selected from the main Italian universities where collections of mechanical models, mainly from 19th century, are still preserved with some attention (poor or not) to their preservation and fruition, very often even with no archival data. The reported examples want to emphasize aspects and problems referring to the valorization of the physical models of mechanisms used in the past, and even recently, starting from an analysis of the state of conservation and consideration of the frames of their current preservation and use.

6.1. An Illustrative Example: The Models at the Engineering School in Rome

The images in Figure 9 show how the models of mechanisms remaining in the department, now called Mechanics and Aeronautics, are preserved, a tradition probably originating from the 1930s. The models, of which the few examples that remain are wooden models, were designed and built by the course teachers on Mechanics Applied to Machines for teaching and explaining both the functional design and the operation of the reproduced mechanisms. The preservation of these mechanisms in the two cabinets shown in the image has been provisional since the 1970s, when Professor Scotto Lavina, the last teacher to use them, retired at the end of the 1980s. In reality, these models were complemented by graphic tables, unfortunately lost, were large in size with construction details and sketches of models for design calculations, and with practical constructions that were jointly explained with the corresponding wooden model. As can be seen from the images in Figure 9, the models today have no use, and are practically forgotten without even having an historical display usefulness for teaching; posters and announcements relating to other departmental initiatives are also often posted, so that they completely obscure the visibility of the models and contents in the cabinets.

Figure 9. Frame of today's storage of wood models at the Department of Mechanics and Aeronautics in Rome: (**a**) the storage wardrobe; (**b**) zoomed view of the storage within the wardrobe.

The examples shown in Figure 10 show two wooden models of excellent workmanship, probably built in the years 1950–1960, to explain the fundamental constructive elements of gear transmission in the two cases of rotary motion and rectilinear motion, coming from a toothed wheel pinion. The running of the models is ensured by a special crank that a user can easily rotate to not only move the transmission components, but also to have a direct experience of mechanics in the operation of gear transmission components, in terms of motion and force. The dimensions of the two specimens with wheels are about 20–30 cm in diameter, while the linear rod with rungs has a size of about 40 cm, to also indicate a realistic mechanical design with organs applicable to industrial-type machines. These two specimens, like others crammed into the cabinets of Figure 9, can be appreciated both at the construction level in terms of manufacturing wooden models for teaching, but also at the level of the history of teaching, with a value that could still be useful today to show students the essential aspects of gear transmission.

The last example of Figure 11 shows a very small-scale model of an industrial crane, made with essential components in terms of both structure, and components for the transmission of power necessary for the application purpose. The model is made with metal elements and with bronze gear wheels to denote the attention to detail of the functionality of the model, which is also completed by suitable ropes for a hoist application. This model appears to have been built in the 1970s, perhaps more for research needs or even better as a scale model for design purposes, even if it seems to have been used for didactic purposes, limited to the use of a few students with thesis work. This example is significant to represent how a combined physical model with metal and wooden elements, even in modern times, has useful and interesting results from various technical-scientific aspects, but also from a historical point of view, it shows how design and research activities were accompanied with functional and detailed physical models.

(a)   (b)

Figure 10. Examples of wood models stored as in Figure 9: (**a**) two-circular gear transmission; (**b**) gear-rack transmission.

Figure 11. An example of an industrial-like crane model stored as in Figure 9.

Very likely, in other departments or warehouses of the engineering faculty of the University of Rome, one could find other physical and paper models used for the explanation and teaching of the mechanisms of the machines, as well as models used in research and applied design for what was taught and investigated for the development of machines, in both more efficient and traditional applications.

6.2. An Illustrative Example: The Models at the Engineering School in Bologna

The Department of Industrial Engineering of the University of Bologna has a collection of 53 models of mechanisms, largely from 20th century, for didactic purposes, that are stored in two cabinets, as shown in Figure 12.

(a)   (b)

Figure 12. The collection of models of mechanisms at the Department of Industrial Engineering at the University of Bologna: (**a**) main wardrobe; (**b**) secondary wardrobe.

At present, the models are no longer used for teaching but are kept in two cabinets with sliding glass doors, arranged in the entrance corridor, so that a student or visitor who enters the section of Mechanics Applied to Machines can look at them. The storage in the cabinets is not functional for museum purpose: the models are arranged in several overlapping rows, due to the limited space, and there are no descriptive plates. The collection is divided into 34 models in wood, nine in plastic and 10 in metal.

The wooden models follow the classification of the Reuleaux mechanisms, but the construction is that of the tables shown in Masi's book, as shown in Figure 2, with the comparison between the real model and the image in his book [35]. We can therefore date the collection to the teaching period of Francesco Masi (1889–1927) at the University of Bologna, or Modesto Prosciutto (1927–1954), the one immediately after. More likely, they date back to the early twentieth century, considering the importance of experimental applications in Masi's teaching [27,45]. The wooden models mainly concern gear wheels, cams, pulleys, chains, and joints.

As an example, Figure 13 shows an eccentric mechanism with several lobes. This mechanism–the most famous is the heart-shaped cam-has the property of having all its diameters of constant length, as evidenced by the diameters drawn on the front, and easily measurable. For this property, the eccentric can be closed within a frame between two wheels, the centers of which are placed at a distance equal to the diameter of the primitive curve of the eccentric. The model is mounted on a 40 × 20 cm base and has a height of approximately 25 cm. On the back there is a crank to rotate the cam manually, and to move the reciprocating rectilinear motion rod. On the back of the cam, *"Scuola d'applicazione per gli ingegneri in Torino. Blotto Giovanni meccanico"* (Application school for engineers in Turin. Blotto Giovanni mechanic) is engraved, but not the production date. Giovanni Blotto, model maker, was an assistant at the University of Turin, and together with Vittorio Canepa he created a large collection of models at the request of Giovanni Curioni, Professor of Construction at the Application School of Turin [46]. Subsequently, Blotto and Canepa opened a model-making company for didactic equipment, to which the models from the Bolognese collection were probably commissioned (other examples of Blotto and Canepa's models can be found at the University of Naples "Federico II" [47]).

Figure 13. Front and rear view of a multilobes cam that transforms the rotation of the input cam into and alternative motion of the rod.

The plastic models were made during the period of activity of Funaioli, in the 1980s, and concern plane kinematics. The different colors highlight the different parts of the mechanism, and engravings on the base highlight the trajectory traveled by the kinematic pairs. Figure 14 shows a four-bar linkage, in which the intermediate connecting rod (green) is the driver and is actuated by means of a gear (red). On the base (blue) there is an engraving depicting the maximum angular excursion of the rocker (white). The base of the model has dimensions 40×30 and there are feet to be able to place it vertically. On the back there is the manufacturer's label: *"Institut für landtechnische grundlagenforschung model. Braunschweig-Völkenrode"*. This label suggests that the models were a gift from a visiting professor (or another meeting occasion between the Italian and German research groups). One of the authors saw similar models in the office of Hanfreid Kerle at the Technical University of Braunschweig, which belonged to the previous Hoecken's collection [48]. It should be emphasized that on both types of model there are consecutive numbers, which suggest a much larger collection than the one currently left.

Figure 14. Four-bar linkage mechanism with an actuated connection rod and the maximum angular displacement of the rocket engraved on the base.

All the pieces are in an excellent state of conservation, and they fascinate anyone who passes in front of the cabinets. The value of these pieces is not only historiographical, but above all, didactic. Even today they can be used to clarify the functioning of complex mechanisms, and help future engineers to assimilate the concepts of machine mechanics.

6.3. An Illustrative Example: The Models at the Engineering School in Turin

The collection of mechanisms in the Polytechnic of Turin is stored in various cabinets in the meeting room of the Department of Mechanical and Aerospace Engineering, Figure 15, for preservation purposes and only recently with an attempt to re-evaluate both teaching and museum valorization, as reported in [49].

The collection is of more than 100 specimens of various sizes and various mechanisms, with constructions of different materials including both acquired models and models designed and built in-house. Unfortunately, as with other collections in academic institutions, there is no compilation of archival data that can uniquely identify the period and authors of many of these models, excluding those that with their own tags give specific indications on the origin and construction of the model. Most of the models have been acquired from the Schröder-Reuleaux commercial collection, such as [6], to which many of the later models actually refer both in terms of structure and design. Each Schröder-Reuleaux model is provided with a small plate with the Schröder company logo, and its corresponding catalog number.

A reusage of some models in teaching activity started in 2015, by planning three hours using a few of those models in combination with numerical computer-assisted procedures in a didactic laboratory, for the second-year course of Automatic Machine Mechanics in the MSc in Mechanical Engineering program, which involves the use of historical models of basic mechanisms.

Figures 16–18 are illustrative examples of the above-mentioned variety. In particular, Figure 16 shows a metallic model of hypoid connections that represents both the concepts of screw (helicoidal) motion and hypoid gear design. The model looks to be built and functioning for theoretical purposes, with a robust design on a wood plate of about 80×40 cm that, very likely made in Turin, was supposed to be movable by lost cranks in the circular wheel extremities.

Figure 15. The collection of models of mechanisms at the Department of Mechanical and Aerospace Engineering at the Polytechnic of Turin.

Figure 16. The iron-made model of the hypoid connections representing the screw motion and hypoid gears.

Figure 17 shows the Watt mechanism, with construction details in its approximate linear guide function using metal elements, being a mechanical model of an improved mechanism from a specimen from the commercial Schröder-Reuleaux collection. Also in this case, the implementation of the mechanism is obtained by means of a small crank, which allows a user not only to manage the functioning of the mechanism but also to have feedback in terms of mechanical efficiency during operation. Finally, of note is the valuable workmanship that makes the model an historical piece of high interest for a museum valorization as it is linked not only to the history of steam engines and the mechanisms used in them, but also for specific interest in the production of these mechanical models.

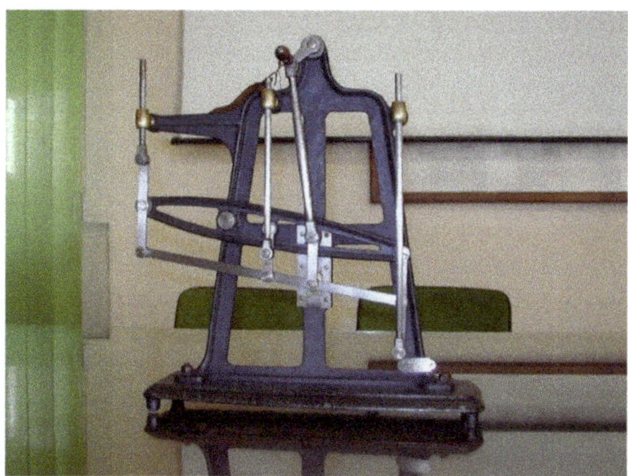

Figure 17. The iron-made model the Watt linkage for approximate-line guide.

Figure 18 shows two characteristic aspects of the preservation of the models in the Italian engineering schools, considering the fact that the models are crammed together without a suitable space between them, which not only prevents possible use, but does not allow complete visibility of each model. In addition, the models built in the offices are considered in the same way as those coming from external purchase, such as those from the Schröder-Reuleaux collection. In particular, Figure 18 shows at its front view the model of a Cardan transmission with torsion rods, with an assembly combining bronze elements with steel rods on cast iron supports, which can be oriented on a wood platform. In addition, in this model the functionality and efficiency of the mechanism is obtained with the action of a user through a crank on the input disk, for the rotation of the initial shaft, which allows a visualization and quantification of the efficiency of the motion transmission by another disk on the last rod of the transmission. It should also be noted that this model hides two other models of exquisite workmanship, and of considerable interest, for the structure of the mechanisms: the first, on the left, is a belt transmission with flywheel, and the second is a gear train, not completely visible even for the structure on which the gears are keyed. The first of these two models look like a model built in-house when you also notice the manufacturing of the operating crank of the mechanism, while the second is a product of the Schröder-Reuleaux collection.

Figure 18. The iron-made model of axle transmission with Cardan joints.

6.4. An Illustrative Example: The Models at the Engineering School in Milan

The collection of models preserved at the Department of Mechanics of the Politecnico di Milano, is an emblematic example of how much the historical value and interest of such models can be also suitable to outline the historical development of the discipline and institutions, as reported and emphasized in the publications of historical celebrations of the department in [50,51]. The collection is preserved mainly with models made in the 1920s–1950s, especially by the work and use of Professor Ottorino Sesini (who retired in 1962) and his successors. Previous acquisitions and constructions remain missing, with some surviving thanks to the historical didactic interest of some teachers, such as those of the publications cited [50,51]. Unfortunately, despite having such recognized historical value, the collection is not usable, but remains preserved as a set of historical memorabilia of the department, especially with regard to past didactic activities. Many of the models which survived the time, were made in the Laboratory of the Society of Encouragement of Arts and Crafts, Milan, following the instruction of teachers for their teaching activities on Kinematics and Design of mechanisms, and in aspects related to the Drawing of machines, in a similar way to what was in use in other Italian academic centers, but also in Europe, at the end of the 19th century. Many of the remaining models, of which examples are shown in Figure 19, are made with steel and bronze elements mounted on a wooden base, with dimensions for easy portability and functionality suitable for use by students for mechanical efficiency with low friction. The examples of Figure 19 refer to the variety of crank-slider mechanisms, useful for the transformation of the continuous rotary motion into an alternative translation, or vice versa. Of note is the essential structure of the models to indicate and realize the functionality of the mechanism, as combined with a practical construction useful for easy portability (both in theory classes and in laboratory sessions), as well as the elegant workmanship that still today makes them attractive both for future re-evaluation in didactic activities, and for a concrete museum valorization.

Figure 19. Examples of the iron-made crank-slider mechanism models from the 1930s' within the collection at the Department of Mechanics, Milan.

6.5. An Illustrative Example: The Models at the Engines Museum in Palermo

The Palermo collection, including models of mechanisms and restored working engines [52], is a valuable example of museum valorization of recognized prestige, even with an international award such as the ASME Landmark prize [53]. The collection is presented with an emphasis on engines, starting with steam engines of the 19th century up to experimental engines in the aeronautical fields that have been developed over time in other locations and in industrial productions, but used in the laboratories of the University of Palermo for research. Remarkable to remember is the commitment and the activity carried out in restoring those engines and the models of mechanisms, for their recovery not only in the structure but mainly in the functionality of a clear museum display, and an efficient presentation from an engineering viewpoint, and for educational purposes, that are limited to the rooms of the museum. The models of the mechanisms refer to a collection started by Professor Ovazza in the early years of the 20th century, with models acquired and built similar to those present at the Polytechnic of Turin, his academic site of origin. They were then enriched with models of local construction in Palermo by his successors, until reaching the collection in the 1970s. After a period of abandonment, both the collection of mechanisms and the engines received deserving renewed interest; the museum was founded and specifically dedicated to this museum valorization.

Figures 20–22 show illustrative examples of the museum arrangement, and the rich collection that the museum houses both in internal combustion and steam engines, and in models of mechanisms. In particular, although the models of the mechanisms are arranged with appropriate open display cabinets, they can be available for interactive use by visitors and students, with an appreciable way of direct experience on the functionality of the models.

In particular, Figure 20 shows a panoramic view of the main museum itinerary, hosting a variety of internal combustion engines and the first steam engines, with a clearly indicated route to facilitate a visitor, according to the historical development of the engines on display, many of which function with the help of a modern electric actuation system.

Figure 20. A view of the main exhibition room of the Museum of Engines and Mechanisms at University of Palermo, with engines form several periods and several applications.

Figure 21 shows the main exhibition frame for the models of the mechanisms, housed in a special room containing about 100 models of mechanisms of various manufacturing, size, and functionality, as collected over time with elements both acquired from commercial and other productions by academic bodies, as well as models built at the laboratory of the University of Palermo.

Figure 21. A cabinet with open-air exhibition of mechanism models in a museum room specifically dedicated to the models for teaching.

In Figure 21, it is possible to see several mechanisms that can be used, and were used, to explain the possibilities of mechanisms in the conversion of motion, and in the structure of the machines as already represented in the collections previously discussed. There are wooden models probably built in-house, according to the needs and instructions of the teachers of the time, as well as by the availability of manufacturing and budget, just as there are metal models of commercial collections, also built in-house to indicate the possibility of design and prototyping of the models themselves. Figure 22 shows an original model of the collection, with a structure that can be recognized as common to commercial-type models, such as the one shown by the Turin collection in Figure 18.

Figure 22 is an example of how the models have been duplicated with constructions in local frames, as the example of the figure represents the model of a rod transmission with Cardan joints, also present in the Turin collection. What is shown in Figure 22 is an example built with a design for compact dimensions, in materials typical of industrial applications for naval transmissions, and placed on a wooden base with one movable support; this is different from what is seen in the model shown in Figure 18, but it somehow has a more evident and appreciable effect.

The experience of the museum valorization in Palermo can be considered an emblematic example of how it is possible to valorize technical and scientific content, and also the historical value of the models of mechanisms, combined with applications of complete and perhaps more significant machines, such as the engines that are indicated as the main purpose of the museum. Moreover, the experience of the Palermo Museum shows how such models can have a value not only as cultural heritage, but as an interest and applicability in explanatory activities for visitors, in addition to having an educational purpose for engineering students.

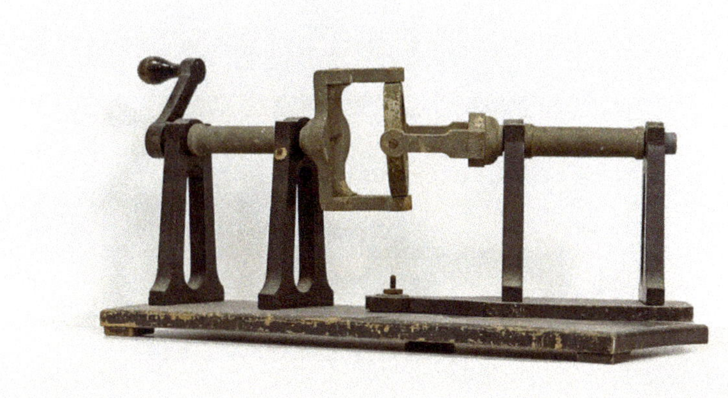

Figure 22. An example of the on-site built mechanism models at University of Palermo.

7. Conclusions

This paper presents aspects and problems for didactic and technical-scientific reconsideration of the models of mechanisms, especially in the format of a scaled mechanical structure for the purpose of an historical valorization for one's own awareness, and use in the formation and tracing of the history of Italian mechanical engineering. The examples that are reported from the major Italian academic sites refer to collections present in some of the major universities, with preservation and archiving solutions and use modalities that give hope for an adequate, especially historical, valorization of cultural heritage. The models of the mechanisms, which historically have evolved from graphic representations to constructions with mechanical structure, right up to today's virtual solutions, have an importance not always recognized both at the didactic and design level with aspects that, however, lead back to the development of mechanical engineering, with content worthy of preservation as cultural heritage. The models in Italian universities have this aspect of cultural heritage, being adequately preserved for historical interest, with models purchased both outside the academic site and produced on site, helping students in research and design activities. The discussed examples show the variety of solutions in terms of mechanisms, and constructive solutions in materials and functionalities, with a richness yet to be fully discovered in Italian academic centers.

Funding: This research received no external funding.

Institutional Review Board Statement: Not applicable.

Informed Consent Statement: Not applicable.

Conflicts of Interest: The authors declare no conflict of interest.

References

1. Singer, C.; Holmyard, E.J.; Hall, A.R.; Williams, T.I. *History of Technology*; Bollati Boringhieri: Turin, Italy, 2012; Volume 2. (In Italian)
2. Capocaccia, A. (Ed.) *History of Technique—From Prehistory to the Year One Thousand*; UTET: Turin, Italy, 1973. (In Italian)
3. Nolle, H. *Linkage Coupler Curve Synthesis: A Historical Review—I and II*; Mechanism and Machine Theory; Elsevier: Oxford, UK, 1974; Volume 9.
4. Kerle, H.; Mauersberger, K.; Ceccarelli, M. Historical Remarks on Past Model Collections of Machines and Mechanisms in Europe. In Proceedings of the 13th World Congress in Mechanism and Machine Science IFToMM 2011, Guanajuato, Mexico, 19–23 June 2011. paper n. A21-279.
5. Reuleaux, F. *Theoretische Kinematik*; F. Viewg und Sohn: Braunschweig, Germany, 1875.
6. Voigt, G. *Kinematische Modelle nach Professor Reuleaux, catalogue I and II*; Voigt: Berlin, UK, 1907.

7. TU Dresden, Collection of Mechanism and ger Models. Available online: https://tu-dresden.de/kustodie/sammlungen-kunstbesitz/ingenieurwissenschaften/getriebemodell-sammlung?set_language=en (accessed on 1 June 2022).

8. Golovin, A.; Tarabarin, V. *Russian Models from the Mechanisms Collection of Bauman University*; Springer: Dordrecht, The Netherlands, 2008.

9. Ceccarelli, M. A historical account on Italian mechanism model. *Elementa. Intersect. Between Philos. Epistemol. Empir. Perspect.* **2021**, *1*, 115–134. [CrossRef]

10. Ceccarelli, M. Figures and achievements in MMS as landmarks in history of MMS for inspiration of IFToMM activity. *Mech. Mach. Theory* **2016**, *105*, 529–539. [CrossRef]

11. ASME History and Heritage Committee (Ed.) *Landmarks in Mechanical Engineering*; Purdue University Press: West Lafayette, India, 1997.

12. Black, J.M. *Machines That Made History—Landmarks in Mechanical Engineering*; ASME Press: New York, NY, USA, 2014.

13. Ceccarelli, M. Are Theory and Procedures for Mechanism Designs Suitable as Goods of Cultural Heritage? *Rev. Int. Estud. Vascos* **2014**, *59*, 36–50.

14. IFToMM. Special Issue: Standardization and Terminology. In *Mechanism and Machine Theory*; No. 7–10; Elsevier: Oxford, UK, 2003; Volume 38.

15. Uicker, J.J.; Pennock, G.R.; Shigley, J.E. *Theory of Machines and Mechanisms*; Oxford University Press Inc.: Oxford, UK, 2017.

16. Lopez-Cajùn, C.S.; Ceccarelli, M. *Mechanisms: Fundamentals of Kinematics for Design and Optimization of Machinery*, 2nd ed.; Trillas: Mexico City, Mexico, 2013. (In Spanish)

17. Ceccarelli, M. Models of mechanisms for teaching and experimental activity. In Proceedings of the Atti del XXXVIII Convegno annuale Società Italiana degli Storici della Fisica e dell'Astronomia, Messina, Italy, 3–6 October 2018; Pavia University Press: Pavia, Italy, 2018; pp. 255–266, ISBN 978-88-6952-7.

18. Ceccarelli, M. An Outline of History of Mechanism Design in servicing Science. In *Physics, Astronomy and Engineering: Critical Problems in the History of Science and Society, Proceedings of the 32nd International Congress of the Italian Society of Historians of Physics and Astronomy, Rome, Italy, 27–29 September 2012*; Scientia Socialis Press: Šiauliai, Lithuania, 2012; pp. 1–10. ISBN 978-609-95513-0-2.

19. Ceccarelli, M.; Cigola, M. On the evolution of graphical representation of gears. In *Proceeding of the Conference MeTrApp 2011 Mechanisms, Transmissions, Applications*; Book Series on Machines and Machine Science; Springer: Dordrecht, The Netherlands, 2011; Volume 1, pp. 3–14.

20. Ceccarelli, M. Short History of Mechanics of Machinery in Italy. In *Proceedings of the 5th Italian Conference on History of Engineering*; Cuzzolin Publ.: Napoli, Italy, 2014; pp. 87–102. (In Italian)

21. Rossi, C.; Ceccarelli, M. Science, Technology and Industry in Southern Italy Before the Unification. In *Essays on the History of Mechanical Engineering, History of Mechanism and Machine Science 31*; Springer: Dordrecht, The Netherlands, 2016; pp. 159–180. [CrossRef]

22. Ceccarelli, M. Giuseppe Antonio Borgnis (1781–1863). In *Distinguished Figures in Mechanism and Machine Science—Part 3, History of Mechanism and Machine Science Volume 26*; Springer: Dordrecht, The Netherlands, 2014; pp. 41–56. [CrossRef]

23. Giulio, C.I. *Lezioni di Meccanica Applicata alle Arti*; Tipografia Pomba: Torino, Italy, 1846.

24. Giorgini, G. *Intorno alle Proprietà Geometriche dei Movimenti di un Sistema di Punti di Forma Invariabile*; Memorie di Matematica e Fisica della Società Italiana delle Scienze; Tomo XXI; Tipografia Camerale: Modena, Italy, 1836; pp. 1–54.

25. Ceccarelli, M.; Sorge, F.; Genchi, G. Elia Ovazza: Professor of TMM in Palermo around the End of the 19th Century. In *Essays on the History of Mechanical Engineering, History of Mechanism and Machine Science 31*; Springer: Dordrecht, The Netherlands, 2016; pp. 47–64. [CrossRef]

26. Fang, Y.; Ceccarelli, M. Medium Size Companies of Mechanical Industry in Northern Italy during the Second Half of the 19th Century. In *Essays on the History of Mechanical Engineering, History of Mechanism and Machine Science 31*; Springer: Dordrecht, The Netherlands, 2016; pp. 181–200. [CrossRef]

27. Ceccarelli, M. 'Francesco Masi (1852–1944)'. In *Distinguished Figures in Mechanism and Machine Science: Their Contributions and Legacies—Part 2, Book Series on History of Machines and Machine Science*; Springer: Dordrecht, The Netherlands, 2010; Volume 7, pp. 141–162. ISBN 978-90-481-2345-2. [CrossRef]

28. Allievi, L. *Cinematica della Biella Piana*; Regia Tipografia Francesco Giannini & Figli: Napoli, Italy, 1895. (In Italian)

29. Ceccarelli, M.; Koetsier, T. Burmester and Allievi: A Theory and Its Application for Mechanism Design at the End of 19th Century. *J. Mech. Des.* **2008**, *130*, 072301. [CrossRef]

30. Ceccarelli, M. Allievi Lorenzo (1856–1941). In *Distinguished Figures in Mechanism and Machine Science—Part 3, History of Mechanism and Machine Science Volume 26*; Springer: Dordrecht, The Netherlands, 2014; pp. 1–17. [CrossRef]

31. Ceccarelli, M.; Teoli, G. Corradino D'Ascanio and His Design of Vespa Scooter. In *Multibody Mechatronic Systems, Mechanisms and Machine Science 25*; Springer: Dordrecht, The Netherlands, 2015; pp. 399–410. [CrossRef]

32. Cocconcelli, M. The Teaching of Applied Mechanics through Textbooks in Italy. *Adv. Hist. Stud.* **2020**, *9*, 358–376. [CrossRef]

33. Reuleaux, F. *Lehrbuch der Kinematic*; F. Vieweg und Sohn: Braunschweig, Germany, 1900.

34. Masi, F. *Manuale di Cinematica Applicata: Nuova Classificazione dei Meccanismi*; Zanichelli: Bologna, Italy, 1883.

35. Masi, F. *La Teoria dei Meccanismi*; Zanichelli: Bologna, Italy, 1897.

36. Ceccarelli, M. Classifications of mechanisms over time. In *Proceedings of International Symposium on History of Machines and Mechanisms HMM2004*; Kluwer: Dordrecht, The Netherlands, 2004; pp. 285–302.

37. *Museum of Engines and Mechanisms, Engine of Innocenti Lambretta 125 C*; University of Palermo: Palermo, Italy. Available online: https://www.museomotori.unipa.it/scheda.php?id=40 (accessed on 1 June 2022).
38. Corazzari, L.; Di Felice, E.; Mucchi, E. Design and Construction of 3-Step Gear by 3D Printing. Master's Thesis, University of Ferrara, Ferrara, Italy, 2019. (In Italian).
39. Bellinazzi, T.; Mucchi, E. Design and Construction of a V8 Engine by 3D Printing. Bachelor's Thesis, University of Ferrara, Ferrara, Italy, 2020. (In Italian).
40. UNESCO. Cultural Heritage-2010. Available online: http://portal.unesco.org/culture/ (accessed on 11 October 2013).
41. Ceccarelli, M. Considerations on Mechanism Designs as Suitable for Cultural Heritage Evaluation. *Adv. Hist. Stud.* **2013**, *2*, 175–184. [CrossRef]
42. ASME History and Heritage Committee (Ed.) About the Landmarks Program. Available online: https://www.asme.org/about-asme/engineering-history/landmarks/about-the-landmarks-program (accessed on 19 May 2022).
43. Ceccarelli, M. Italian Landmarks in the History of MMS. In Proceedings of the 2015 IFToMM Workshop on History of Mechanism and Machine Science, St-Petersburg, Russia, 26–28 May 2015. paper no. 20-03.
44. ASME. *A Historic Mechanical Engineering Collection: The Collection of Engines at the Museum of Engines and Mechanism of University of Palermo, Palermo*; Museum Brochure; ASME: New York, NY, USA, 2017.
45. Ceccarelli, M.; Molari, P.G. Come insegnare la meccanica. Il metodo di Francesco Masi all'Università di Bologna e all'Istituto Aldini Valeriani. *Riv. Sc. Off.* **2016**, *2*, 4–11.
46. Gianasso, E. Cultura e formazione degli ingegneri. Studi ottocenteschi intorno a Leonardo da Vinci. In Proceedings of the 4th International Conference on History of Engineering, Naples, Italy, 11 December 2020; pp. 511–522.
47. Università Degli Studi di Napoli "Federico II", I Modelli Didattici Conservati Presso il Centro Interdipartimentale di Ingegneria per i Beni Culturali. Available online: http://www.nordsud.unina.it/storia_ing/interna11-b.html (accessed on 1 June 2022).
48. Kerle, H. About Karl Hoecken and some of his works on mechanisms. In *Explorations in the History of Machines and Mechanisms: Proceedings of HMM2012*; Springer: Dordrecht, The Netherlands, 2012; pp. 123–134.
49. Franco, W.; Trivella, A.; Quaglia, G. Mechanism and Machine Science Educational Workshop Based on Schröder-Reuleaux Ancient Models of Politecnico di Torino. *Adv. Hist. Stud.* **2020**, *9*, 295–311. [CrossRef]
50. De Alberti, L.; Rovida, E. (Eds.) *Historical Heritage of Mechanics Department*; Politecnico di Milano: Milan, Italy, 1999. (In Italian)
51. Curami, A.; Rovida, E.; Zappa, E. *Mechanics Teachers Since 1863: History of Mechanics Department*; Politecnico di Milano: Milan, Italy, 2015. (In Italian)
52. Monastero, R.; Genchi, G. The Museum of Engines and Mechanisms. More Than a Century of History of Technology. In *Essays on the History of Mechanical Engineering. History of Mechanism and Machine Science, Volume 31*; Sorge, F., Genchi, G., Eds.; Springer: Cham, Switzerland, 2016; pp. 201–225. [CrossRef]
53. Genchi, G. (Ed.) *The Collection of Engines at the Museum of Engines and Mechanisms University of Palermo—A Mechanical Engineering Heritage Collection for ASME Landmark*; Museum of Engines and Mechanisms, University of Palermo: Palermo, Italy, 2017.

 *machines*

MDPI

Review

The Role of Instant Centers in Kinematics and Dynamics of Planar Mechanisms: Review of LaMaViP's Contributions

Raffaele Di Gregorio

LaMaViP, Department of Engineering, University of Ferrara, 44122 Ferrara, Italy; raffaele.digregorio@unife.it; Tel.: +39-0532-974828

Abstract: Theoretical kinematics and dynamics is one of the research fields where LaMaViP (Laboratory of Mechatronics and Virtual Prototyping) operates, which is the lab leaded by the author at the University of Ferrara. In the last two decades, this research activity at LaMaViP has produced, among others, many novel results that highlight how instant centers' (ICs) locations condition the kinetostatic and dynamic behaviors of planar mechanisms, and that provide tools suitable for design purposes. This paper reviews/summarizes the tools devised at LaMaViP for PM analysis and synthesis through ICs, and shows that they are a complete set of tools, which make the full description of PMs' kinematics and dynamics possible, and that the new IC features, identified while setting up these tools, are relevant in machine design.

Keywords: kinematics; planar mechanism; instant center; singular configuration; machine design

Citation: Di Gregorio, R. The Role of Instant Centers in Kinematics and Dynamics of Planar Mechanisms: Review of LaMaViP's Contributions. *Machines* **2022**, *10*, 732. https://doi.org/10.3390/machines10090732

Academic Editors: Marco Ceccarelli, Giuseppe Carbone and Alessandro Gasparetto

Received: 30 July 2022
Accepted: 25 August 2022
Published: 26 August 2022

Publisher's Note: MDPI stays neutral with regard to jurisdictional claims in published maps and institutional affiliations.

1. Introduction

Planar mechanisms (PMs) enter in many even spatial machines. That is why ample literature has been devoted to them (e.g., [1–16]). Instantaneous kinematics and statics of PMs can be fully described [1–3] by using instant centers (ICs), which are also relevant in many design problems. For instance, vehicle-suspension design [17], lower-limb prostheses for amputees [18], and Remote Center of Compliance (RCC) systems [19] are only some of these problems.

The use of ICs in PMs' analysis and synthesis requires the availability of techniques for determining their positions on the motion plane. In single-DOF PMs, the IC positions depend only on the mechanism configuration. Common methods to locate the ICs in single-DOF PMs, are based on the direct application of the Aronhold–Kennedy (AK) theorem. Unfortunately, when the mechanism architecture becomes complex, these methods fail [1,20]. In the literature, the PMs whose ICs cannot be located by directly applying the AK theorem are called "indeterminate" [5].

Graphic methods to locate the ICs of indeterminate PMs have been proposed in the literature [1,5–7]. In [1], Klein proposed a graphic trial-and-error method based on the fact that (a) some theorems of projective geometry guarantee the existence of an alternative geometric locus for each secondary IC, not locatable through the AK theorem, and that (b) such a locus can be determined by arbitrarily assuming a number of positions of another secondary IC, for which the AK theorem gives only one straight line, and, then, by coherently determining the positions of the sought-after IC through the AK theorem. So doing, a number of points of the alternative geometric locus are determined, which are sufficient to draw the locus; then, the intersection of the so-determined alternative locus with another locus of the same IC, provided by the AK theorem, gives the correct IC position. In [7], Foster and Pennock proposed another graphic method that can be applied to all the indeterminate PMs. Their work also presents analytical relationships that could be used to formulate an analytical method. Their relationships contain kinematic coefficients whose analytical determination would require the computation of the derivative of the mechanism closure equations. Previously, Yan and Hsu [4] presented another analytic approach, applicable also to multi-DOF PMs, based on a similar principle, which, in practice, requires the solution of the velocity analysis to determine the IC positions. Such analytical approaches, when applied to single-DOF PMs, take no advantage from the fact that the IC positions of these PMs depend only on the mechanism configuration.

In this context, at the Laboratory of Mechatronics and Virtual Prototyping (LaMaViP) [21], an exhaustive analytic technique for the IC-positions' determination of single-DOF PMs has been conceived which uses only the mechanism-configuration data. Then, in [22], the IC determination in multi-DOF PMs has been addressed and some theoretical results have been proved which make their systematic determination possible.

The availability of analytic techniques for determining the IC positions open to the possibility of devising algorithms that fully describe the instantaneous kinematics of PMs through the ICs. PMs' singularity analysis is a relevant problem to address when studying PMs' instantaneous kinematics. Indeed, it is central in the design of parallel PMs (PPMs) since it deeply affects their kinetostatic performances [23]. Singularities are mechanism configurations where mechanism's instantaneous kinematics becomes indeterminate [24–28]. Hunt [28] identified two special types of mechanism configurations where the instantaneous kinematic behavior is singular: the stationary configurations and the uncertainty configurations. Stationary configurations are mechanism configurations where the rate of a joint variable is instantaneously equal to zero (i.e., the joint is instantaneously inactive which can be alternatively said as follows: "the joint is at a dead center position"). Uncertainty configurations are configurations where the mechanism locally gains extra additional DOFs (transitory mobility). With reference to single-DOF PMs, Hunt [28] highlighted that some of the ICs coincide when the mechanism assumes a stationary configuration; whereas the same ICs fall on particular straight lines when the mechanism assumes an uncertainty configuration. Then, Yan and Wu [29,30] gave a geometric criterion to identify which ICs coincide at a stationary configuration [29] and developed a geometric methodology to generate single-DOF PMs at dead center positions [30].

In this context, at LaMaViP, firstly [31,32], an exhaustive geometric and analytic technique for identifying the singular configurations of single-DOF PMs has been devised which is based on ICs' positions. Then [22], new theoretical results, which relate ICs positions of multi-DOF PMs with the positions of the same ICs in the single-DOF PMs generated from the multi-DOF one by locking all the actuated joints but one, have been used to set up a novel technique for identifying singularities of multi-DOF PMs.

The IC positions can be also used to compute the velocity (influence) coefficients (VCs) [2,33] of single-DOF PMs and the entries of velocity-coefficient-vectors (VCVs) [34] of multi-DOF PMs, which are indeed VCs of the single-DOF PMs generated from the multi-DOF PM by locking all the generalized coordinates but one. VCs are by definition the ratios between the first-time derivatives (rates) of any two motion variables. In single-DOF PMs, they only depend on the mechanism configuration and enter in both the kinematics [2,33] and statics analyses [32] of these mechanisms. By exploiting this fact with reference to PMs, at LaMaViP, new dynamic formulations [35–38] have been conceived which fully highlight the role of ICs in PMs' dynamics.

This paper reviews the tools devised at LaMaViP for PM analysis and synthesis through ICs and shows that they are a complete set of tools, which make the full description of PMs' kinematics and dynamics possible, and that the new IC features, identified while setting up these tools, are relevant in machine design.

The paper is organized as follows. Section 2, over reminding some background concepts, illustrates the tools devised for ICs' position determinations and PMs' singularity analysis. Section 3 illustrates the dynamic formulations. Eventually, Section 4 draws the conclusions.

2. Planar Kinematics Revisited through Instant Centers (ICs)

The instant center, $\mathbf{C}_{ji}$, of the relative (planar) motion between links j and i (Figure 1) is the point of the motion plane that has the same velocity, no matter whether it is considered fixed to one or the other of the two links, whatever be the third link, link k in Figure 1, fixed to the observer that measures its velocity.

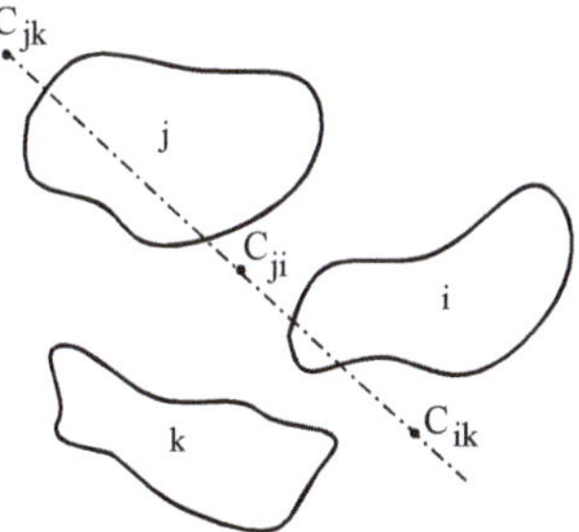

Figure 1. Diagram of the Aronhold-Kennedy theorem.

Prismatic (P)/revolute (R) pairs and rolling (C_r)/slipping (C_s) contacts are the only kinematic pairs that appear in PMs. P, R and C_r are single-DOF kinematic pairs; whereas, C_s is a two-DOF kinematic pair. In a PM, if links j and i are joined by any of these pairs (see Figure 2), the instant center, $\mathbf{C}_{ji}$, of their relative motion has a known position for the three single-DOF pairs, that is, P, R and C_r (see Figure 2a–c); whereas, it must lie on the common normal at the contact point for C_s contacts (Figure 2d). These four kinematic pairs generate only holonomic and time-independent (scleronomic) constraints [33,34]. Therefore, in PMs, a possible time-dependent (rheonomic) constraint can only be generated by a mobile frame, and, even in this special case, the relative motions between links are scleronomic. Accordingly, in PMs internal motion (i.e., the one with respect to the frame), any motion variable not chosen as generalized coordinate depends only on the chosen generalized coordinates. This fact involves that in l-DOF PMs, the VCs, over depending on the mechanism configuration, depend also on ($l - 1$) ratios between generalized-coordinate rates (see [22] for details).

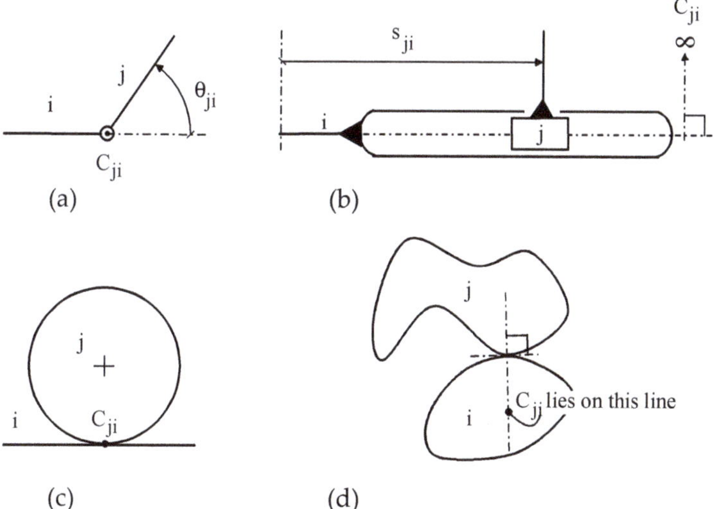

Figure 2. Position of the IC, $\mathbf{C}_{ji}$, in the four kinematic pairs that can appear in PMs: (**a**) revolute pair, (**b**) prismatic pair, (**c**) rolling contact, and (**d**) slipping contact.

In a PM with m links, m(m − 1)/2 relative motions can be identified and as many are the ICs (i.e., one for each relative motion). The relative motion theorems [39] bring the conclusion that only (m − 1) of such motions (e.g., the ones of the mobile links with respect to the frame) are independent (i.e., are sufficient to determine all the remaining ones). Accordingly, determining the positions of (m − 1) ICs is sufficient to solve the velocity analysis of any PM and to fully identify its instantaneous-kinematics features. The ICs whose positions can be determined through a simple inspection of the PM (i.e., the ones related to the single-DOF kinematic pairs) are named "primary" ICs; all the other ICs are named "secondary" ICs [5].

The Aronhold-Kennedy (A-K) theorem states that, in planar motion, the ICs ($\mathbf{C}_{ji}$, $\mathbf{C}_{ik}$, and $\mathbf{C}_{jk}$ in Figure 1) of the three relative motions definable among three rigid bodies (links i, j, and k in Figure 1) must lie on the same line. This condition together with the IC definition brings one to compute the following VCs' explicit expressions (see Figures 1 and 3).

$$\frac{\dot{\theta}_{ji}}{\dot{\theta}_{ki}} = \frac{(\mathbf{C}_{kj} - \mathbf{C}_{ki}) \cdot (\mathbf{C}_{kj} - \mathbf{C}_{ji})}{\|\mathbf{C}_{kj} - \mathbf{C}_{ji}\|^2}; \quad \frac{\dot{s}_{ji}}{\dot{\theta}_{ki}} = -(\mathbf{C}_{kj} - \mathbf{C}_{ki}) \cdot \mathbf{u}_{ji}; \quad \frac{\dot{s}_{ji}}{\dot{s}_{ri}} = \frac{(\mathbf{C}_{kj} - \mathbf{C}_{ki}) \cdot \mathbf{u}_{ji}}{(\mathbf{C}_{kr} - \mathbf{C}_{ki}) \cdot \mathbf{u}_{ri}} \quad (1)$$

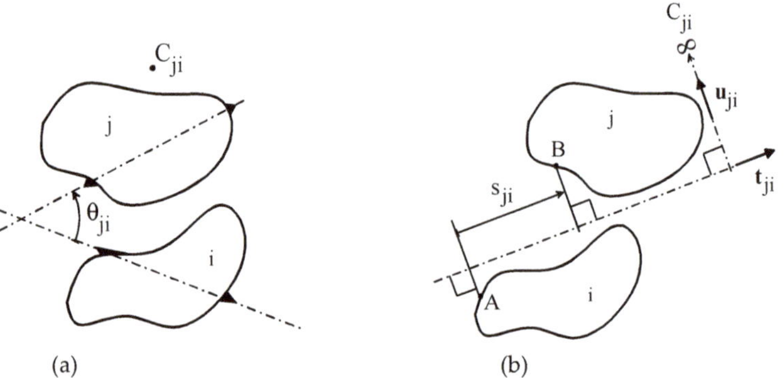

Figure 3. Relative (planar) motion between any two links, denoted j and i: (**a**) instantaneous rotation (the IC $\mathbf{C}_{ji}$ is a finite point), (**b**) instantaneous translation along the direction of the unit vector $\mathbf{t}_{ji}$ (the IC $\mathbf{C}_{ji}$ is the point at infinity the unit vector $\mathbf{u}_{ji}$ points to).

In Equation (1), a bold capital letter denotes the position vector of the point, the capital letter refers to, in a reference system fixed to the motion plane. The angle θ_{ji} (θ_{ki}), positive if counterclockwise, identifies the orientation of link j (link k) with respect to link i (see Figure 3a). The unit vector $\mathbf{u}_{ji}$ ($\mathbf{u}_{ri}$) is obtained through a counterclockwise rotation of 90° from the unit vectors $\mathbf{t}_{ji}$ ($\mathbf{t}_{ri}$) (Figure 3b), which gives the positive directions of instantaneous translation, in a relative motion where the IC $\mathbf{C}_{ji}$ ($\mathbf{C}_{ri}$) is a point at infinity, and $\dot{s}_{ji}$ ($\dot{s}_{ri}$) is the signed magnitude of the instantaneous-translation velocity, $\dot{s}_{ji}\mathbf{t}_{ji}$ ($\dot{s}_{ri}\mathbf{t}_{ri}$).

Equation (1) and the fact that, in *l*-DOF PMs, VCs depend on (*l* − 1) ratios between generalized-coordinate rates bring the conclusion that, in single-DOF PMs, the IC positions depend uniquely on the mechanism configuration; whereas, in *l*-DOF PMs, they, over depending on the mechanism configuration, depend also on (*l* − 1) ratios between generalized-coordinate rates [22].

2.1. Instant Center Determination

The analysis of Figure 2 reveals that, in a reference system fixed to PM's frame, the coordinates of all the primary ICs and, in C_s contacts, the pose parameters of all the normal lines at the contact points depend only on PM's configuration. Therefore, they can be analytically determined by solving the constraint-equation system once the generalized coordinates of the PM are assigned.

2.1.1. Single-DOF PMs

In single-DOF PMs [21], firstly, the coordinates of all the primary ICs and, in C_s contacts, the pose parameters of all the normal lines at the contact points must be computed as functions of the assigned generalized coordinate. Then, these data must be used together with the AK theorem (Figure 1) for determining the pose parameters of two lines where one secondary IC must lie on. In addition, the equations of the two lines must be written and the system constituted by these two linear equations must be solved to determine the coordinates of the secondary IC that lie on the two lines. Successively, the coordinates of the just-determined secondary IC together with the previously determined data must be used to repeat this algorithm to compute the coordinates of another secondary IC and so on until the coordinates of all the secondary ICs are determined. The central issues to solve for automatically using such a procedure are

(i) the determination of the sequence, hereafter called S_0, with which the secondary ICs must be computed;

(ii) the alternative algorithm to use when the determined data are not sufficient to identify two lines where another secondary IC must lie on, which is the case of indeterminate linkages [1,5–7,20,21].

If the PM is not indeterminate, issue (i) (i.e., the determination of S_0) can be solved by implementing an algorithm easy to deduce from the "circle diagrams" [20,40]. Differently, for indeterminate PMs, the solution of issue (i) is much more cumbersome [1,5–7,21] since it is connected to the solution of issue (ii). In [21], both the issues have been solved by using the "table" shown in Figure 4, and a filling procedure for the proposed "table", which identifies both the sequence S_0 in the case of not-indeterminate PMs and the alternative algorithm to use for indeterminate PMs. Such a general technique is illustrated below.

	1	2	...	i	...	j	...	m	1_{sc}	2_{sc}	...	q_{sc}
1	■	X	...	X	...		...		X		...	X
2	X	■	...		...		...	X		X	...	
⋮	⋮	⋮	⋮	⋮	⋮	⋮	⋮	⋮	⋮	⋮	⋮	⋮
i	X		...	■	...		...		X		...	
⋮	⋮	⋮	⋮	⋮	⋮	⋮	⋮	⋮	⋮	⋮	⋮	⋮
j			...		...	■	...	X		X	...	
⋮	⋮	⋮	⋮	⋮	⋮	⋮	⋮	⋮	⋮	⋮	⋮	⋮
m		X	...		...	X	...	■			...	X

Figure 4. Table proposed in [21] to determine all the secondary ICs of a single-DOF PM with m links and q C_s-contacts. The first row (column) just contains the column (row) index and it is just a heading row (column).

The table to use (Figure 4) for a single-DOF PM with m links and q C_s-contacts has m rows, which one-to-one correspond to the m links, and (m + q) columns with the first m columns corresponding one-to-one to the m links and the last q columns corresponding one-to-one to the q C_s-contacts. In the first m columns, the cells with the row index equal to the column index are black; whereas, all the remaining cells are empty at the beginning of the filling procedure. In these columns, the cell indices identify the IC with the same lower-right indices, that is, the cell (i, j) corresponds to the IC $\mathbf{C}_{ij}$. Since $\mathbf{C}_{ij}$ and $\mathbf{C}_{ji}$ are the same IC, there are two cells for each IC, which are symmetrically disposed with respect to the diagonal black cells. In the last q columns, only the two cells whose row indices correspond to one or the other of the two links joined by the C_s-contact the column refers to are filled with the symbol "X" and the remaining cells are left empty. In these other columns the two filled cells indicate that the pose parameters of a line, which the IC with indices given by the two row indices of the two filled cells lies on, are known. The central filling rule is that, in the first m column, an empty cell can be filled only when the data necessary to compute the corresponding IC are known. So doing, in the first m column, two filled cells mean that two ICs with one common index (i.e., the one of the column) have known coordinates and identify the pose parameters of a line (i.e., the one passing through the two ICs) on which the IC with indices given by the two row indices of the two filled cells lies. Therefore, the following filling procedure/computation algorithm can be conceived (see Figure 5).

(a) in the first m columns the cells corresponding to the primary ICs are filled with the symbol "X" and the coordinates of these ICs are computed by solving the constraint-equation system;

(b) the rows with at least two filled cells are selected and compared to identify all the couple of rows that have two filled cells in the same two columns;

(c) for each couple of rows identified in the previous step, the two linear equations are written which correspond to the lines identified by the two columns with filled cells. The so-obtained system of two linear equations is solved to determine the coordinates of the secondary IC with indices given by the indices of the two rows; then, the corresponding cells are filled with the roman number "I";

(d) focusing only on the row couples whose indices correspond to the ones of the still empty cells of the first m columns, the steps (b) and (c) are repeated until either all the cells of the first m columns are filled or at step (b) is not possible to identify any couple of rows (i.e., the mechanism is indeterminate). At each repetition of the steps (b) and (c) the roman number used to fill the cells is increased of one unit (see, Figure 5d);

(e) if step (d) brings to fill all the cells of the first m columns, the coordinates of all the secondary ICs have been computed and the algorithm is stopped; otherwise (i.e., in the case of indeterminate mechanisms) the following steps are implemented:

(e.1) focusing only on the row couples whose indices correspond to the ones of the still empty cells of the first m columns, a row couple that has two filled cells in the same column is selected. Moreover, the two cells with indices coincident with the row indices of the selected row couple are filled with the starred roman number "I*";

(e.2) the coordinates of the secondary IC, whose indices coincide with the two row indices of the row couple selected in the previous step, are written as the ones of a point lying on the line passing through the two IC identified by the two filled cells located in the above-mentioned same column. That is, a line parameter, say λ, is introduced and the two IC coordinates are explicitly written as linear functions of λ;

(e.3) focusing only on the row couples whose indices correspond to the ones of the still empty cells of the first m columns, the steps (b) and (c) are repeated by taking into account also the cells filled with "I*" and analytically solving the two-linear-equations system of step (c) still to identify a secondary IC that must lie on three lines. During this step, the cells corresponding to the located ICs are filled with "I*" and the analytic solutions of the above-mentioned equation systems bring to explicitly write the coordinates of these ICs as functions of the parameter λ introduced in the previous step;

(e.4) the equations of the three lines, which the secondary IC identified in the previous step lies on, are written. Such equations constitute a system of three equations in three unknowns, the two coordinates of the IC and the parameter λ, and the three equations are all linear in the two coordinates of the IC;

(e.5) the coordinates of the above-mentioned IC are explicitly expressed as functions of λ by solving the first two equations of the system deduced in the previous step. Then, the so-obtained expressions are introduced in the third equation of the same system to obtain one equation in the unique unknown λ;

(e.6) the equation in λ deduced in step (e.5) is solved and the computed value of λ is back substituted in the explicit expressions of the coordinates of all the ICs deduced in the previous steps to compute their numeric values;

(e.7) jump to step (d)

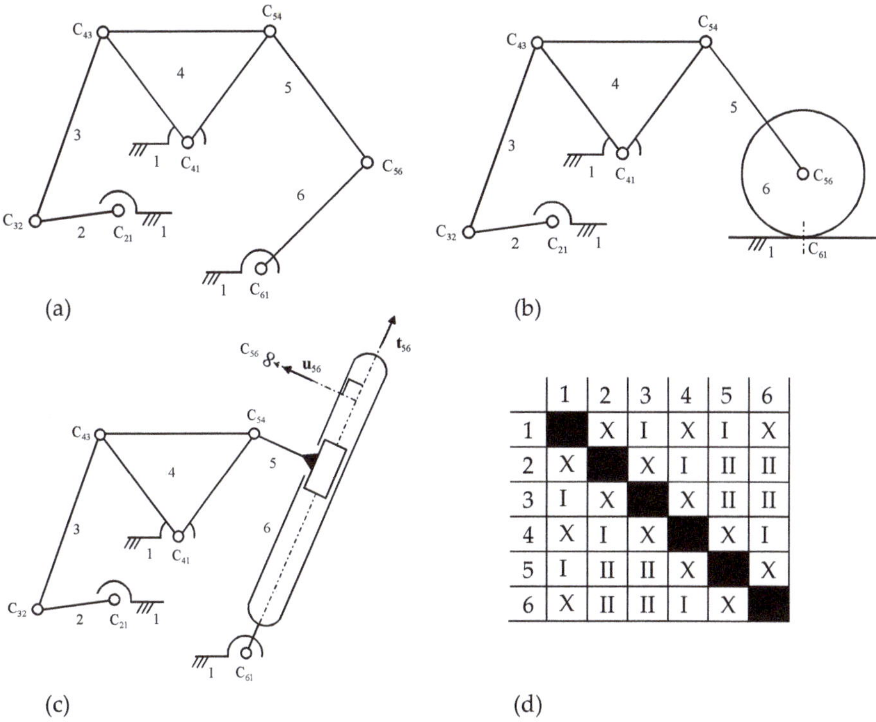

Figure 5. The three PMs (**a–c**) share the same table (**d**).

It is worth noting that the coordinates of the ICs with the same roman number (see Figure 5d) in the table can be simultaneously computed by using the coordinates of the primary ICs and/or of the secondary ICs with a lower roman number. Accordingly, once the table of Figure 4 has been filled, the roman numbers that appear in it give the sequence S_0 with which the secondary-ICs' positions must be computed.

The table of Figure 4 reveals the following pieces of information about the PM it refers to: the number and the type (binary, ternary, etc.) of links; how many two-DOF and one-DOF pairs (i.e., C_s and primary ICs) there are, and which links they join. Such pieces of information are sufficient to compute the DOF number with the Kutzbach criterion [2]. Since the table does not provide any piece of information about the types of one-DOF pairs the primary ICs refer to, it does not change if a R-pair is replaced by either a P-pair or a C_r contact and vice versa. Consequently, different PMs can be solved together by using the same table. For instance, the table of Figure 5d holds for the three PMs of Figure 5a–c and for many others. Consequently, once the table has been filled for one PM, it gives the IC sequence to use for computing the IC positions of all the PMs that share that table and the computation algorithm is much faster for all the other PMs of this family since the sequence S_0 is already known for all of them.

In addition, the proposed table is able to identify both the presence of an over con-straint (i.e., a structure or a substructure) and the DOF number of a multi-DOF PM. In-deed, if, during the filling procedure without assuming the position of any secondary IC (i.e., without inserting any starred roman number in the table), either of the following condition is found

(i)   a primary IC must lie on one or more known lines,

(ii)   a still-unknown secondary IC is found which must lie on three or more lines,

particular geometric conditions must be satisfied to make the primary IC lies on the known line(s) (case (i)) or to make the lines have a common intersection (case (ii)), that is, an over constraint is present. In this case, the order of the over constraint is given by the number of lines in case (i) and by the number of lines, which must have a common intersection, minus two in case (ii). Moreover, if, for determining the positions of all the secondary ICs, the positions of a number, say n, of secondary ICs must be assumed on known lines without reaching any over-constraint condition, the PM has n + 1 DOF.

2.1.2. Multi-DOF PMs

In the motion of a generic link j relative to another generic link i (see Figure 3), both belonging to one *l*-DOF PM, let $\omega_{ji}\,(=\dot{\theta}_{ji})$ and $^i\mathbf{v}_{O|j}$ denote the signed magnitude, positive if counterclockwise, of the angular velocity and the velocity, measured from link i, of a generic point, O, considered fixed to link j, respectively. The relationship between the 3-tuple $(\omega_{ji},\,^i\mathbf{v}_{O|j})$ and the rates, $\dot{q}_r$ for r = 1, … , *l*, of the generalized coordinates, q_r for r = 1, … , *l*, of the *l*-DOF PM is linear and homogeneous. Such a property makes the superposition principle applicable. In [22], the superposition principle has been exploited to deduce the instantaneous-kinematics relationships that relate an *l*-DOF PM to the *l* single-DOF PMs generated from the *l*-DOF PM by locking all the generalized coordinates but one. Hereafter, for the sake of brevity, the phrase "the r-th single-DOF PM" will mean "the single-DOF PM generated from the *l*-DOF PM by locking all the generalized coordinates, but the r-th one, that is, with $\dot{q}_k = 0$ for any $k\in\{1, … , l \mid k \neq r\}$."

Indeed, the application of this principle yields the following relationships (see Figure 6)

$$\omega_{ji} = \sum_{r=1,l} \omega_{ji,r} \tag{2a}$$

$$^i\mathbf{v}_{O|j} = \sum_{r=1,l} {}^i\mathbf{v}_{O|j}^{(r)} \tag{2b}$$

where $\omega_{ji,r}$ $({}^i\mathbf{v}_{O|j}^{(r)})$ is ω_{ji} $({}^i\mathbf{v}_{O|j})$ computed by locking all the generalized coordinates but q_r, that is, measured in the r-th single-DOF PM. From now on, $(\cdot)_{ji,r}$ will denote the quantity $(\cdot)_{ji}$ evaluated in the r-th single-DOF PM. By using complex numbers to represent planar vectors in the Cartesian/Argand reference Oxy (see Figure 6), the following relationships hold (hereafter, a bold letter will denote a planar vector and/or the corresponding complex number according to the context and $i = \sqrt{-1}$):

$$^i\mathbf{v}_{O|j} = -i\dot{h}_{ji}\mathbf{d}_{ji} \tag{3a}$$

$$^i\mathbf{v}_{O|j}^{(r)} = -i\dot{h}_{ji,r}\mathbf{d}_{ji,r} \tag{3b}$$

where, with reference to Figures 3 and 6, if $\omega_{ji} \neq 0$ ($\omega_{ji,r} \neq 0$), then $\dot{h}_{ji} = \omega_{ji} = \dot{\theta}_{ji}$ $(\dot{h}_{ji,r} = \omega_{ji,r} = \dot{\theta}_{ji,r})$ and $\mathbf{d}_{ji} = \mathbf{C}_{ji}$ $(\mathbf{d}_{ji,r} = \mathbf{C}_{ji,r})$, else $\dot{h}_{ji} = \dot{s}_{ji}$ $(\dot{h}_{ji,r} = \dot{s}_{ji,r})$ and $\mathbf{d}_{ji} = \mathbf{u}_{ji}$ $(\mathbf{d}_{ji,r} = \mathbf{u}_{ji,r})$. The introduction of Equations (3a) and (3b) into Equation (2b) transform system (2) as follows

$$\omega_{ji} = \sum_{r=1,l} \omega_{ji,r} \tag{4a}$$

$$\dot{h}_{ji}\mathbf{d}_{ji} = \sum_{r=1,l} \dot{h}_{ji,r}\mathbf{d}_{ji,r} \tag{4b}$$

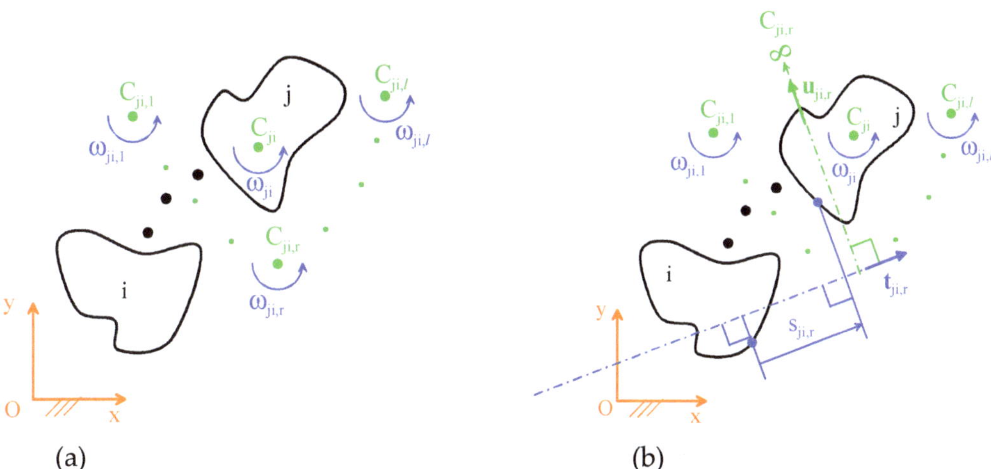

Figure 6. Instantaneous relative motion between two links, j and i, of an *l*-DOF PM ($\mathbf{C}_{ji,r}$ denotes the IC of this relative motion in the r-th single-DOF PM) when (**a**) $\mathbf{C}_{ji,r}$ is a finite point (rotation around $\mathbf{C}_{ji,r}$), and (**b**) $\mathbf{C}_{ji,r}$ is a point at infinity located by the unit vector $\mathbf{u}_{ji,r}$ (translation with the direction of the unit vector $\mathbf{t}_{ji,r}$).

Equations (4a) and (4b) make it possible to determine the position of $\mathbf{C}_{ji}$ in the *l*-DOF PM once the positions of all the $\mathbf{C}_{ji,r}$ for $r = 1, \ldots, l$ (i.e., of the same IC in all the single-DOF PMs generated from the *l*-DOF PM) are known together with the PM configuration and the generalized-coordinate rates $\dot{q}_r$ for $r = 1, \ldots, l$. It is worth stressing that this determination can be conducted even in the case $\mathbf{C}_{ji}$ is a point at infinity. This result, deduced in [22], together with the algorithm for the determination of the IC positions in single-DOF PM, presented in [21] and summarized in the previous subsection, makes it possible to determine analytically the IC positions in any PM.

In the case that $\omega_{ji} \neq 0$ and all the $\mathbf{C}_{ji,r}$ for $r = 1, \ldots, l$ are finite points of the motion plane, Equation (4) becomes

$$\omega_{ji} = \sum_{r=1,l} \omega_{ji,r} \tag{5a}$$

$$\omega_{ji}\mathbf{C}_{ji} = \sum_{r=1,l} \omega_{ji,r}\mathbf{C}_{ji,r} \tag{5b}$$

which bring the conclusion that, in this case, the following statement is true:

Statement 1. *From a geometric point of view, if the points $C_{ji,r}$ for $r = 1, \ldots, l$ are considered heavy points whose signed weight is $\omega_{ji,r}$, C_{ji} is the centroid of these heavy points.*

This conclusion makes it possible to employ the simple geometric rules of mass point geometry [41], which, for instance, yield that, in a 2-DOF PM, $\mathbf{C}_{ji}$ must lie on the line passing through $\mathbf{C}_{ji,1}$ and $\mathbf{C}_{ji,2}$.

2.2. Singularity Analysis

The instantaneous kinematics (IK) of a mechanism can be described by considering it as an input/output (I/O) device [24–26] where the input variables are the actuated-joint rates and the output variables are the ones that uniquely identify the twist of one link chosen as output link. In PMs and, in general, in all the mechanisms with holonomic and time-independent constraints, the analytic relationship (I/O instantaneous relationship) between actuated-joint rates and output-link twist is a linear and homogeneous system where the two coefficient matrices (Jacobians) that multiply the actuated-joint rates and the output-link twist, respectively, depend only on the mechanism configuration. The

I/O instantaneous relationship brings one to define two IK problems: the forward IK (FIK) problem and the inverse IK (IIK) problem. The FIK problem is the determination of the output-link twist for assigned actuated-joint rates; vice versa, the IIK problem is the determination of the actuated-joint rates for an assigned output-link twist.

In this context, the mechanism configurations where one or the other or both of the two Jacobians are rank deficient make the FIK or the IIK or both the IK problems undetermined and are the mechanism's singularities. Accordingly [24], three types of mechanisms' singularity are identifiable: (I) those that make the IIK problem undetermined (serial singularities), (II) those that make the FIK problem undetermined, (parallel singularities), and (III) those that make both the IIK and the FIK problems undetermined.

At a serial singularity, the output link has a limitation on the possible twists that it can assume, that is, it has a reduction in its instantaneous mobility. Such a condition identifies the workspace boundary. Consequently, serial singularities are configurations where the output link reaches the borders of its motion range, which, in single-DOF PMs, correspond to the extreme values (dead center positions) of one or more variables related to the output-link pose.

At a parallel singularity, the output-link twist can be different from zero even though all the actuated joints are locked, that is, the actuated joints are not able to control the output-link pose any longer and, somehow, the instantaneous mobility of the output link increases. Such singularity can occur inside the output-link workspace and must be avoided during the mechanism motion. In addition, parallel singularities are configurations where at least one of the input links (i.e., those links related to the actuated joints) reaches the borders of its motion range, which, in single-DOF PMs, correspond to the extreme values (dead center positions) of one or more input variables. According to Hunt's definitions [28], in single-DOF PMs, both serial and parallel singularities are stationary configurations; whereas, type-(III) singularities are uncertainty configurations.

The I/O instantaneous relationship of a single-DOF PM can be written in the following canonical form [31]:

$$b\dot{z} = a\dot{q} \tag{6}$$

where q is the generalized coordinate (input variable), z is any possible output variable; whereas, a and b are two coefficients uniquely depending on the PM configuration (i.e., on q). In Equation (6), the input (output) variable q (z) can be either an angle, θ_{ji} (Figure 3a), related to a relative rotation (Rot) or a linear variable, s_{ji} (Figure 3b), related to a relative translation (Tra). Therefore, according to the types of input and output variables, there are only four possible types of Equation (6) (the first (second) acronym refers to the type of input (output) variable): Rot-Rot, Rot-Tra, Tra-Rot, and Tra-Tra. Let "i", "o", "f", and "k" denote the indices of the input link, of the output link, of the reference link that is used to evaluate the rate of the input variable, and of the reference link that is used to evaluate the rate of the output variable, respectively. The positions of the six ICs corresponding to all the possible relative motions among these four links are geometrically related by the AK theorem (Figure 1) as shown in Figure 7. The analytic expressions (see [31] for demonstration and further details) of the coefficients a and b of Equation (6) are reported in Table 1 where $\mathbf{C}_{mn}$ ($\mathbf{t}_{mn}$) with mn∈{if, ok, ik, kf, oi, of} is the complex number that gives the IC position (the unit vector of the positive translation direction (Figure 3b) in the relative motion of link m with respect to link n. Also, as indicated in the note of Table 1, if, for special choices of the indices "i", "o", "f", and "k", a and b share common factors such factors must be simplified.

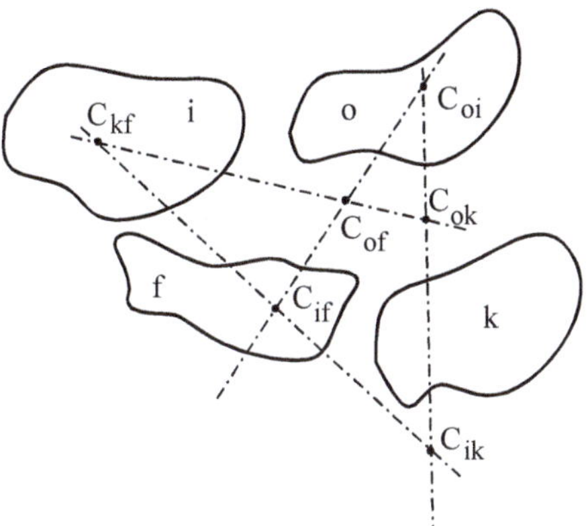

Figure 7. Geometric relationship among the positions of the six ICs corresponding to all the possible relative motions among the links i, o, f, and k.

Table 1. Analytic expressions of the coefficients a and b appearing in the I/O instantaneous relationships (Equation (6)) of single-DOF PMs deduced in [31] (C_{mn} (t_{mn}) with $mn \in \{if, ok, ik, kf, oi, of\}$ is the complex number that gives the IC position (the unit vector of the positive-translation direction (Figure 3b) of the relative motion of link m with respect to link n).

Input-Output	$\dot{q}$	$\dot{z}$	a [*]	b [*]
Rot-Rot	$\omega_{if} = \dot{\theta}_{if}$	$\omega_{ok} = \dot{\theta}_{ok}$	$(C_{of} - C_{if})(C_{oi} - C_{ik})$	$(C_{ok} - C_{ik})(C_{oi} - C_{of})$
Rot-Tra	$\omega_{if} = \dot{\theta}_{if}$	$\dot{s}_{ok}$	$(C_{of} - C_{if})(C_{oi} - C_{ik})$	$i\, t_{ok}\,(C_{oi} - C_{of})$
Tra-Rot	$\dot{s}_{if}$	$\omega_{ok} = \dot{\theta}_{ok}$	$i\, t_{if}\,(C_{oi} - C_{ik})$	$(C_{ok} - C_{ik})(C_{of} - C_{oi})$
Tra-Tra	$\dot{s}_{if}$	$\dot{s}_{ok}$	$i\, t_{if}\,(C_{oi} - C_{ik})$	$i\, t_{ok}\,(C_{of} - C_{oi})$

[*] If, for special choices of the indices "i", "o", "f", and "k", a and b share common factors such factors must be simplified.

Equation (6) reveals that, in single-DOF PMs, a serial (a parallel) singularity occurs at PM configurations for which the coefficient a (the coefficient b) is equal to zero and, accordingly, a type-III singularity occurs when both the coefficients a and b are equal to zero. These conditions and Table 1 yield the geometric conditions to check for identifying single-DOF PMs' singularities, through the IC positions, reported in Table 2.

Table 2. Geometric conditions on the IC positions that identify the singularities of single-DOF PMs.

Input-Output	$\dot{q}$	$\dot{z}$	a = 0 (Serial Singularity)	b = 0 (Parallel Singularity)
Rot-Rot	$\omega_{if} = \dot{\theta}_{if}$	$\omega_{ok} = \dot{\theta}_{ok}$	$C_{of} = C_{if}$ or $C_{oi} = C_{ik}$	$C_{ok} = C_{ik}$ or $C_{oi} = C_{of}$
Rot-Tra	$\omega_{if} = \dot{\theta}_{if}$	$\dot{s}_{ok}$	$C_{of} = C_{if}$ or $C_{oi} = C_{ik}$	$C_{oi} = C_{of}$
Tra-Rot	$\dot{s}_{if}$	$\omega_{ok} = \dot{\theta}_{ok}$	$C_{oi} = C_{ik}$	$C_{ok} = C_{ik}$ or $C_{of} = C_{oi}$
Tra-Tra	$\dot{s}_{if}$	$\dot{s}_{ok}$	$C_{oi} = C_{ik}$	$C_{of} = C_{oi}$

In an *l*-DOF PM, Equations (4a) and (4b) can be exploited to deduce the general I/O instantaneous relationship [22]. Indeed, the replacement of the subscript "ji" with "ok" in Equation (4) transforms them as follows:

$$\omega_{ok} = \sum_{r=1,l} \omega_{ok,r} \qquad (7a)$$

$$\dot{h}_{ok}\mathbf{d}_{ok} = \sum_{r=1,l} \dot{h}_{ok,r}\mathbf{d}_{ok,r} \tag{7b}$$

where, for each $\dot{h}_{ok,r}$, the following relationship, deduced from Equation (6), holds

$$b_r\dot{h}_{ok,r} = a_r\dot{q}_r \qquad r = 1, \ldots, l \tag{8}$$

with a_r and b_r that are the coefficients a and b of Table 1 when referred to the r-th single-DOF PM, that is, they are the ones reported in Table 3. It is worth noting that, in Table 3, the values of the indices "i" and "f" depend on the particular r-th single-DOF PM.

Table 3. Analytic expressions of the coefficients a_r and b_r appearing in the I/O instantaneous relationship (Equation (8)) of the r-th single-DOF PM ($C_{mn,r}$ ($t_{mn,r}$) with mn$\in${if, ok, ik, kf, oi, of} is the complex number that gives the IC position (the unit vector of the positive-translation direction (Figure 3b) of the relative motion of link m with respect to link n in the r-th single-DOF PM).

Input-Output	$\dot{q}_r$	$\dot{h}_{ok,r}$	a_r (*)	b_r (*)
Rot-Rot	$\omega_{if,r} = \dot{\theta}_{if,r}$	$\omega_{ok,r} = \dot{\theta}_{ok,r}$	$(C_{of,r} - C_{if,r})(C_{oi,r} - C_{ik,r})$	$(C_{ok,r} - C_{ik,r})(C_{oi,r} - C_{of,r})$
Rot-Tra	$\omega_{if,r} = \dot{\theta}_{if,r}$	$\dot{s}_{ok,r}$	$(C_{of,r} - C_{if,r})(C_{oi,r} - C_{ik,r})$	$i\,t_{ok,r}\,(C_{oi,r} - C_{of,r})$
Tra-Rot	$\dot{s}_{if,r}$	$\omega_{ok,r} = \dot{\theta}_{ok,r}$	$i\,t_{if,r}\,(C_{oi,r} - C_{ik,r})$	$(C_{ok,r} - C_{ik,r})(C_{of,r} - C_{oi,r})$
Tra-Tra	$\dot{s}_{if,r}$	$\dot{s}_{ok,r}$	$i\,t_{if,r}\,(C_{oi,r} - C_{ik,r})$	$i\,t_{ok,r}\,(C_{of,r} - C_{oi,r})$

(*) If, for special choices of the indices "i", "o", "f", and "k", a_r and b_r share common factors such factors must be simplified.

The introduction of the analytic expressions of $\dot{h}_{ok,r}$, for r = 1, ... , l, obtained from Equation (8) into Equations (7a) and (7b) transforms system (7) as follows

$$\omega_{ok} = \sum_{r=1,l} \frac{a_r}{b_r}\delta_r\dot{q}_r \tag{9a}$$

$$\dot{h}_{ok}\mathbf{d}_{ok} = \sum_{r=1,l} \frac{a_r}{b_r}\mathbf{d}_{ok,r}\dot{q}_r \tag{9b}$$

where δ_r is a binary digit that, if $\omega_{ok,r} \neq 0$, is equal to 1, else (i.e., if $\omega_{ok,r} = 0$) is equal to 0. System (9), with simple algebraic manipulations, can be put in the following form

$$\left(\prod_{\substack{p=1,l \\ \delta_p \neq 0}} b_p \right)\omega_{ok} = \sum_{r=1,l} a_r \left(\prod_{\substack{p=1,l \\ \delta_p \neq 0 \,\&\, p \neq r}} b_p \right)\delta_r\dot{q}_r \tag{10a}$$

$$\left(\prod_{p=1,l} b_p \right)\dot{h}_{ok}\mathbf{d}_{ok} = \sum_{r=1,l} a_r \left(\prod_{\substack{p=1,l \\ p \neq r}} b_p \right)\mathbf{d}_{ok,r}\dot{q}_r \tag{10b}$$

which constitute the I/O instantaneous relationship of the *l*-DOF PM expressed through the ICs. In this relationship, the output-link twist is given through the three-tuple (ω_{ok}, $\dot{h}_{ok}\mathbf{d}_{ok}$).

In [22], the analysis of Equation (10) brought this author to demonstrate the following statements and theorems (see [22] for the demonstrations):

Statement 2. *The union of the serial-singularity sets of all the l single-DOF PMs generated from an l-DOF PM is a subset of the set of all the serial singularities of that l-DOF PM.*

Theorem 1. *For $l \geq 3$ and provided that all the a_r and b_r coefficients are different from zero, a serial singularity of the l-DOF PM occurs if and only if three ICs, $C_{ok,r}$, are either aligned or all ideal points.*

Statement 3. *The set of all the parallel singularities of an l-DOF PM is the union of the l sets of parallel singularities of the single-DOF PMs generated from the l-DOF PM.*

Theorem 2. *The coincidence of all the $C_{ok,r}$, $r = 1, \ldots , l$, (the $C_{ok,r}$ that are point at infinity are included) identifies a particular parallel singularity where all the b_r coefficients vanish.*

It is worth noting that, since Theorem 1 identifies all the serial singularities that are not serial singularities of any single-DOF PM generated from the l-DOF PM, in two-DOF PMs, the serial-singularity set is the union of the two serial-singularity sets of the two single-DOF PMs generated from the two-DOF PM.

Eventually, both the IIK and the FIK problems are unsolvable (i.e., a type-III singularity occurs) when, at a given configuration, at least one serial-singularity condition together with at least one parallel-singularity condition occur. In particular, if at least one out of the b_r coefficients together with at least one out of the a_r coefficients are equal to zero, a type-(III) singularity will occur. Therefore, the following statement holds [22]:

Statement 4. *The union of the l sets of type-(III) singularities of the single-DOF PMs generated from an l-DOF PM is a subset of the set of all the type-(III) singularities of the l-DOF PM.*

The above-reported statements and theorems demonstrate that the singularities of an l-DOF PM can be collected into two classes: (a) configurations that are singular for at least one single-DOF PM generated from the l-DOF PM, and (b) singularities of the l-DOF PM that are not singularities of any single-DOF PM generated from that l-DOF PM (i.e., those identified by Theorem 1). In addition, all the above-listed singularity conditions relate l-DOF PM's singularities to the IC positions in the single-DOF PMs generated from the l-DOF PM. By exploiting the fact that all the ICs of single-DOF PMs can be analytically/geometrically determined [21] and that the coefficients a_r and b_r together with their vanishing conditions are expressible through the IC of single-DOF PMs (see Tables 1–3), an analytic/geometric algorithm for determining l-DOF PM's singularities has been proposed in [22]. Such an algorithm systematically uses IC positions of single-DOF PMs for identifying the singularities of any PM.

The above-reported singularity conditions are illustrated in Figure 8 through the 3-$\underline{R}$RR parallel PM, which is a 3-DOF PM. In this PM (see Figure 8a), links 7 and 0 are chosen as output link (i.e., o = 7) and reference link for all the variables (i.e., k = f = 0); whereas, q_r for r = 1, 2, 3 are the actuated-joint (input) variables, and, in the r-th single-DOF PM for r = 1, 2, 3, the input link is link r. So, the three $C_{ok,r}$ for r = 1, 2, 3 are $C_{70,1}$, $C_{70,2}$, and $C_{70,3}$ (see Figure 8). With these choices, the analysis of Figure 8 brings to the following conclusions. The configuration of Figure 8a does not match any singular condition (i.e., it is a non-singular configuration). The configuration of Figure 8b satisfies the singularity condition $C_{oi,1} = C_{ik,1}$ ($C_{71,1} = C_{10,1}$) of Table 2, that is, it is a serial singularity of the 3-DOF PM, which is also a serial singularity of the 1st single-DOF PM. The configuration of Figure 8c satisfies the geometric condition of Theorem 2 (i.e., the coincidence of all the $C_{ok,r}$) thus, it is a parallel singularity both of the 3-DOF PM and of all the single-DOF PMs. The configuration of Figure 8d satisfies the geometric condition of Theorem 2 (i.e., the coincidence of all the $C_{ok,r}$) and the serial-singularity condition $C_{oi,1} = C_{ik,1}$ ($C_{71,1} = C_{10,1}$) of Table 2, that is, it is a type-(III) singularity.

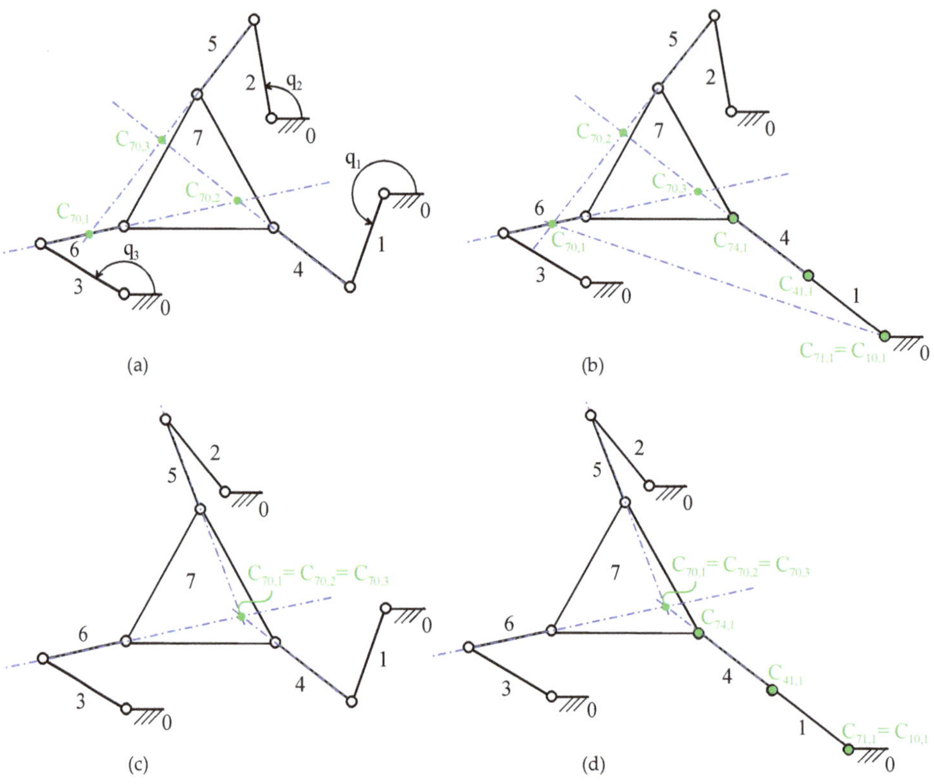

Figure 8. The 3-<u>R</u>RR parallel PM: (**a**) non-singular configuration, (**b**) serial singularity that is also a serial singularity of the 1st single-DOF PM (i.e., the one obtained by locking q_2 and q_3), (**c**) parallel singularity that satisfy Theorem 2, (**d**) type-(III) singularity that satisfies Theorem 2 and, simultaneously, is a serial singularity of the 1st single-DOF PM.

3. Influence of IC Locations on Planar-Mechanism Dynamics

At LaMaViP, the research on ICs properties, over PMs' kinematics, addressed also PMs' dynamics and has brought to provide an exhaustive set of tools for addressing/modelling PMs' dynamics [35–38,42]. Since a complete presentation of these tools would require much more room in term of pages than what would be reasonable for a journal paper, the present review was intended to be mainly focused on PMs' kinematics. Nevertheless, a short summary of the results obtained on PMs' dynamic is necessary and is presented in this section.

3.1. Single-DOF PMs

The analytic determination of ICs' positions [21] together with Equation (1) makes the analytic determination of all the VCs in single-DOF PMs possible. The VCs together with their derivative with respect to the generalized coordinate of the single-DOF PM are the only terms necessary to build the dynamic model of a single-DOF PM by using the Eksergian's equation [43]. This approach has been adopted in [35] to deduce a general dynamic model for single-DOF PMs based on ICs' positions, which explicitly appear in it. In [35], over the model, the algorithms for systematically solving the two main dynamics problems (i.e., inverse and direct dynamics problems [15]) with this approach have been presented. Then, in [37], by expressing the acceleration of each link's barycenter through the acceleration of the IC of the relative motion between the same link and the frame, single-DOF PMs' dynamic models deduced both from Newton-Euler formulation and from D'Alembert's principle have been built which systematically use ICs' position. All the dynamic approaches proposed in [35,37] relate single-DOF PMs' dynamic behavior to ICs' positions, but are mainly analytic.

From a different point of view, in [38], D' Alembert's principle has been revisited and by using a new type of diagram, named "active-load diagram", a novel geometric and analytic technique for studying single-DOF PMs' dynamics has been presented which refers PM's dynamic behavior to ICs' positions. An active load is, by definition, any internal or external load (force or moment) that provides a non-null virtual work when the mechanism changes its configuration. Accordingly, the active-load diagram of a single-DOF PM is a sketch of the PM, at a given configuration, that also show the active internal loads in the joints and, for each mobile link, all the pieces of information (active loads, C_{j1} or $\mathbf{u}_{j1}$, etc.) reported in Figure 9, which are necessary to systematically write the terms appearing in D' Alembert's principle. Such a diagram makes it possible to write immediately the dynamic model of the single-DOF PM without any analytic consideration. Indeed, for instance, Figure 10 shows the active-load diagram of a shaper mechanism where the active loads are the external forces $\mathbf{F}_4$ and $\mathbf{F}_6$, and the moment M_{12}, applied inside the actuated joint, and where, for the sake of simplicity, the hypothesis that inertia forces are negligible is introduced. With reference to Figure 10, from a geometric point of view, the following model of the shaper mechanism can be immediately deduced (here, $F_4 = |\mathbf{F}_4|$ and $F_6 = |\mathbf{F}_6|$):

$$M_{12} = F_4 b_4 \frac{\overline{C_{42}C_{21}}}{\overline{C_{42}C_{41}}} + F_6 b_6 \tag{11}$$

which gives the following analytic model (here, (x_A, y_A) are the coordinates of a generic point, A, and $(F_{j,x}, F_{j,x})$ are the components of the force $\mathbf{F}_j$ measured in the Cartesian reference $C_{21}xy$ fixed to link 1 (see Figure 10)

$$M_{12} = -\left[(x_{P_4} - x_{C_{41}})F_{4,y} - (y_{P_4} - y_{C_{41}})F_{4,x}\right]\frac{\overline{C_{42}C_{21}}}{\overline{C_{42}C_{41}}} - \left[(x_{C_{62}} - x_{C_{21}})F_{6,y} - (y_{C_{62}} - y_{C_{21}})F_{6,x}\right] \tag{12}$$

where the ratio $\frac{\overline{C_{42}C_{21}}}{\overline{C_{42}C_{41}}}$ is the VC $\frac{\dot{\theta}_{41}}{\dot{\theta}_{21}}$ and can be analytically expressed through the first formula of Equation (1).

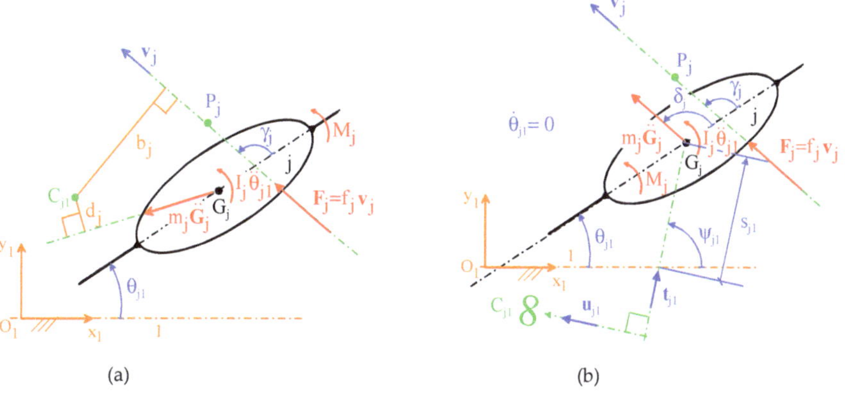

(a)   (b)

Figure 9. Active-load diagram of the j-th link of a single-DOF PM (link 1 is the frame and $O_{1}x_1y_1$ is a Cartesian reference fixed to link 1; $\mathbf{F}_j$ (M_j) is a generic external force (moment) applied to link j; m_j, G_j, and I_j are mass, barycenter and mass moment of inertia about G_j of the j-th link, respectively): (**a**) instantaneous rotation of the j-th link, (**b**) instantaneous translation of the j-th link.

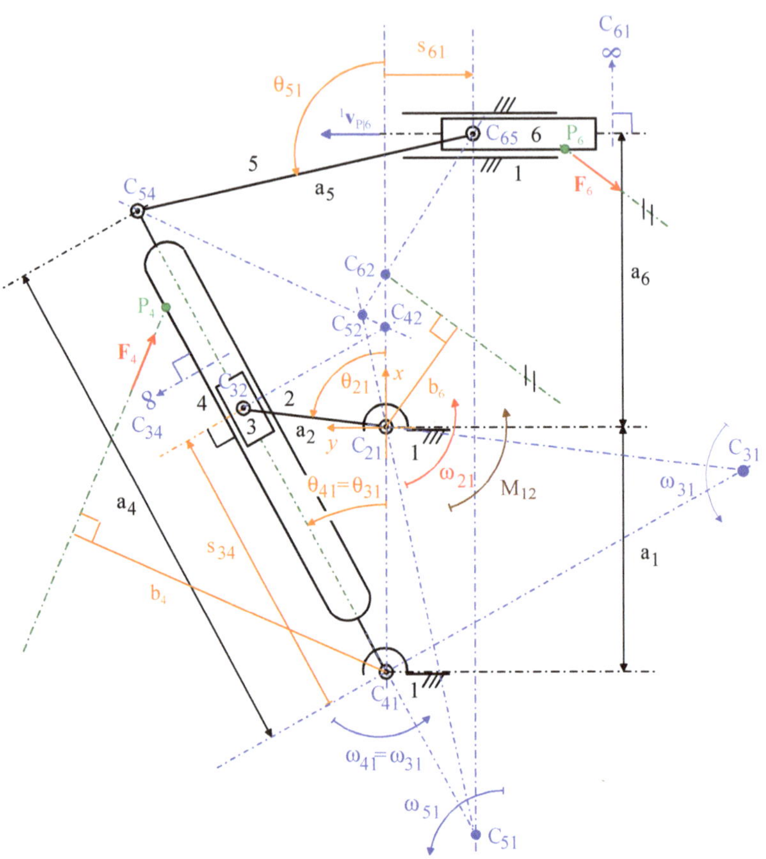

Figure 10. Active-load diagram of the shaper mechanism.

With reference to Figure 10 and the above-reported relationships, the necessary ICs that are not primary ICs are only two: C_{31} and C_{51}. Indeed, in this case, since $\theta_{31} = \theta_{41}$, the relationship $\frac{\overline{C_{42}C_{21}}}{\overline{C_{42}C_{41}}} = \frac{\overline{C_{32}C_{21}}}{\overline{C_{32}C_{31}}}$ holds. Nevertheless, it is worth noting that, when the method is applied to solve only one configuration, it is possible to determine some additional IC (in this case C_{62}) to write also the term associated to the active forces applied to a translating link (link 6) in the form of a moment about an IC. Indeed, in this case, the virtual work of force $\mathbf{F}_6$ can be computed as follows:

$$\delta L_{\mathbf{F}_6} = \mathbf{F}_6 \cdot \delta \mathbf{P}_6 = \delta \theta_{21} \mathbf{F}_6 \cdot [\mathbf{k} \times (C_{62} - C_{21})] = \mathbf{k} \cdot [(C_{62} - C_{21}) \times \mathbf{F}_6] \delta \theta_{21} = -F_6 b_6 \delta \theta_{21} \tag{13}$$

where $\mathbf{k}$ is the unit vector perpendicular to the motion plane that points toward the reader and $F_6 = |\mathbf{F}_6|$.

For a single-DOF PM with m links where link 1 is the frame, all the approaches presented in [35,37,38] require the determination of the positions of all the C_{j1} for j = 2, . . . , m (i.e., of (m − 1) ICs). Nevertheless, only a few of these ICs are secondary ICs and really require additional computations. In addition, as it is shown in the case studies presented in [35,38], these additional computations bring to determine VCs that are necessary to write the model; therefore, they are not additional in practice.

In [38], how to write the model in an analytical and systematic form is presented together with the algorithms for solving the two main dynamics problems by using the proposed approach. Eventually, the active-load diagrams, presented in [38], (see Figures 9 and 10, and

Equation (11)) immediately reveal that an external force $\mathbf{F}_j$, applied to link j, whose line of action passes through C_{j1} (i.e., with $b_j = 0$ (see Figure 9a), if the link rotates, or with $\mathbf{v}_j = \mathbf{u}_{j1}$ (see Figure 9b), if the link translates) is inactive. That is, it is directly equilibrated by the mechanism structure without requiring the application of a generalized torque in the actuated joint. This result is relevant in PMs' design since it provides a criterion for reducing the generalized torque in the actuated joint.

3.2. Multi-DOF PMs

At LaMaViP, two formulations [36,42] have been developed which relate multi-DOF PMs' dynamics to ICs' position. The first one [36] introduces, for planar motion, rigid-body's (link's) configuration space (c-space), which is a three-dimensional space whose coordinates are the "pose coordinates" of the link, that is the two coordinates, say (x, y), of a point fixed to the link plus the slope angle, say $\theta \in\,]-\pi, +\pi]$, which identifies the link orientation. Then, it gives the structure of an affine space to the c-space by defining three-dimensional vectors, named planar-pose vectors. Eventually, by using the planar wrench, it formulates the "motion laws" of c-space's point. Such laws make it possible to study PMs' dynamics as a number of c-space points that interact one another and stand external planar wrenches. This approach, over providing a coherent, autonomous and self-contained set of tools and laws that model PMs' kinematics and dynamics, has brought to confirm the above-reported statement 1 and to demonstrate the following new statement on ICs (see [36] for details and examples):

Statement 5. *A planar passive kinematic chain with l-DOF can transmit between links j and i only a force system whose central axis passes through all the instant centers $C_{ji,r}$ for r = 1, ... , l.*

Statement 5 implies the following corollaries:

(5.a) a single-DOF planar passive kinematic chain can transmit only force systems whose central axis belongs to the pencil of lines passing through the unique C_{ji};

(5.b) a two-DOF planar passive kinematic chain can transmit only one force with line of action that passes through $C_{ji,1}$ and $C_{ji,2}$;

(5.c) an l-DOF planar passive kinematic chain with $l \geq 3$ cannot transmit any force system unless the ICs $C_{ji,r}$ for r = 1, ... , l are all aligned (i.e., only in particular configurations).

The second formulation [42], by reinterpreting D' Alembert's principle, relates the dynamics behavior of an l-DOF PM to the ones of the l single-DOF PMs generated from the l-DOF PM and uses the "active-load diagrams" of these single-DOF PMs to provide a geometric interpretation that fully discloses the role of ICs in multi-DOF PMs' dynamics. Such diagrams can also be used to write immediately the dynamic model of the l-DOF PM without any analytic consideration. Indeed, for instance, Figure 11a–c show the active-load diagrams of the three single-DOF PMs generated from the 3-$\underline{R}$RR parallel PM (Figure 8) with link 7 (mobile platform) loaded by a force system whose resultant force, located on its central axis, is force $\mathbf{F}_7$ of Figure 11 and by the generalized torques, τ_r for r = 1, 2, 3, applied in the actuated joints. With reference to Figure 11, from a geometric point of view, according to this formulation (see [42] for demonstrations) the following model of the 3-$\underline{R}$RR parallel PM can be immediately deduced (the signs refer to the configuration of Figure 11 and $F_7 = |\mathbf{F}_7|$):

$$\begin{cases} \tau_1 = F_7 b_{7,1}\dfrac{\dot{\theta}_{70,1}}{\dot{\theta}_{10,1}} = -F_7 b_{7,1}\dfrac{\overline{C_{71,1}C_{10,1}}}{\overline{C_{71,1}C_{70,1}}} \\[4mm] \tau_2 = F_7 b_{7,2}\dfrac{\dot{\theta}_{70,2}}{\dot{\theta}_{20,2}} = F_7 b_{7,2}\dfrac{\overline{C_{72,2}C_{20,2}}}{\overline{C_{72,2}C_{70,2}}} \\[4mm] \tau_3 = -F_7 b_{7,3}\dfrac{\dot{\theta}_{70,3}}{\dot{\theta}_{30,3}} = F_7 b_{7,3}\dfrac{\overline{C_{73,3}C_{30,3}}}{\overline{C_{73,3}C_{70,3}}} \end{cases} \qquad (14)$$

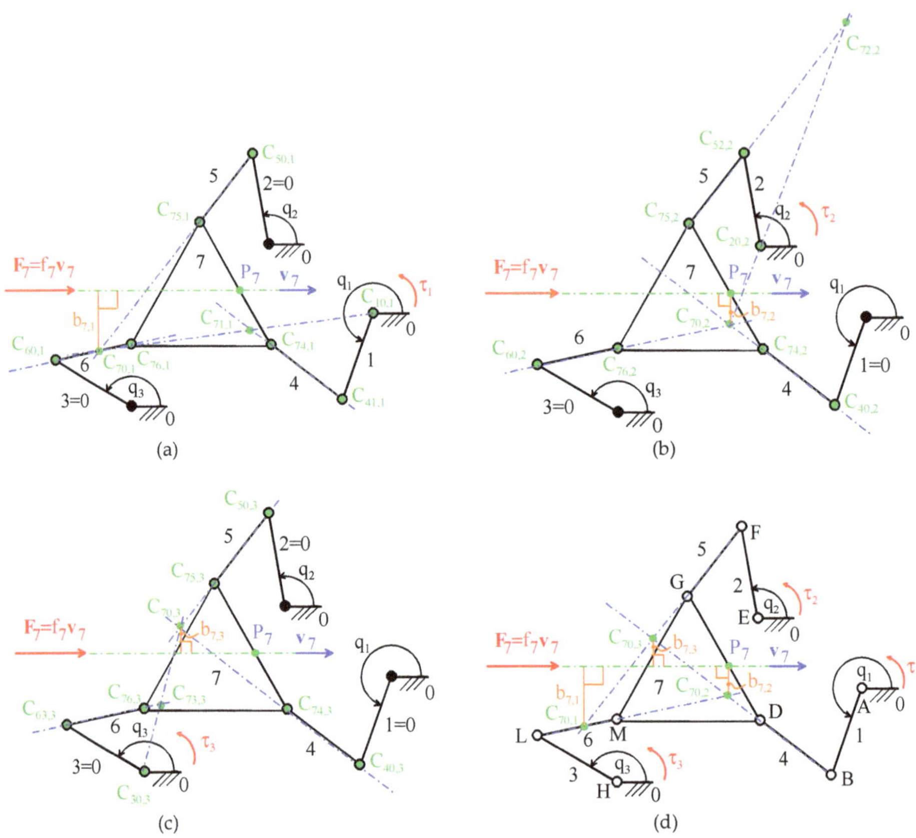

Figure 11. Active-load diagrams of the three single-DOF PMs generated from the 3-RRR parallel PM: (**a**) r = 1 (i.e., q_2 and q_3 are locked), (**b**) r = 2 (i.e., q_1 and q_3 are locked), (**c**) r = 3 (i.e., q_1 and q_2 are locked), (**d**) extract of the previous diagrams that is useful to visualize how an active force is redistributed among the actuators.

Moreover, in [42], how to write the model in an analytical and systematic form is presented together with the algorithms for solving the two main dynamics problems by using the proposed formulation. Eventually, let link 0 be the frame, this formulation immediately reveals (see Figure 11 and Equation (14)) that an external force $\mathbf{F}_j$, applied to link j, whose line of action passes through $C_{j0,r}$ (i.e., with $b_{j,r} = 0$ (see Figure 11)) has no effect on the r-th generalized torque τ_r. This consideration can be further detailed to give a clear graphical idea on how an active load is redistributed among the actuators. Indeed, diagrams such as the ones of Figure 11d where the distances $b_{j,r}$, of $\mathbf{F}_j$ from the ICs $C_{j0,r}$, for r = 1, . . . , l, are drawn all together can be used during design to optimize the load distribution among the actuators.

4. Conclusions

At LaMaViP (Laboratory of Mechatronics and Virtual Prototyping) of Ferrara University, Italy, the research on planar mechanisms' (PMs) kinematics and dynamics of the last two decades has brought to highlight how the instant-centers' (ICs) positions can be exploited to identify relevant properties of PMs' behavior. The results of this research have been summarized/reviewed here to give the reader a guide to use them in PMs' analysis and design.

In particular, an exhaustive technique for analytically determining IC positions, even for indeterminate PMs, has been obtained and the relationship between one IC in a multi-DOF PM and the same IC in the single-DOF PMs generated from that multi-DOF PM has been identified.

The geometric conditions that ICs' positions satisfy when a PMs is at a singular configuration have been exhaustively enumerated and analytic techniques that exploit such conditions to implement PMs' singularity analysis have been developed.

The role that ICs' positions play to determine PMs' dynamic behavior has been fully disclosed and PM-dynamics formulations that explicitly contain ICs' positions have been developed.

Such findings provide a complete set of tools that can be easily used for PMs' analysis and design.

Funding: This research was funded by University of Ferrara (UNIFE), FAR2020 and developed at the Laboratory of Mechatronics and Virtual Prototyping (LaMaViP), Department of Engineering, UNIFE.

Institutional Review Board Statement: Not applicable.

Informed Consent Statement: Not applicable.

Data Availability Statement: Not applicable.

Conflicts of Interest: The authors declare no conflict of interest. The funders had no role in the design of the study; in the collection, analyses, or interpretation of data; in the writing of the manuscript, or in the decision to publish the results.

References

1. Klein, A.W. *Kinematics of Machinery*; McGraw-Hill Book Company, Inc.: New York, NY, USA, 1917.
2. Paul, B. *Kinematics and Dynamics of Planar Machinery*; Prentice-Hall, Inc.: Englewood Cliffs, NJ, USA, 1987.
3. Gans, R.F. *Analytical Kinematics: Analysis and Synthesis of Planar Mechanisms*; Butterworth-Heinemann Inc.: Boston, MA, USA, 1991.
4. Yan, H.-S.; Hsu, M.-H. An analytical method for locating instantaneous velocity centers. In Proceedings of the 22nd ASME Biennial Mechanisms Conference, Scottsdale, AZ, USA, 13–16 September 1992; Volume 47, pp. 353–359.
5. Foster, D.E.; Pennock, G.R. A Graphical Method to Find the Secondary Instantaneous Centers of Zero Velocity for the Double Butterfly Linkage. *ASME J. Mech. Des.* **2003**, *125*, 268–274. [CrossRef]
6. Pennock, G.R.; Kinzel, E.C. Path curvature of the single flier eight-bar linkage. *ASME J. Mech. Des.* **2004**, *126*, 470–477. [CrossRef]
7. Foster, D.E.; Pennock, G.R. Graphical methods to locate the secondary instantaneous centers of single-degree-of-freedom indeterminate linkages. *ASME J. Mech. Des.* **2005**, *127*, 249–256. [CrossRef]
8. Butcher, E.A.; Hartman, C. Efficient enumeration and hierarchical classification of planar simple-jointed kinematic chains: Application to 12- and 14-bar single degree-of-freedom chains. *Mech. Mach. Theory* **2005**, *40*, 1030–1050. [CrossRef]
9. Cardona, A.; Pucheta, M. An automated method for type synthesis of planar linkages based on a constrained subgraph isomorphism detection. *Multibody Syst. Dyn.* **2007**, *18*, 233–258.
10. Murray, A.P.; Schmiedeler, J.P.; Korte, B.M. Kinematic Synthesis of Planar, Shape-Changing Rigid-Body Mechanisms. *ASME J. Mech. Des.* **2008**, *130*, 032302. [CrossRef]
11. Kong, X.; Huang, C. Type synthesis of single-DOF single-loop mechanisms with two operation modes. In Proceedings of the ASME/IFToMM International Conference on Reconfigurable Mechanisms and Robots, ReMAR 2009, London, UK, 22–24 June 2009; pp. 136–141.
12. Wolbrecht, E.T.; Reinkensmeyer, D.J.; Perez-Gracia, A. Single degree-of-freedom exoskeleton mechanism design for finger rehabilitation. In Proceedings of the 2011 IEEE International Conference on Rehabilitation Robotics, Zurich, Switzerland, 29 June–1 July 2011; pp. 1–6. [CrossRef]
13. Wang, J.; Ting, K.-L.; Zhao, D. Equivalent Linkages and Dead Center Positions of Planar Single-Degree-of-Freedom Complex Linkages. *ASME J. Mech. Robot.* **2015**, *7*, 044501. [CrossRef]
14. St-Onge, D.; Gosselin, C.M. Synthesis and Design of a One Degree-of-Freedom Planar Deployable Mechanism with a Large Expansion Ratio. *ASME J. Mech. Robot.* **2016**, *8*, 021025. [CrossRef]
15. Nikravesh, P.E. *Planar Multibody Dynamics: Formulation, Programming, and Applications*; CRC Press: Boca Raton, FL, USA, 2008.
16. Flores, P.; Seabra, E. *Dynamics of Planar Multibody Systems*; VDM: Saarbrucken, Germany, 2010.
17. Dixon, J.C. *Suspension Geometry and Computation*; John Wiley & Sons Ltd.: Chichester, UK, 2009.
18. Windrich, M.; Grimmer, M.; Christ, O.; Rinderknecht, S.; Beckerle, P. Active lower limb prosthetics: A systematic review of design issues and solutions. *Biomed. Eng. Online* **2016**, *15*, 5–19. [CrossRef]
19. Watson, P.C. Remote Center Compliant System. U.S. Patent US4098001, 4 July 1978.
20. Kumar Mallik, A.; Ghosh, A.; Dittirich, G. *Kinematic Analysis and Synthesis of Mechanisms*; CRC: Boca Raton, FL, USA, 1994.

21. Di Gregorio, R. An Algorithm for Analytically Calculating the Positions of the Secondary Instant Centers of Indeterminate Linkages. *ASME J. Mech. Des.* **2008**, *130*, 042303. [CrossRef]
22. Di Gregorio, R. A novel method for the singularity analysis of planar mechanisms with more than one degree of freedom. *Mech. Mach. Theory* **2009**, *44*, 83–102. [CrossRef]
23. Gosselin, C.; Angeles, J. A Global Performance Index for the Kinematic Optimization of Robotic Manipulators. *ASME J. Mech. Des.* **1991**, *113*, 220–226. [CrossRef]
24. Gosselin, C.M.; Angeles, J. Singularity analysis of closed-loop kinematic chains. *IEEE Trans. Robot. Autom.* **1990**, *6*, 281–290. [CrossRef]
25. Ma, O.; Angeles, J. Architecture singularities of platform manipulators. In Proceedings of the 1991 IEEE International Conference on Robotics and Automation (ICRA1991), Sacramento, CA, USA, 9–11 April 1991; pp. 1542–1547.
26. Zlatanov, D.; Fenton, R.G.; Benhabib, B. A Unifying Framework for Classification and Interpretation of Mechanism Singularities. *ASME J. Mech. Des.* **1995**, *117*, 566–572. [CrossRef]
27. Bonev, I.A.; Zlatanov, D.; Gosselin, C.M.M. Singularity Analysis of 3-DOF Planar Parallel Mechanisms via Screw Theory. *ASME J. Mech. Des.* **2003**, *125*, 573–581. [CrossRef]
28. Hunt, K.H. *Kinematic Geometry of Mechanisms*; Oxford University Press: Oxford, UK, 1978; ISBN 0-19-856124-5.
29. Yan, H.-S.; Wu, L.-I. The stationary configurations of planar six-bar kinematic chains. *Mech. Mach. Theory* **1988**, *23*, 287–293. [CrossRef]
30. Yan, H.-S.; Wu, L.-I. On the dead-center positions of planar linkage mechanisms. *ASME J. Mech. Transm. Automat. Des.* **1989**, *111*, 40–46. [CrossRef]
31. Di Gregorio, R. A novel geometric and analytic technique for the singularity analysis of one-dof planar mechanisms. *Mech. Mach. Theory* **2007**, *42*, 1462–1483. [CrossRef]
32. Simionescu, P.A.; Talpasanu, I.; Di Gregorio, R. Instant-Center Based Force Transmissivity and Singularity Analysis of Planar Linkages. *ASME J. Mech. Robot.* **2010**, *2*, 021011. [CrossRef]
33. Di Gregorio, R. Systematic use of velocity and acceleration coefficients in the kinematic analysis of single-DOF planar mechanisms. *Mech. Mach. Theory* **2019**, *139*, 310–328. [CrossRef]
34. Di Gregorio, R. Kinematic analysis of multi-DOF planar mechanisms via velocity-coefficient vectors and acceleration-coefficient Jacobians. *Mech. Mach. Theory* **2019**, *142*, 103617. [CrossRef]
35. Di Gregorio, R. A Novel Dynamic Model for Single Degree-of-Freedom Planar Mechanisms Based on Instant Centers. *ASME J. Mech. Robot.* **2016**, *8*, 011013. [CrossRef]
36. Di Gregorio, R. Kinematics and dynamics of planar mechanisms reinterpreted in rigid-body's configuration space. *Meccanica* **2015**, *51*, 993–1005. [CrossRef]
37. Di Gregorio, R. On the Use of Instant Centers to Build Dynamic Models of Single-dof Planar Mechanisms. In *Advances in Robot Kinematics 2018 (ARK2018), Proceedings of the 16th International Symposium on Advances in Robot Kinematics, Bologna, Italy, 1–5 July 2018*; Lenarcic, J., Parenti-Castelli, V., Eds.; Springer: Cham, Switzerland, 2018; Volume 8, pp. 242–249.
38. Di Gregorio, R. A geometric and analytic technique for studying single-DOF planar mechanisms' dynamics. *Mech. Mach. Theory* **2021**, *168*, 104609. [CrossRef]
39. Ardema, M.D. *Newton-Euler Dynamics*; Springer: New York, NY, USA, 2005.
40. Barton, L.O. *Mechanism Analysis: Simplified Graphical and Analytical Techniques*; Marcel Dekker Inc.: New York, NY, USA, 1993; ISBN 9780429102493.
41. Coxeter, H.S.M. *Introduction to Geometry*, 2nd ed.; John Wiley & Sons Inc.: New York, NY, USA, 1969; ISBN 9780471504580.
42. Di Gregorio, R. A geometric and analytic technique for studying multi-DOF planar mechanisms' dynamics. *Mech. Mach. Theory* **2022**, *176*, 104975. [CrossRef]
43. Eksergian, R. Dynamical Analysis of Machines. Ph.D. Thesis, Clark University, Worcester, MA, USA, 1928.

Article

Kinematic Modeling and Motion Planning of the Mobile Manipulator Agri.Q for Precision Agriculture

Giovanni Colucci *, Andrea Botta, Luigi Tagliavini, Paride Cavallone, Lorenzo Baglieri and Giuseppe Quaglia

Department of Mechanical and Aerospace Engineering, Politecnico di Torino, 10129 Torino, Italy;
andrea.botta@polito.it (A.B.); luigi.tagliavini@polito.it (L.T.); paride.cavallone@polito.it (P.C.);
lorenzo.baglieri@studenti.polito.it (L.B.); giuseppe.quaglia@polito.it (G.Q.)
* Correspondence: giovanni_colucci@polito.it

Abstract: In recent years, the study of robotic systems for agriculture, a modern research field often shortened as "precision agriculture", has become highly relevant, especially for those repetitive actions that can be automated thanks to innovative robotic solutions. This paper presents the kinematic model and a motion planning pipeline for a mobile manipulator specifically designed for precision agriculture applications, such as crop sampling and monitoring, formed by a novel articulated mobile base and a commercial collaborative manipulator with seven degrees of freedom. Starting from the models of the two subsystems, characterized by an adjustable position and orientation of the manipulator with respect to the mobile base, the linear mapping that describes the differential kinematics of the whole custom system is expressed as a function of the input commands. To perform pick–and–place tasks, a motion planning algorithm, based on the manipulator manipulability index mapping and a closed form inverse kinematics solution is presented. The motion of the system is based on the decoupling of the base and the arm mobility, and the paper discusses how the base can be properly used for manipulator positioning purposes. The closed form inverse kinematics solution is also provided as an open-source Matlab code.

Keywords: precision agriculture; mobile manipulation; motion planning; analytic Jacobian; inverse kinematics; manipulability

Citation: Colucci, G.; Botta, A.; Tagliavini, L.; Cavallone, P.; Baglieri, L.; Quaglia, G. Kinematic Modeling and Motion Planning of the Mobile Manipulator Agri.Q for Precision Agriculture. *Machines* **2022**, *10*, 321. https://doi.org/10.3390/machines10050321

Academic Editors: Marco Ceccarelli, Giuseppe Carbone and Alessandro Gasparetto

Received: 22 March 2022
Accepted: 23 April 2022
Published: 29 April 2022

Publisher's Note: MDPI stays neutral with regard to jurisdictional claims in published maps and institutional affiliations.

1. Introduction

The development of mobile manipulators, defined as complex robotic systems composed by a mobile base and a robotic manipulator mounted upon it, has become a relevant research field inside the service robotics world. They were first invented at the end of 1990s for industrial applications, and with the development of the so-called Autonomous Industrial Mobile Manipulators (AIMM) [1], their relevance is strictly related to the evolution of industrial production processes, currently characterized by the introduction of digital technologies, e.g., the integration of autonomous robotic systems into the traditional production plants for massive and repetitive tasks. To meet the current demand of product customization and differentiation, researchers and manufacturers from all the world have created several AIMM solutions, that integrate autonomous mobility, exteroceptive and proprioceptive sensing, and dexterous manipulation [2–4]. Such systems can autonomously navigate inside a production plant, perform several tasks, e.g., packaging, painting, and logistics, and can be easily set up and programmed to change their main purpose.

Moreover, service robotics can take advantage of such complex systems and their related functionality. Indeed, recent years have seen the birth of mobile manipulator solutions for non-industrial objectives, e.g., assistance, precision agriculture, and search and rescue, as briefly depicted in Figure 1.

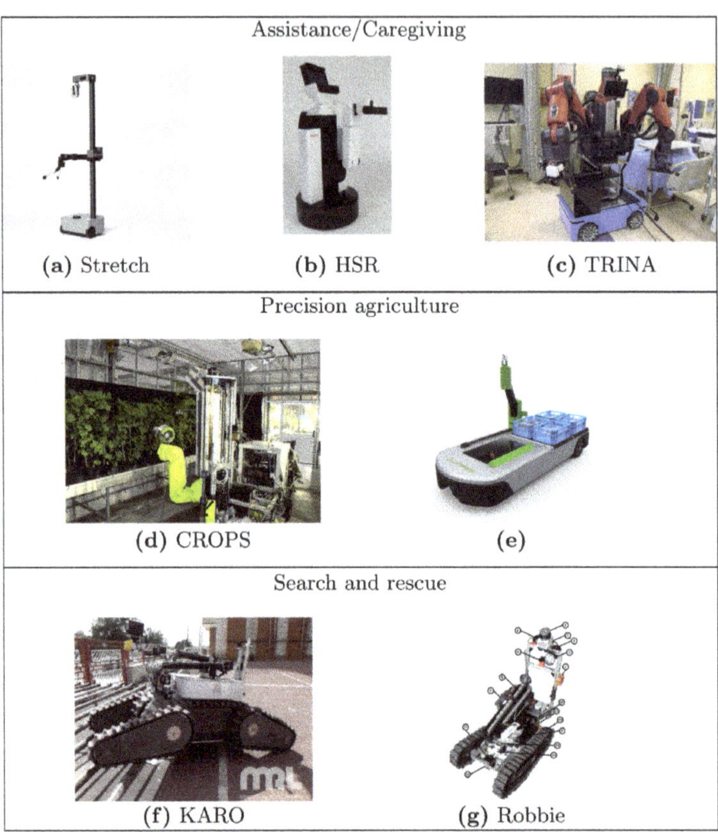

Figure 1. (a) Stretch, from hello robot company [5,6], (b) Human Support Robot (HSR), from Toyota company [7,8], (c) TRINA, developed by the researchers of Worcester Polytechnic Institute and Duke University [9,10], (d) CROPS, developed by the researchers of Università degli Studi di Milano, University of Ljubljana and Technische Universitat Munchen [11], (e) robot for strawberry harvesting, from Octinion [12,13], (f) KARO, developed by the researchers of the Advanced Mobile Robotics Lab, Qazvin Azad University [14], and (g) Robbie, developed by the researchers of Fachhochschule Technikum Wien [15].

In 2021, Kemp et al. presented Stretch [5], specifically designed for indoor human environments and telepresence. The key design goal was the reduction in the number of actuators, and it ended up with a conventional differential drive mobile base and a manipulator with three degrees of freedom (d.o.f.). For motion planning purposes, the authors considered two separate operation modes, the navigation and manipulation mode, underlining how the correct position of the base is fundamental for a correct task execution. Again for assistive and collaborative applications, since 2006, researchers from Toyota company have developed HSR (Human Support Robot) [8], a domestic mobile manipulator with a dual wheel caster mechanism and a seven d.o.f. robotic arm. To pick objects from the floor or tables, the first manipulator d.o.f. enables the motion along the vertical axis of the structure. Focusing on humanoid systems, Li et al. [9] implemented an off-the-shelf dual armed humanoid manipulator and an omnidirectional base to create TRINA (Tele-Robotic Intelligent Nursing Assistant) a complex system with 26 d.o.f. commanded by an operator console for telepresence in hospitals.

The development and implementation of such systems can be effective also in the agriculture field, where all the tasks have been traditionally carried out by manual labor.

In 2017, Silwal et al. [16] presented an autonomous mobile apple harvester formed by a six d.o.f. arm mounted upon a generic mobile base, successfully tested in a commercial apple orchard. Even though the system was incredibly innovative, there was no cooperation between the mobile base and manipulator motion, since the mobile base is fixed during the apple harvesting. A similar work was presented by De Preter et al. in 2018 [12], where a commercial robot from Octinion company was used for strawberry harvesting. The integration of mobile manipulators in the agriculture field can be useful also for selective spraying, thus reducing the production cost and the use of pesticides on crops. To this end, in 2016 Oberti et al. [11] presented a novel integration of a selective spaying system on the modular agriculture robot CROPS, with a multispectral imaging system and an associated image recognition to select the diseased target area.

In all the cited precision agriculture cases, the development of the mobile manipulator systems was substantially done by the implementation of a custom manipulator and/or sensing system upon a generic and multi-purpose mobile base, and the mounting position of the arm with respect to the base was fixed. This paper presents the kinematic model and a simplified motion planning method for the novel Agri.Q mobile manipulator, designed by the researchers from Politecnico di Torino for precision agriculture applications, where the robotic arm mounting position and orientation with respect to the mobile base is adjustable thanks to the mobile base mobility characteristics.

The main contributions to the subject investigated in this work are:

- The differential kinematic model of the whole custom system, described by a linear mapping from the velocity input commands to the system velocities. The kinematic model for the planar motion of the base was already completed by the authors in [17], where the base was treated as a mobile rover and the pitch mobility was not considered; so, the work is here significantly extended considering the pitch motion, which translates and rotates the manipulator base, and the manipulator mobility itself.
- The description of a decoupled motion planning algorithm for sampling and pick/place tasks, where the base mobility is used to properly reach the target and also take advantage of the manipulator dexterity.
- Manipulator inverse kinematics formulation with the use of the elbow or swivel angle, which is a closed form analytic solution of the inverse kinematics problem. An open-source algorithm written in Matlab code is provided (https://github.com/giocolucci/Jaco2SwivelIK; https://it.mathworks.com/matlabcentral/fileexchange/108419-jaco2swivelik, accessed in 21 March 2022).

The next sections will be organized as follows. First, a brief description of the Agri.Q system is provided in Section 1.1. Next, Section 2 focuses on the kinematic model, and Section 3 focuses on the simplified motion planning pipeline.

1.1. Agri.Q Mobile Manipulator

The Agri.Q prototype was formed by a custom mobile base [18], specifically designed for precision agriculture applications, and a commercial seven d.o.f. collaborative arm Kinova Jaco2. The mobile base belongs to the articulated type, with two modules, a front module and a back module; each of these is provided by two locomotion units, one for each side, formed by two traction wheels. Each locomotion unit is commanded by a single electric motor, and a chain drive provides the mechanical power to the two wheels, which are connected to the main module with a rocker with a passive revolute joint, to enhance the obstacle overcoming abilities (Figure 2). Since the base has to navigate inside loose and irregular soil, the use of eight wheels ensures a wide contact surface between the vehicle and the ground, as occurs with a track system. Moreover, the traction control can decide to move the system using a front-wheel drive or all-wheel drive, enabling navigation inside quite steep rows of vines. The base is also provided with a solar panel, designed for drone landing and the recharge of the 24 V lithium-ion battery that feeds the required electric power to the traction elements, the robotic arm, and to the electronic components. Thanks to a pitch mechanism [19], the orientation of the panel is adjustable in order to to take full advantage of the recharging phase in different sunlight conditions [20].

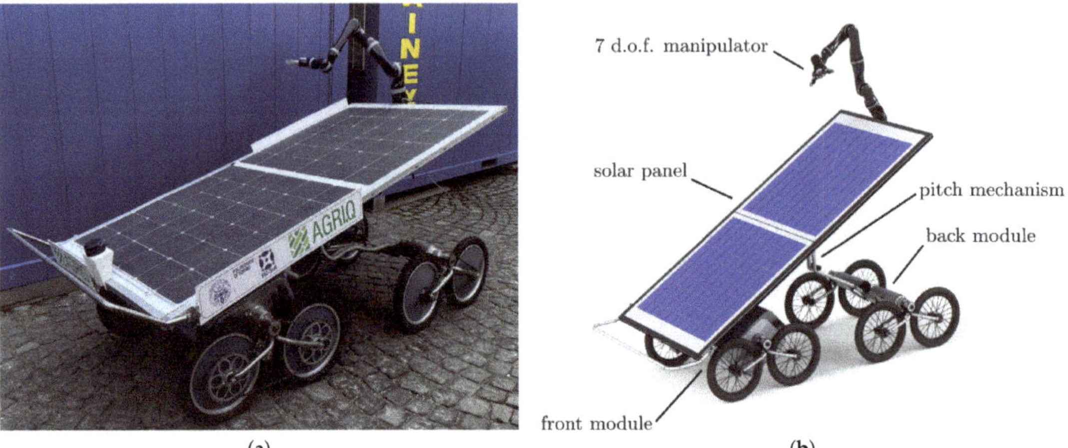

Figure 2. (**a**) The Agri.Q mobile manipulator prototype developed at Politecnico di Torino. (**b**) Render of the prototype to highlight its fundamental components.

The robotic arm was the Jaco2 model from Kinova company, a collaborative serial manipulator with seven d.o.f. designed for assistance and other collaborative tasks. Since the d.o.f. of the arm is larger than the dimension of its workspace, it is defined as intrinsically redundant, which makes it able to perform difficult tasks and use the kinematic redundancy to avoid obstacles or reach a target inside a narrow space. To take advantage of the base mobility, the manipulator was fixed to the solar panel, so the pitch motion of the base also results in a modification of the mounting position and orientation of the arm with respect to the mobile base, as illustrated in Figure 2.

2. Kinematic Model

2.1. Hypotheses and Representation of the System

To study the 2-D motion of the articulated mobile base, the rover platform was modeled as two modules, a front module and back module, connected by a revolute joint centered in the origin of $\{F\}$, that enabled the yaw motion in the $\{\hat{i}_O, \hat{j}_O\}$ plane, as described in Figure 3a. The traction power was provided by the front wheels, that were commanded by ω_L and ω_R, the angular velocity of the left front wheels and angular velocity of the right front wheels, respectively. It is worth pointing out that the angular velocity was the same for the two right or left wheels, since each locomotion unit is commanded by a single motor and the chain drive does not cause velocity disparities.

The following assumptions were made:

- The traction wheels are subject to pure rolling conditions;
- No lateral slip of the front and back modules is enabled;
- The surface is flat and no out-of-plane motion are considered;
- Each couple of wheels can be reduced to a single virtual wheel with the rotation axis aligned with the module body, as showed in Figure 3b.

The latter assumption, that treats the front module as a differential drive system, was already presented and discussed by the authors in [17].

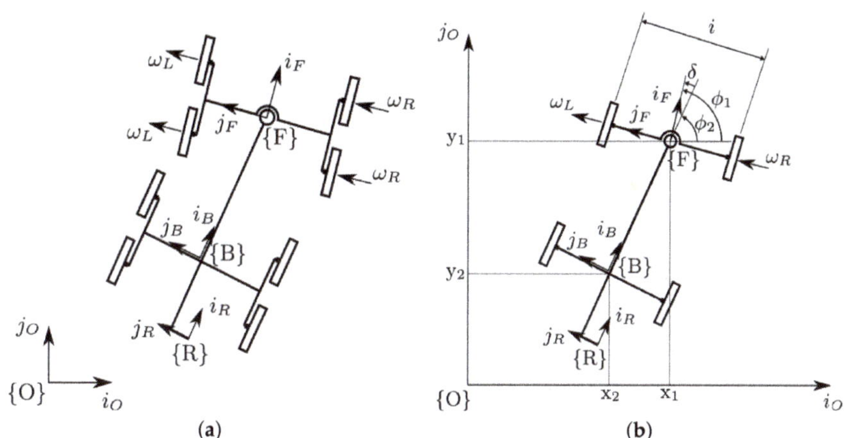

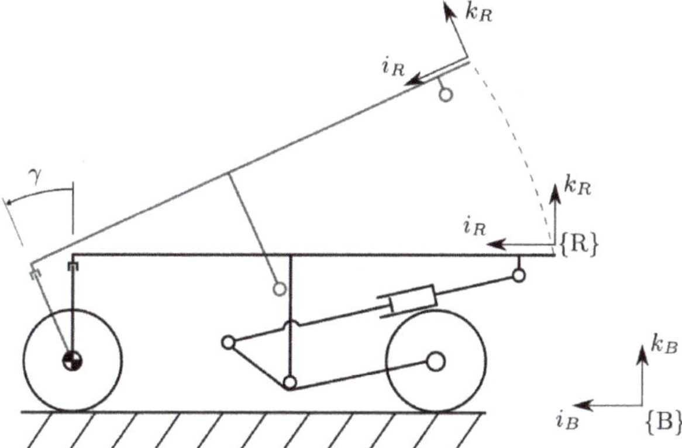

Figure 3. (**a**) Model of Agri.Q articulated mobile base on $\{\hat{i}_O, \hat{j}_O\}$ plane. The system is described by the position and orientation of the three mobile reference frames, where $\{F\}$ is the front module r.f., $\{B\}$ is the back module r.f., and $\{R\}$ refers to the manipulator. In (**b**) the reduction of each couple of wheels to a single one is represented.

The mobile base was also provided with a pitch mechanism [19], that can translate and rotate $\{R\}$, the robotic arm base reference frame, in the $\{\hat{i}_B, \hat{k}_B\}$ plane, i.e., it can translate and rotate the serial manipulator, as shown in Figure 4, by the angular velocity command $\dot{\gamma}$.

Figure 4. Pitch motion of the mobile base that causes the motion of $\{R\}$, described, for simplicity, with a relative yaw angle $\delta = 0$. In black is the start position of the base, in gray is the new configuration. It is worth noting that it is a planar motion in the $\{\hat{i}_B, \hat{k}_B\}$ plane.

2.2. Kinematic Model of the Mobile Base

The number of degrees of freedom (dof) of the system, composed by the front module, back module, and pitch system, is equal to five, that corresponsd to the four dof in $\{\hat{i}_O, \hat{j}_O\}$ plane and the pitch motion in $\{\hat{i}_B, \hat{k}_B\}$. The chosen set of generalized coordinates of the system is the following:

$$\underline{q}_P = \{x_1, y_1, \phi_1, \delta, \gamma\}, \tag{1}$$

where x_1 and y_1 are the Cartesian coordinates of the origin of $\{F\}$ with respect to $\{O\}$, ϕ_1 is the yaw angle of $\{F\}$ with respect to $\{O\}$, δ is the relative yaw between front and back module defined as:

$$\delta = \phi_1 - \phi_2, \tag{2}$$

and γ is the pitch angle of the panel.

The relationship between the command vector $\underline{u}$ and q can be found in the form:

$$\dot{q}_P = G(q_P)\underline{u}_P, \tag{3}$$

where $\underline{u} = \{\omega_L, \omega_R, \dot{\gamma}\}$, as explained before, and the 5×3 *Author : The change is correct* matrix $G(q_P)$ expresses the linear mapping between the time derivative of the generalized coordinates of the system and the system input $\underline{u}$. For this, the Cartesian coordinates $\{x_2, y_2\}$ of the origin of $\{B\}$, which describe the Cartesian coordinates of the back module of the system, can be written as a function of $\{x_1, y_1\}$:

$$x_2 = x_1 - b\cos(\phi_2) \tag{4}$$

$$y_2 = y_1 - b\sin(\phi_2) \tag{5}$$

Moreover, a non-holonomic equation constraint for the back module can be introduced, which expresses the absence of lateral slip of the module:

$$\underline{v_2} \cdot \hat{j}_B = 0 \tag{6}$$

where $\underline{v_2} = \{\dot{x}_2, \dot{y}_2\}$ is the velocity of the origin of $\{B\}$ expressed with respect to $\{O\}$, and $\hat{j}_B = \{-\sin\phi_2, \cos\phi_2\}$ is the unit vector of the reference frame $\{B\} = \{\hat{i}_B, \hat{j}_B, \hat{k}_B\}$.

By deriving Equations (4) and (5) with respect to time and replacing them into (6), the following result can be obtained:

$$\dot{\delta} = \dot{\phi}_1 - \frac{\sin(\delta)}{b}|\underline{v_1}|. \tag{7}$$

Thus, the linear mapping between $\{|\underline{v_1}|, \dot{\phi}_1, \gamma\}$ and $\dot{q}$ can be found:

$$\dot{q}_P = \left\{ \begin{array}{c} \dot{x}_1 \\ \dot{y}_1 \\ \dot{\phi}_1 \\ \dot{\delta} \\ \dot{\gamma} \end{array} \right\} = \left[\begin{array}{ccc} \cos(\phi_1) & 0 & 0 \\ \sin(\phi_1) & 0 & 0 \\ 0 & 1 & 0 \\ -\dfrac{\sin(\delta)}{b} & 1 & 0 \\ 0 & 0 & 1 \end{array} \right] \left\{ \begin{array}{c} |\underline{v_1}| \\ \dot{\phi}_1 \\ \dot{\gamma} \end{array} \right\} \tag{8}$$

To obtain the linear mapping from $\underline{u}$ to $\dot{q}$, the differential-drive model of the front module must be taken into account. Thus, the instantaneous center of rotation icr_F of the front module must lie on the $\hat{j}_F$ axis, and the following relationship can be written:

$$|\underline{v_1}| = |v_{1,R}| + \frac{|v_{1,L}| - |v_{1,R}|}{2} = \frac{1}{2}(|v_{1,L}| + |v_{1,R}|), \tag{9}$$

$$\dot{\phi}_1 = \frac{|v_{1,R}| + |v_{1,L}|}{i}. \tag{10}$$

By substituting Equations (9) and (10) into (8), and taking into account the pure rolling hypothesis, the following final result is obtained:

$$\dot{q}_P = \begin{bmatrix} \dfrac{R\cos(\phi_1)}{2} & \dfrac{R\cos(\phi_1)}{2} & 0 \\[2mm] \dfrac{R\sin(\phi_1)}{2} & \dfrac{R\sin(\phi_1)}{2} & 0 \\[2mm] -\dfrac{R}{i} & \dfrac{R}{i} & 0 \\[2mm] -\dfrac{R}{i} - \dfrac{R\sin(\delta)}{2b} & \dfrac{R}{i} - \dfrac{R\sin(\delta)}{2b} & 0 \\[2mm] 0 & 0 & 1 \end{bmatrix} \begin{Bmatrix} \omega_L \\ \omega_R \\ \dot{\gamma} \end{Bmatrix} \tag{11}$$

2.3. Analytic Jacobian of the System

To combine the mobility of the mobile base and the manipulator and to plan the motion of the end effector of the mobile manipulator in $\mathbb{R}^6$ (position and orientation), the analytic Jacobian matrix of the system must be computed. The total set of generalized coordinates of the system $q \in \mathbb{R}^{p+r}$ is composed of both the generalized coordinates of the platform $q_P \in \mathbb{R}^p$ and the manipulator ones $q_R \in \mathbb{R}^r$:

$$q = \{q_P, \ q_R\}, \tag{12}$$

Although this general formulation can also be used with other mobile manipulator systems, it is worth recalling that for the Agri.Q prototype, $p = 5$ and $r = 7$.

The position and orientation of the end-effector (ee) of the mobile manipulator with respect to the fixed frame $\{O\}$ can be described by the 4×4 homogeneous transformation matrix T_{ee}^0 defined as follows:

$$T_{ee}^0(q) = \begin{bmatrix} R_{ee}^O(q) & p_{ee}^O(q) \\ 0 & 1 \end{bmatrix} \tag{13}$$

where R_{ee}^O is the 3×3 rotation matrix describing the orientation of the end-effector wrt $\{O\}$, and p_{ee}^O is the associated 3×1 position vector. T_{ee}^O can be computed as the product of three matrices, so that the mobility of the system is decomposed and studied individually:

$$T_{ee}^O(q) = T_F^O \ T_R^F \ T_{ee}^R \tag{14}$$

where:

- $T_F^O = T_F^O(x_1, y_1, \phi_1)$ represents the transformation matrix from the fixed frame $\{O\}$ to $\{F\}$, fixed to the front module. It depends on the three degrees of freedom of the front module motion in the $\{\hat{i}_O, \hat{j}_O\}$ plane;
- $T_R^F = T_R^F(\delta, \gamma)$ represents the transformation from $\{F\}$ to $\{R\}$, which is the reference frame fixed to the manipulator base, and it depends on the relative yaw angle δ between the front and back modules and the pitch angle γ. It also contains information about the mounting parameters x_0, y_0, and z_0;
- $T_{ee}^R = T_{ee}^R(q_R)$ describes the forward kinematics of the serial kinematic chain of the manipulator.

The analytic Jacobian of the system is formed by the $3 \times (p+r)$ matrix J_v, which contains the linear mapping between the joint space velocity and the end effector linear twist space, and the $3 \times (p+r)$ matrix J_ω, which transforms $\dot{q}$ into the end-effector angular twists:

$$\begin{Bmatrix} v_{ee} \\ \omega_{ee} \end{Bmatrix} = J(q)\dot{q} = \begin{bmatrix} J_v \\ J_\omega \end{bmatrix} \dot{q} \tag{15}$$

The analytic expressions of $\underline{v}_{ee}$ and $\underline{\omega}_{ee}$ can be easily derived from T_{ee}^O:

$$\underline{v}_{ee} = \frac{\mathrm{d}\underline{p}_{ee}^O}{\mathrm{d}t}, \tag{16}$$

$$\underline{\omega}_{ee} = \{-S_{23}, S_{13}, -S_{12}\}, \tag{17}$$

where S is the 3×3 skew-symmetric matrix defined as:

$$S = \dot{R}_{ee}^O \, R_{ee}^O{}^T \tag{18}$$

As explained by De Luca et al. in [21], for motion planning purposes it could be useful to express the Equation (15) in terms of the command vector $\underline{u} = \{\underline{u}_P, \underline{u}_R\}$, where $\underline{u}_R$, i.e., the command vector of the manipulator, coincides with the joint space velocity vector:

$$\underline{u}_R = \underline{\dot{q}}_R \tag{19}$$

Thus, the $6 \times (p + r)$ Jacobian matrix can be decomposed into the $6 \times p$ matrix J_P, which describes the analytical Jacobian of the mobile platform and the $6 \times r$ Jacobian J_R of the manipulator. Thus, Equation (15) can be modified into:

$$\left\{ \begin{array}{c} \underline{v}_{ee} \\ \underline{\omega}_{ee} \end{array} \right\} = J_P \underline{\dot{q}}_P + J_R \underline{\dot{q}}_R = J_P G(\underline{q}_P)\underline{u}_P + J_R \underline{\dot{q}}_R, \tag{20}$$

where the matrix $G(\underline{q}_P)$ was defined in Equation (3) and calculated in (11).

3. Motion Planning Pipeline for Decoupled Motion

Even though the analytical Jacobian of the mobile manipulator can be used for motion planning using both the mobile base and manipulator mobility, exploiting the augmented capability to reach a target point, perform a task, or avoid obstacles, a simpler but still interesting way to move the system lies in the motion decoupling of the mobile base and the manipulator [22], motivated by the general higher position accuracy of the manipulator with respect to the base. Although such an approach does not allow the simultaneous motion of the base and the arm, it provides significant advantages in terms of motion planning, since it reduces the degrees of freedom that are involved during the two motions. A decoupled motion planning technique provides the great advantage of using the mobile base for a gross positioning of the manipulator, then exploiting the accuracy of the arm for grasping and manipulation.

Thus, a novel motion planning method that exploits the base and arm mobility of Agri.Q prototype for a pick and place task is presented and discussed.

In a generic scenario where Agri.Q moves inside the rows of vineyards for sampling crops or soil, the motion planning pipeline can be described as follows:

- Start phase: a high-level command requires the execution of a given task, such as a pick-and-place execution, which corresponds to a crop sampling task;
- First perception phase: according to the specifications indicated in the start phase, a perception system, e.g., a depth camera, recognizes the target and evaluates its position and orientation with respect to the manipulator base frame $\{R\}$;
- Mobile base motion planning and execution: while the arm is still fixed, the mobile base is moved to enable the arm to reach the target and, moreover, to place the manipulator in a proper way;
- Second perception phase: the perception system recomputes the new position and orientation of the target with respect to the arm base frame;
- Arm motion planning and execution: while the base is now fixed, the arm performs the task taking care to not collide with the mobile base and the rest of the environment.

In the next subsections, the discussion is focused on the mobile base motion planning and execution phase, which contains the main contribution of the authors to the depicted pipeline. As mentioned before, the mobile base is moved in order to enable the arm to reach the target, so that it can also exploit its best manipulability area to perform the task. Then, the definition of the best manipulability area is presented, also with a closed form inverse kinematics technique that is able to evaluate the manipulability index of the arm. In the last subsection, we illustrate how the mobile base can properly position the arm.

3.1. Modified Manipulability Index of the Manipulator

A common approach to evaluate the dexterity of a robotic system is the use of the manipulability index defined as [23]:

$$c = \frac{1}{\kappa},\tag{21}$$

where κ is the 2-norm condition number of Jacobian matrix J_R, calculated as the ratio between the highest and lowest eigenvalues of J_R, respectively, $\lambda_{max,min}$:

$$\kappa = \frac{\lambda_{max}}{\lambda_{min}}\tag{22}$$

In order to take care of the joint limits constraints, where the arm may have good dexterity even though it is close to the limit of its motion, Chan and Dubey proposed in [24] a gradient function that is used to penalize the Jacobian matrix J_R in a given configuration of the arm, already implemented by Vahrenkamp et al. in [25] and by Chen et al. in [26]. The gradient function $\nabla h(\theta_j)_j$ for the j-th joint is defined as:

$$\nabla(\theta_j)_j = \frac{(\theta_{j,max} - \theta_{j,min})^2(2\theta_j - \theta_{j,max} - \theta_{j,min})^2}{\alpha(\theta_{j,max} - \theta_j)^2(\theta_j - \theta_{j,min})^2},\tag{23}$$

where $\theta_{j,max}, \theta_{j,min}$ are, respectively, the upper and lower j-th joint limit, and α is a parameter that accentuates the closeness to the limit. The penalization factor for each column of J_R is then defined as:

$$p_j = \frac{1}{\sqrt{1 + |\nabla h(\theta_j)_j|}}.\tag{24}$$

In Figure 5 the behavior of p_j for different values of α is presented. As mentioned before, a larger value of α causes a relaxation of the joint limits closeness. For a small value of α, $p_j = 1$ in a small area near the value $(q_{j,max} - q_{j,min})/2$. According to [25], the value of the parameter $\alpha = 4$ was selected.

Thus, the modified manipulability index c_{mod} is evaluated as the 2-norm condition number of the modified Jacobian $J_{R,mod}$, where each entry $J_{i,j}$ of J_R is modified according to the corresponding penalization factor:

$$J_{ij\,R,mod} = p_j J_{ij} \qquad i = 1,\ldots,6 \quad j = 1,\ldots,7\tag{25}$$

$$c_{mod} = \frac{1}{\kappa(J_{R,mod}(\underline{q}_R))}\tag{26}$$

It is worth noting that $j = 1,\ldots,7$, because $r = 7$ for the Agri.Q prototype.

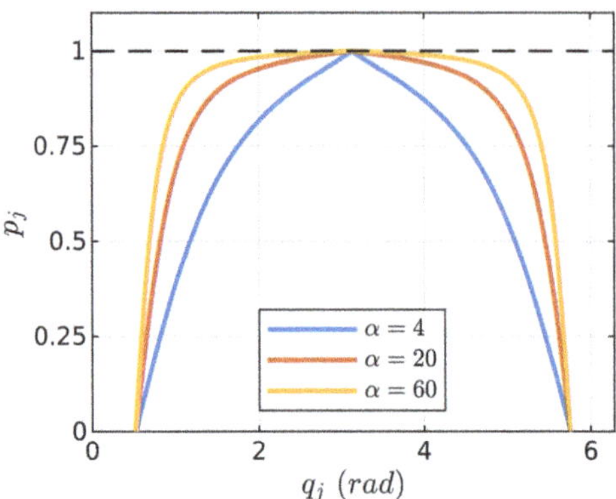

Figure 5. Penalization factor as a function of a generic j-th joint. For this case, $q_{j,min}$ = 0.5 rad, $q_{j,max}$ = 5.8 rad.

3.2. Manipulator Manipulability Mapping

In order to properly move the mobile base so that the arm can take advantage of its best manipulability properties, the value of c_{mod} must be mapped inside the manipulator workspace. To this aim, the authors have implemented a closed form inverse kinematics algorithm that provides all the possible solutions as a function of a parameter ϕ_R [27], which is the swivel or elbow angle of the manipulator [28,29]. The presented formulation treats the seven d.o.f. manipulator as a human-like arm, where, for a fixed shoulder and wrist position, the elbow can be rotated, while the orientation of the hand remains fixed. Thus, the kinematic structure of the manipulator can be modeled with a first spherical joint, centered in the shoulder, and a second similar joint, centered in the wrist, and the two points are connected by a revolute joint that represents the elbow mobility. For more details about the closed form inverse kinematics formulation, please refer to the related work presented by the author in [27].

An open-source copy of the inverse kinematics algorithm is available online (https://github.com/giocolucci/Jaco2SwivelIK; https://it.mathworks.com/matlabcentral/fileexchange/108419-jaco2swivelik, accessed in 21 March 2022). It was tested on a Dell XPS machine, with Ubuntu 20.04 LTS, and an Intel i7-10510U@1.80 GHz processor. The total execution time was approximately 1 ms when used with a single fixed elbow angle, and 100 ms when the solution with the highest manipulability index was extracted.

Thus, since the robotic arm has seven degrees of freedom, there are ∞^1 solutions for a given pose (position and orientation), and the algorithm enables evaluating the dexterity of the arm for every possible solution, eventually extracting the solution with the highest manipulability index.

An example of a manipulability map is presented in Figure 6, where the value of c and c_{mod} was evaluated inside the yellow domain w for a fixed end-effector orientation, illustrated in the figure. Both the area where $c_{mod} > 0.5$, in blue, and $c_{mod} > 0.6$, in red, were significantly reduced due to the joint limit constraints, introduced by the penalization factor p_j described in (24).

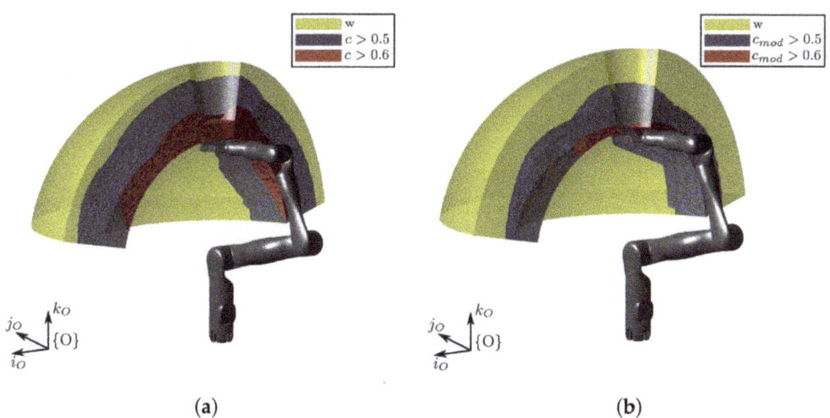

(a)   (b)

Figure 6. (**a**) Manipulability map for the Kinova Gen2 7 d.o.f. arm in terms of c, (**b**) Manipulability map for the Kinova Gen2 7 d.o.f. arm in terms of c_{mod}. The position and orientation of the end-effector during the mapping process is showed in both figures. w is the total explored workspace, c and c_{mod} are the 2-norm condition number and the modified version of it, respectively, as presented in Equations (21) and (26).

3.3. Mobile Base Motion Planning

Since it was designed for crop sampling applications inside a vineyard, the mobile base of Agri.Q prototype can be used to position the manipulator where it can actually reach the target and take advantage of its improved dexterity. To develop the proposed simplified motion planning pipeline, the assumption of no motion in $\hat{i}_O$ direction was made, because it represents the perpendicular direction to the vineyard rows, as described in Figure 7. While navigating inside a vineyard, the distance between the manipulator and a target, e.g., a grape, is fixed and equal to the distance between the manipulator and the row. Even though the arm will take advantage of better dexterity if it moves closer to the target, this implies that also the mobile base will move closer to the rows, leading to potential collisions. Thus, the navigation of the mobile manipulator Agri.Q inside the vineyard rows can be simplified, and the base motion can be described by a reduced set of generalized coordinates:

$$q_P = \{y_1, \gamma\}. \tag{27}$$

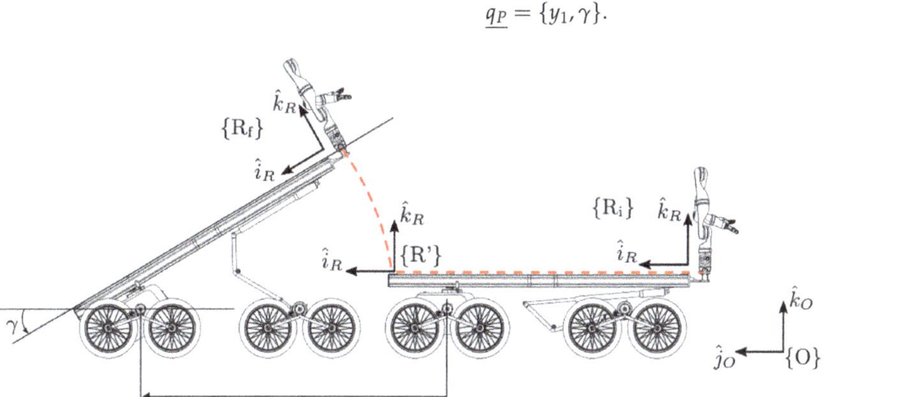

Figure 7. Effect of s and γ on the position and orientation of the manipulator. $\{R_i\}$ is the initial configuration of the manipulator, $\{R'\}$ takes into account the effect of the linear motion s along $\hat{j}_O$, and $\{R_f\}$ is the final configuration of the manipulator, also with the contribution of the pitch angle γ.

Thus, the reference frame fixed to the manipulator $\{R\}$ is moved by the value s defined as the difference between the final and initial Cartesian coordinates y_1:

$$s = y_{1,f} - y_{1,i},\tag{28}$$

then, the new r.f. $\{R'\}$ is transformed into $\{R_f\}$ due to the pitch angle γ (Figure 7).

Since the manipulability map is fixed to $\{R\}$, the motion planning pipeline described in the following is proposed by the authors:

- Starting from a desired goal pose, described by the homogeneous transformation matrix T^O_{goal} with respect to $\{O\}$, a slice of the entire workspace, parallel to the $\{\hat{j}_O, \hat{k}_O\}$ plane and passing through the P target point, is extracted, depicted in Figure 8a as the circle in black.
- Inside the extracted domain, one can evaluate what portion of the workspace can be actually reached with the desired end-effector orientation. It is worth pointing out that, in general, the domain of the reachable points with a desired pose does not coincide with the manipulator workspace, which, instead, considers all the possible reachable points without an orientation specification. The boundaries of the modified workspace are described in Figure 8b with the red line.
- One can evaluate whether the point P lies inside the modified workspace boundaries. If it does, the manipulator can actually reach the target, and no motion of the mobile base is requested. If it does not, the mobile base must move to reach the target point.
- If a mobile base motion is requested, the reduced workspace domain is mapped in terms of the modified manipulability index c_{mod}, defined in Equation (26).
- An optimal area where $c_{mod} > c_{mod,min}$, represented in the figure inside the gray line, is calculated as a portion of the modified workspace. Its manipulability barycenter coordinates C are calculated with respect to $\{O\}$:

$$\underline{r_C} = \frac{\sum_{i=1}^N c_{mod,i}\underline{r_i}}{\sum_{i=1}^N c_{mod,i}},\tag{29}$$

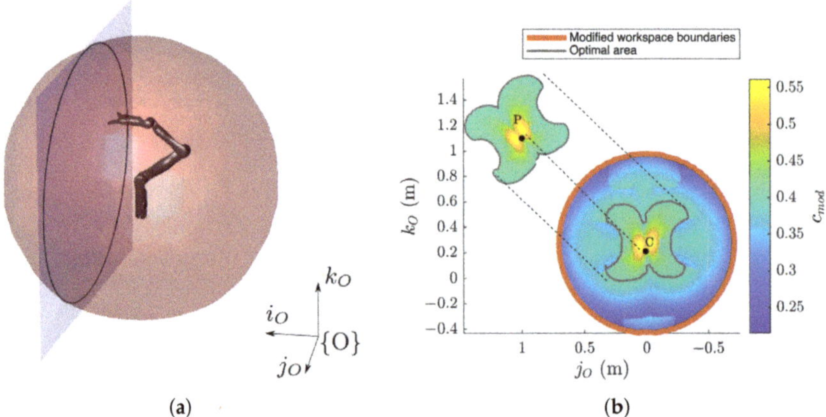

Figure 8. (**a**) Manipulator workspace representation (red volume) and extraction of the plane of interest (blue). Since no motion is enabled along the i_O axis, the x off-set of the plane with respect to $\{O\}$ is the x coordinates of the target point P. (**b**) Optimal manipulability area identification (inside the gray line). The total workspace is reduced due to the orientation specification, thus obtaining the area inside the red circle that, in general, does not coincide with the black circle in (**a**). The optimal area is then moved and rotated according to Figure 7.

where $r_i\ i = 1, \ldots, N$ is the position vector of the i-th point inside the optimal area, and $c_{mod,i}$ is its modified manipulability index.

In the reported example, $c_{mod,min} = 0.4$.

- $\{y_1, \gamma\}$ are calculated in order to move C to the target point P. In general, the procedure implies the rotation of the optimal area due to γ.

4. Conclusions

The paper presented the kinematic modeling of the Agri.Q mobile manipulator prototype, designed for precision agriculture purposes. Starting from the subsystems' mobility, the linear mapping from the input velocity commands to the linear and angular twists of the whole system was presented. The differential kinematic model, that is based on the no lateral wheel slipping assumption, is fundamental for singularity analysis and to implement traditional redundancy resolution methods, e.g., the Extended Jacobian or the Reduced Gradient. To perform useful pick-and-place tasks while navigating inside vine rows, a simplified motion planning algorithm, based on the motion decoupling was presented. Thus, the mobile base, less accurate than the manipulator, is used to position the arm, also taking advantage of its best manipulability area. The presented method lends itself to a simple and quick implementation on the real prototype, since the motion decoupling techniques significantly reduce the computational cost of the motion planning phase.

Author Contributions: Conceptualization, G.C. and A.B.; methodology, G.C. and A.B.; software, G.C.; validation, L.B., L.T. and P.C.; formal analysis, G.C. and A.B.; investigation, G.C. and A.B.; resources, G.Q.; data curation, G.C.; writing—original draft preparation, G.C.; writing—review and editing, A.B., L.B., P.C., L.T. and G.Q.; visualization, G.C. and A.B.; supervision, G.Q.; project administration, G.Q.; funding acquisition, G.Q. All authors have read and agreed to the published version of the manuscript.

Funding: This research received no external funding.

Institutional Review Board Statement: Not applicable.

Informed Consent Statement: Not applicable.

Data Availability Statement: The data presented in this study are available on request from the corresponding author.

Conflicts of Interest: The authors declare no conflict of interest.

Abbreviations

The following abbreviations are used in this manuscript:

d.o.f.	degree of freedom
r.f.	reference frame

References

1. Hvilshøj, M.; Bøgh, S.; Skov Nielsen, O.; Madsen, O. Autonomous industrial mobile manipulation (AIMM): Past, present and future. *Ind. Robot. Int. J.* **2012**, *39*, 120–135. [CrossRef]
2. Outón, J.L.; Villaverde, I.; Herrero, H.; Esnaola, U.; Sierra, B. Innovative Mobile Manipulator Solution for Modern Flexible Manufacturing Processes. *Sensors* **2019**, *19*, 5414. [CrossRef] [PubMed]
3. RB Kairos + Mobile Manipulator. Available online: https://robotnik.eu/products/mobile-manipulators/rb-kairos/ (accessed on 17 April 2022).
4. KMR Iiwa. Available online: https://www.kuka.com/en-se/products/mobility/mobile-robots/kmr-iiwa (accessed on 17 April 2022).
5. Kemp, C.C.; Edsinger, A.; Clever, H.M.; Matulevich, B. The Design of Stretch: A Compact, Lightweight Mobile Manipulator for Indoor Human Environments. *arXiv* **2021**, arXiv:2109.10892.
6. Hello Robot Website. Available online: https://hello-robot.com/product (accessed on 15 March 2022).
7. Toyota Global Site | Frontier Research. Available online: https://www.toyota-global.com/innovation/partner_robot/index.html (accessed on 15 March 2022).
8. Yamamoto, T.; Terada, K.; Ochiai, A.; Saito, F.; Asahara, Y.; Murase, K. Development of Human Support Robot as the research platform of a domestic mobile manipulator. *ROBOMECH J.* **2019**, *6*, 4. [CrossRef]

9. Li, Z.; Moran, P.; Dong, Q.; Shaw, R.J.; Hauser, K. Development of a tele-nursing mobile manipulator for remote care-giving in quarantine areas. In Proceedings of the 2017 IEEE International Conference on Robotics and Automation (ICRA), Singapore, 29 May–3 June 2017; pp. 3581–3586. [CrossRef]

10. Intelligent Motion Laboratory at Duke University. Available online: http://motion.pratt.duke.edu/nursing/index.html (accessed on 15 March 2022).

11. Oberti, R.; Marchi, M.; Tirelli, P.; Calcante, A.; Iriti, M.; Tona, E.; Hočevar, M.; Baur, J.; Pfaff, J.; Schütz, C.; et al. Selective spraying of grapevines for disease control using a modular agricultural robot. *Biosyst. Eng.* **2016**, *146*, 203–215. [CrossRef]

12. De Preter, A.; Anthonis, J.; De Baerdemaeker, J. Development of a Robot for Harvesting Strawberries. *IFAC-PapersOnLine* **2018**, *51*, 14–19. [CrossRef]

13. Strawberry Picker. Available online: http://www.octinion.com/strawberry-picker (accessed on 15 March 2022).

14. Habibian, S.; Dadvar, M.; Peykari, B.; Hosseini, A.; Salehzadeh, M.H.; Hosseini, A.H.M.; Najafi, F. Design and Implementation of a Maxi-Sized Mobile Robot (Karo) for Rescue Missions. *ROBOMECH J.* **2021**, *8*, 1. [CrossRef]

15. Novotny, G.; Emsenhuber, S.; Klammer, P.; Pöschko, C.; Voglsinger, F.; Kubinger, W. A Mobile Robot Platform for Search and Rescue Applications. In Proceedings of the 30th DAAAM International Symposium, Zadar, Croatia, 23–26 October 2019; pp. 0945–0954. [CrossRef]

16. Silwal, A.; Davidson, J.R.; Karkee, M.; Mo, C.; Zhang, Q.; Lewis, K. Design, integration, and field evaluation of a robotic apple harvester. *J. Field Robot.* **2017**, *34*, 1140–1159. [CrossRef]

17. Botta, A.; Cavallone, P.; Tagliavini, L.; Colucci, G.; Carbonari, L.; Quaglia, G. Modelling and simulation of articulated mobile robots. *Int. J. Mech. Control* **2021**, *22*, 15–26.

18. Quaglia, G.; Visconte, C.; Scimmi, L.S.; Melchiorre, M.; Cavallone, P.; Pastorelli, S. Design of a UGV Powered by Solar Energy for Precision Agriculture. *Robotics* **2020**, *9*, 13. [CrossRef]

19. Quaglia, G.; Visconte, C.; Scimmi, L.S.; Melchiorre, M.; Cavallone, P.; Pastorelli, S. Design of the positioning mechanism of an unmanned ground vehicle for precision agriculture. In *Advances in Mechanism and Machine Science*; Uhl, T., Ed.; Mechanisms and Machine Science; Springer International Publishing: Cham, Switzerland, 2019; Volume 73; pp. 3531–3540. [CrossRef]

20. Botta, A.; Cavallone, P. Robotics Applied to Precision Agriculture: The Sustainable Agri.q Rover Case Study. In Proceedings of the I4SDG Workshop 2021, Online, 25–26 November 2021; Quaglia, G., Gasparetto, A., Petuya, V., Carbone, G., Eds.; Springer International Publishing: Cham, Switzerland, 2022; pp. 41–50.

21. De Luca, A.; Oriolo, G.; Giordano, P. Kinematic modeling and redundancy resolution for nonholonomic mobile manipulators. In Proceedings of the 2006 IEEE International Conference on Robotics and Automation (ICRA 2006), Orlando, FL, USA, 15–19 May 2006; pp. 1867–1873. [CrossRef]

22. Lynch, K.M.; Park, F.C. *Modern Robotics*; Cambridge University Press: Cambridge, UK, 2017; pp. 475–476.

23. Merlet, J.P. Jacobian, Manipulability, Condition Number, and Accuracy of Parallel Robots. *J. Mech. Des.* **2005**, *128*, 199–206. [CrossRef]

24. Chan, T.F.; Dubey, R. A weighted least-norm solution based scheme for avoiding joint limits for redundant joint manipulators. *IEEE Trans. Robot. Autom.* **1995**, *11*, 286–292. [CrossRef]

25. Vahrenkamp, N.; Asfour, T.; Metta, G.; Sandini, G.; Dillmann, R. Manipulability analysis. In Proceedings of the 2012 12th IEEE-RAS International Conference on Humanoid Robots (Humanoids 2012), Osaka, Japan, 29 November–1 December 2012; pp. 568–573. [CrossRef]

26. Chen, F.; Selvaggio, M.; Caldwell, D.G. Dexterous Grasping by Manipulability Selection for Mobile Manipulator With Visual Guidance. *IEEE Trans. Ind. Inform.* **2019**, *15*, 1202–1210. [CrossRef]

27. Colucci, G.; Baglieri, L.; Botta, A.; Cavallone, P.; Quaglia, G. Optimal Positioning of Mobile Manipulators Using Closed Form Inverse Kinematics. In *Advances in Service and Industrial Robotics*; Springer International Publishing: Cham, Switzerland, 2022. [CrossRef]

28. Artemiadis, P. Closed-form Inverse Kinematic Solution for Anthropomorphic Motion in Redundant Robot Arms. In Proceedings of the ICRA 2013, Karlsruhe, Germany, 6–10 May 2013. [CrossRef]

29. Faria, C.; Ferreira, F.; Erlhagen, W.; Monteiro, S.; Bicho, E. Position-based kinematics for 7-DoF serial manipulators with global configuration control, joint limit and singularity avoidance. *Mech. Mach. Theory* **2018**, *121*, 317–334. [CrossRef]

Article

Embedded Payload Solutions in UAVs for Medium and Small Package Delivery

Matteo Saponi [1,2], Alberto Borboni [1], Riccardo Adamini [1], Rodolfo Faglia [1] and Cinzia Amici [1,*]

[1] Department of Mechanical and Industrial Engineering, University of Brescia, Via Branze, 38, 25123 Brescia, Italy
[2] Imbal Carton s.r.l., Via Gardesana, 54, 25080 Prevalle, Italy
* Correspondence: cinzia.amici@unibs.it

Abstract: Investigations about the feasibility of delivery systems with unmanned aerial vehicles (UAVs) or drones have been recently expanded, owing to the exponential demand for goods to be delivered in the recent years, which has been further increased by the COVID-19 pandemic. UAV delivery can provide new contactless delivery strategies, in addition to applications for medical items, such as blood, medicines, or vaccines. The safe delivery of goods is paramount for such applications, which is facilitated if the payload is embedded in the main drone body. In this paper, we investigate payload solutions for medium and small package delivery (up to 5 kg) with a medium-sized UAV (maximum takeoff of less than 25 kg), focusing on (i) embedded solutions (packaging hosted in the drone fuselage), (ii) compatibility with transportation of medical items, and (iii) user-oriented design (usability and safety). We evaluate the design process for possible payload solutions, from an analysis of the package design (material selection, shape definition, and product industrialization) to package integration with the drone fuselage (possible solutions and comparison of quick-release systems). We present a prototype for an industrialized package, a right prism with an octagonal section made of high-performance double-wall cardboard, and introduce a set of concepts for a quick-release system, which are compared with a set of six functional parameters (mass, realization, accessibility, locking, protection, and resistance). Further analyses are already ongoing, with the aim of integrating monitoring and control capabilities into the package design to assess the condition of the delivered goods during transportation.

Keywords: drone-based package delivery; UAV transportation; embedded payload; packaging; quick-release system

Citation: Saponi, M.; Borboni, A.; Adamini, R.; Faglia, R.; Amici, C. Embedded Payload Solutions in UAVs for Medium and Small Package Delivery. *Machines* **2022**, *10*, 737. https://doi.org/10.3390/machines10090737

Academic Editors: Marco Ceccarelli, Giuseppe Carbone and Alessandro Gasparetto

Received: 31 July 2022
Accepted: 25 August 2022
Published: 27 August 2022

Publisher's Note: MDPI stays neutral with regard to jurisdictional claims in published maps and institutional affiliations.

1. Introduction

In recent years, exponential growth in the volume of packages to be delivered has occurred in the logistics industry and transportation field. The COVID-19 pandemic further intensified this trend, increasing the demand from the e-commerce market and eliciting the awareness of consumers and stakeholders of the need for new logistics solutions, such as contactless delivery strategies. The world health crisis not only affected consumer purchase habits but also introduced modifications in the kind of goods to be delivered, for instance, increasing the demand for delivery of medicines, vaccines, and biological tests.

In this scenario, investigations about the feasibility of delivery systems based on the use of unmanned aerial vehicles (UAVs) or aerial drones (hereafter drones for the sake of simplicity) has been expanded. As described by Benarbia and Kyamakya in a review [1], much literature on this topic focuses on last-mile delivery (or the "last-mile challenge"), as it represents the most critical step of the delivery process. Transportation costs increase as packages approach their final destination—even more so in rural areas. For such applications, drone-based systems could help to reduce the carbon footprint of the delivery process, as well as traffic jams. On the other hand, commonly identified constraints include short flight ranges and low payloads, which are mainly related to the limited autonomy of UAV batteries [2,3]. To solve this issue, drones would need to either reduce their roundtrip extension or lean on a distributed network of recharging stations, which are generally not yet available [1,4]. In addition, drone performance can be affected by hostile weather conditions, which can reduce aerodynamics and induce communication issues or limit operator effectiveness [5,6]. Developing robust and efficient UAVs requires considerable designer competence, consequently increasing the final production and operating costs of the drone [1].

Slightly different evaluations should be considered when dealing with the delivery of medical items, from first aid and medical supplies [7–9], such as automatic external defibrillators (AEDs) [10–15]; to medicines and pharmaceuticals [16–18], such as insulin [19] or vaccines [20–23]; or biological samples, such as blood [24], tissues [25], or microbiological and laboratory samples [26]. In a 2019 paper, Hii et al. [19] expanded the concept of feasibility analysis of drone-based delivery, highlighting the need to focus on the delivery process itself, the transported good, and in particular, on the potential impact of UAV transportation on the ultimate quality of the delivered medicine. In their work, the authors evaluated the effects of flight conditions on insulin samples, which are very sensitive to environmental stresses, such as high temperatures and exposure to agitation or vibrations. Data revealed that the insulin quality was maintained after delivery, and the authors recommended the application of five tests to assess the feasibility of UAVs for delivery of medicines, including a post-delivery quality test and the on-board monitoring of the pharmaceuticals' environment. Similar studies were presented by Amukele et al. in 2017 [27], Yakushiji et al. in 2020 [28], and Mohd et al. in 2021 [24] but with respect to blood delivery for emergency cases, e.g., transfusions [28] or postpartum hemorrhaging [24]. All the above-mentioned studies, both under simplified [27] and actual operating conditions [24,28], reported positive results. The described tests are not easily compared, as the flight times varied between 9 and 35.52 min (Hii et al. [19] and Yakushiji et al. [28], respectively), the adopted UAVs were of varying sizes (from a few grams [19] to a maximum declared takeoff mass of 24.9 kg [28]), and different operative conditions were applied (such as payload and travelled distance).

Nonetheless, package deliveries reported in the literature are generally characterized by payloads applied externally to the drone main body, either rigidly connected with gimbals or suspended with ropes. These configurations allow for the use of standard components, such as ordinary refrigerators and supply bags [28,29] for transportation but also enable the use of custom packaging systems, providing additional features to fulfill peculiar requirements. Maity presented a lightweight polyisocyanurate thermal insulation system and a comparison with different materials and solutions [30]. Kostin and Silin [31] proposed an isothermal container for medical delivery, whereas Amicone et al. [25] proposed "Smart Capture", i.e., a packaging solution integrating an artificial intelligence (AI) system that can measure relevant parameters, such as temperature or vibrations during flight, and monitor or control them. With respect to blood transportation, insulated containers or shippers provided with ice packs are most commonly used [29], but specific requirements and packing instructions are defined by regulations, such as the United Nations Agreement Concerning the International Carriage of Dangerous Goods by Road (ADR) [32,33]. Biological substances classified as dangerous goods according to UN guideline UN3373 require in particular transportation conditions that vary depending on the specific material to be transported, such as liquid or solid substances, and according to the delivery scenario, such as by road or air, for example, ADR packing instruction 650, category b, or the corresponding IATA (International Air Transport Association) guidelines.

Stephan et al. (2022) [17] described two additional issues associated with medical drone delivery: (i) users interact with UAVs, but little attention is paid to this interaction in the design phases of drones and delivery processes; and (ii) little scientific evidence is available with respect to the effectiveness, user experience, and acceptance of medical delivery with UAVs, as information about the use of drones to deliver medicines is often disseminated through media rather than empirical studies. The acceptance and reputation of drones have proven to be non-trivial aspects with respect to the successful adoption of such devices for everyday delivery in civilian contexts. Zailani et al. [6] suggested that the potential and prior use of UAVs for military purposes affects societal perceptions of drones. de Miguel Molina et al. compiled a list of drone applications considered acceptable by society [34], including medical purposes [27] and for support in emergency scenarios [35,36], as well as applications that enable proactive behaviors for environmental protection, such as forest monitoring [37], surveying marine fauna [38], and mapping coral reefs [39] or orangutan habitats [40]. Within this framework, user perception of safety and security of drone delivery is also fundamental [41] with respect to UAVs themselves, for example, concerning the risk of accidents or failures or the security of the payload, as the good should be reliably delivered without causing harm.

With these considerations in mind, in this paper, we investigate payload solutions for medium and small package delivery with a medium-sized UAV (i.e., with a maximum takeoff mass $\leq$ 25 kg) under the following conditions:

(i)   Embedded solution: the packaging must be hosted in the cargo bay within the drone fuselage;
(ii)   Medical transportation compatibility: the packaging must allow for the delivery of medical items; and
(iii) User-oriented design: the packaging must make voluntary interaction with the user as simple as possible.

The simultaneous focus on these three aspects represents the main innovation of the present study with respect to the currently available literature. Furthermore, we describe, in detail, the methodological steps adopted in the development process, in addition to focusing on design aspects of the payload system, with particular attention to geometry and the material of the package itself, from the collection of the requirements to the concept of a custom box designed to fit within the vehicle fuselage and product industrialization. Interaction with the user is carefully considered throughout the design process, especially in the preliminary investigation of a quick-release system for package protection, and a set of qualitative solutions is proposed.

2. Materials and Methods

In this section, we describe the main steps of the applied design process, including identification of the set of requirements for the packaging system, investigation of the geometry and material of the packaging, and a preliminary optimization for product industrialization. Finally, the framework of the quick-release system is described.

2.1. Requirements Identification

The realization of an embedded payload solution involves a set of requirements, including mandatory specifications and desiderata, as well as constraints and possible additional elements arising from untold needs that should be considered from the early stages of the design process [42,43].

Figure 1 is a qualitative schematic of the proposed UAV cargo bay. The cargo bay is accessible from the top of the UAV body, with the drone on the ground. The drone was designed to ensure a maximum payload of 5 kg, with a maximum takeoff mass $\leq$ 25 kg. We do not provide any technical details about the UAV performance or design, as such parameters fall outside the scope of the present paper.

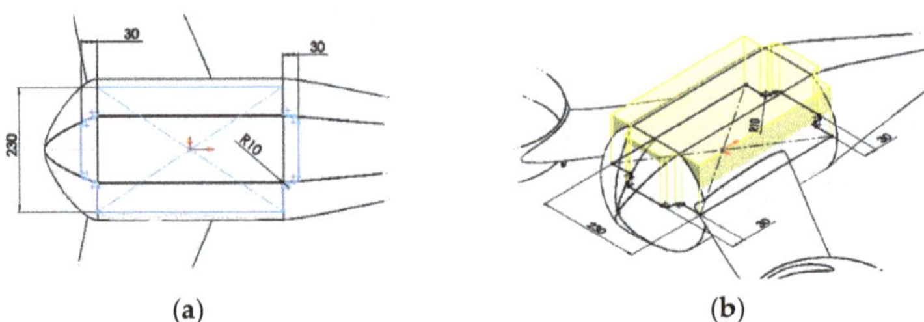

Figure 1. Qualitative schematic of the proposed cargo bay (dimensions in mm): (**a**) top view and (**b**) three-dimensional view.

In particular, specific requirements emerge from the set conditions:

(i)   Embedded solution:

    a.   The available volume within the drone fuselage that can be occupied by the package must be optimized;

    b.   Access to the cargo bay should be as simple as possible for the user, but the package should be protected as much as possible from external undesired interactions and threats.

(ii)   Medical transportation compatibility:

    a.   The package must have an adequate minimum internal volume and be composed of appropriate materials that do not cause danger or harm to the good or the user. The package must also allow for sensor integration for delivery tracking and monitoring of package conditions, with additional space to house dry ice units;

    b.   The release system must work in emergency scenarios, with the user acting under potentially high-stress conditions. Therefore, access to the cargo bay by the user must be as simple and effective as possible.

(iii)   User-oriented design:

    a.   The package must be easy to use, i.e., simple and quick to assemble and open. The box should include user instructions for assembly and opening operations;

    b.   The release system must effectively ensure the safety and security of the good but be intuitive to use. Access to the cargo bay must be safe for the user and as inclusive as possible (e.g., wheelchair-friendly).

Additional packaging constraints are related to the need for industrialization, which must be cost-effective; therefore, the package geometry should be as simple as possible, made of cardboard folded from a plane and generated by a single machining process comprising die cutting and etching.

2.2. Packaging Geometry

The proposed cargo bay is a cylinder with an elliptic section (major axis, 252 mm; minor axis, 242 mm; length, 400 mm).

To identify the optimal package geometry, keeping in mind the requirement of packaging material folded from a plane, only solutions derived from solids of revolution were considered, in addition to an ordinary parallelepipedon and, in particular, right prisms with hexagonal and octagonal sections. Cylindric structures were excluded, as they cannot be feasibly constructed from a plane cardboard sheet.

Maximization of volume was considered as a selection criterion.

2.3. Packaging Material

Given the purpose of the packaging, corrugated papers were evaluated, as well as multilayer containerboard generally composed of at least three sheets of paper: outside and inside liners sandwiching the fluting, i.e., a wave-shaped sheet of paper. The mechanical performance of the final cardboard design is characterized by equal materials, number of layers, and flute dimensions (e.g., height and pitch of the flutes and number of flutes per meter). Table 1 lists the main types of flutes and their most distinguishing features.

Table 1. Main kinds of flutes, general height ranges, and most distinguishing features.

Flute Type	Flute Height Range (h, in mm)	Distinguishing Features
K—high flute	$h \geq 5$	Particularly suitable for heavy cardboards, with high flutes and double- or triple-wall boards
A—high flute	$h \geq 4.5$	High resistance to vertical compression and good dumping performance but low resistance to transversal compression and low linearity of the liners; therefore, low quality is expected for printings on liners
C—mean flute	$3.5 < h \leq 4.4$	Good compromise between mechanical performance (resistance to external forces) and paper consumption
B—low flute	$2.5 < h \leq 3.4$	High linearity of the liners, with a high quality expected for printings on liners, but high paper consumptionGood resistance to transversal compression but limited to vertical compression
E—micro flute	$1.2 < h \leq 2.4$	
F—micro flute	$0.8 < h \leq 0.9$	
G—micro flute	$0.5 < h \leq 0.6$	

Three elements were evaluated in the selection of a grade (or composition) of cardboard most suitable for the delivery of medical items under severe weather conditions:

(e.1) Outside liner: The external liner of the package must be made of virgin fiber paper, as it provides resistance to impacts and absorption of liquids. These characteristics can be assessed through burst tests and Cobb tests, respectively, as described in Test Standard DIN EN ISO 535; the former involves the use of a Mullen tester to measure the pressure force required to puncture or rupture the face of the corrugated board (in kPa), and the latter measures the amount of water absorbed by a defined area of cardboard within a defined amount of time (typically 60, 180, or 1800 s) resulting from one-sided contact with water (in g/m^2). Holding other characteristics equal, the selected grade must maximize the results of the burst test and minimize those of the Cobb test;

(e.2) ECT/cardboard density ratio: The edge compression test (ECT) measures the strength of corrugated board in response to edgewise compression. In the test, a small segment of cardboard is compressed on its edge between two rigid plates orthogonal to the direction of the flutes, and the force required to establish a peak load is measured (in kN per linear meter of load-bearing edge (kN/m), although often reported as an ECT value). The chosen grade must provide the best ECT result with equal mass;

(e.3) ECT_{min}: for the selected grade, a minimum value of 7 kN/m is required.

The following materials were evaluated:

- KLW (white kraftliner), a virgin fiber paper characterized by high resistance to humidity and high mechanical performance;
- WS (Wellenstoff), a recycled paper typically used in flutes and middle-layer paper; and
- TL (test liner), a recycled paper characterized by low cost and low performance.

Low thickness of the grade and low total mass of the package were considered as additional preferences, and environmental friendliness of the material, production costs, and process sustainability were also evaluated as selection criteria for the packaging composition.

2.4. Industrialization Process

The realization process for the package comprises three main phases: the cardboard cut, which includes both the silhouette die cut and the creation of creases; scrap removal; and the final product collection.

The silhouette of the unfolded box must be optimized to allow for quick and simple package folding, optimal operational conditions for the machining process, and maximization of the cardboard sheet yield.

The final package should minimize the number of actions required by the user for folding; this can be achieved by presetting the optimal arrangement of flaps to facilitate the procedure. Specific convenient expedients can be implemented to this end: for example, smart shaping of the flaps could enable auto-centering and plug-in closure or hook-locking, avoiding the need for glues or adhesive tapes.

With respect to realization of prototypes, the employed professional machines (provided by Imbal Carton S.r.l., Prevalle BS, Italy) require a minimum width of 5 mm for die cutting of the windows. When die cutting, non-idealities should be also considered, such as the anisotropy of the cardboard properties along the sheet directions (width and length) resulting from the presence of one or more fluting liners. The position and the maximum extension of the flaps should also be evaluated, as particularly unbalanced ratios of length and width of the flaps could translate to a loss of resistance, in addition to flute orientation with respect to fold lines and the direction of maximum extension of the flaps. The esthetics and equilibrium of shapes should also be considered as relevant factors with respect to the definition of the final package geometry.

2.5. Quick-Release System

The quick-release system is expected to fulfill a double role: to allow for prompt access to the package for the final user on one hand, as well as to protect the package from physical constraints, preventing access by unauthorized personnel.

To lock the relative movement between package and cargo bay, only simple support constraints have been taken into consideration. The constraint is bilateral, considering that the box is supported by the fuselage on the opposite side. This surface, together with a set of clamps, prevents the package from moving, constraining the six degrees of freedom (DoFs) of the rigid-body box.

A set of preliminary parameters were identified to enable a first comparison among possible solutions. The following aspects were evaluated:

(a.1) Mass: overall mass of the system (could be substituted by volume or material cost, as all the proposed solutions are intended to be realized with the same material);

(a.2) Realization: quantification of the realization costs of the system; assessed in terms of the number of elements that require precision machining;

(a.3) Accessibility: ease of use by the user during the release and detachment of the package; quantified in terms of the number of elementary operations required of the user to retrieve the package;

(a.4) Locking: the ability of the system to constrain the package in terms of contact surfaces or redundancy; evaluated in terms of the area of the system at a minimum distance from the package;

(a.5) Protection: provides an indication about the level of protection that the system can assure; is quantified as the area of the package free from fuselage protection, but covered by the system;

(a.6) Resistance: the capacity of the system to provide functional support to the resistance of the fuselage; evaluated in terms of the minimum transversal section of the system along the package extension.

Because the proposed solutions are presented at the concept level and the aim of evaluating the proposed parameters was to rank preferences, the values assigned to the six parameters were normalized between 0 and 100% to facilitate comparison. Following normalization, a value of 100% represents the ideal condition, i.e., low mass, low realization cost, few actions required for accessibility, wide locking and protection surfaces, and a wide resistance section.

3. Results

3.1. Industrialized Packaging

A right prism geometry with an octagonal section was selected for the final package design. The industrialized design was slightly modified with respect to the original concept, making the section prism equilateral. Although this reduces the actual internal volume available for storage, it also simplifies the package realization and folding procedure.

Figure 2 shows the folded prototype of the industrialized packaging. The final dimensions of the unfolded package are 320 × 83 × 83 mm, whereas the dimensions of the cardboard sheet are (1291 × 745) mm. Figure 3 shows a technical drawing of the sheet layout for die cutting of the package, revealing that the optimal solution allows for a yield of two boxes per sheet.

Figure 2. Prototype of the final packaging design in folded configuration.

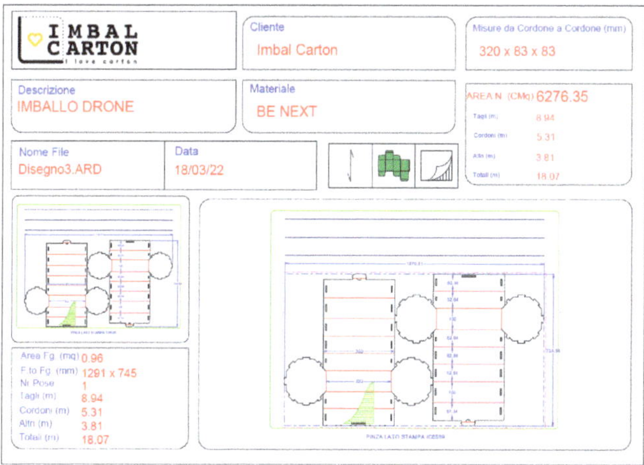

Figure 3. Technical drawing of the sheet layout for die-cutting machining.

A synthesis of the results of the performed tests and of the main features of the grade selection is reported in Table 2. KLW achieved the best results in the Cobb test (lower values of liquid absorption), so was selected as the material for the outside liner; the last column of Table 2 describes whether this specification is met.

Table 2. Extract of the analyzed grades: top ten grades according to ECT/Density ratio. The total box weight was computed based on a sample area of 0.4134 m^2. The first three columns list geometry-related values, and the results of the Burst test and ECT are reported. The final columns report whether (x) or not the checked condition is met.

Grade	Density (g/m^2)	Thickness (mm)	Total Box Weight (kg)	Burst (kPa)	ECT (kN/m)	ECT/Density	External Paper	ECT$_{min}$ Check	External Paper Check
2.40 BE X6	595	3.7	0.246	892.6	8.8	14.790	PLK 140	x	
1.24 B X2	379	2.8	0.157	841.0	5.4	14.248	KLW 115		x
1.22 E X6	319	1.5	0.132	874.8	4.5	14.107	KL 115		x
2.28 BC Q9 *	624	5.7	0.258	1095.8	8.7	13.942	KLW 115	x	x
2.28 BE X7 *	583	3.7	0.241	984.0	8.1	13.894	KLW 115	x	x
1.21 B Q7	411	2.3	0.170	746.0	5.7	13.869	PLK 120		
1.24 B Q7	411	2.3	0.170	856.6	5.7	13.869	KLW 115		x
2.47 BC N9 *	694	5.8	0.287	1252.9	9.6	13.833	KLW 135	x	x
1.40 C X2	502	3.8	0.208	1006.8	6.9	13.745	PLK 175		
2.32 BC N2	543	5.7	0.224	959.3	7.4	13.628	PLK 120	x	

* Grades satisfying both ECT$_{min}$ and external paper check.

Three grades (2.28 BC Q9, 2.28 BE X7, and 2.47 BC N9) satisfy both conditions of both an ECT value higher than 7 kN/m and KLW as the external paper. Among them, 2.28 BE X7 presented the lowest thickness and was therefore chosen as the cardboard material. This grade, also known as "Next generation board$^®$", is a double-wall board, the composition of which is listed in Table 3, with an overall thickness ranging between 3.7 and 3.9 mm, which is considerably lower than the typical values for comparable BE compositions (4.5 mm).

Table 3. Composition of the chosen cardboard.

Layer	Material	Paper Density (g/m^2)
Outside liner	KLW	115
Fluting	WS	115
Middle liner	WS	90
Fluting	WS	80
Inside liner	TL	135

The procedure for packaging folding includes the following steps:

(s.1) The good to be delivered, equipped with its standard box, if necessary, is positioned on the package rectangular face connected to the two lateral octagonal faces;

(s.2) The octagonal faces are folded perpendicularly to the support surface toward the inner side of the box (see Figure 4a);

(s.3) Grasping the box by the octagonal faces, the box is roto-translated on the support surface so that the lateral flaps of the octagonal face lock with the windows of each rectangular face (see Figure 4b);

(s.4) Once all the rectangular faces are locked to the octagonal bases, the final flap on the longitudinal dimension of the last rectangular face is hook-locked to the corresponding window, closing the box (see Figure 4d);

(s.5) Additional elements, such as a seal of warranty, can be affixed to the package, although not necessary for packaging functionality.

If the good is sensitive to orientation, s.4) can be performed inversely, i.e., bringing the rectangular faces to wrap the good and to lock with the octagonal bases (see Figure 4c).

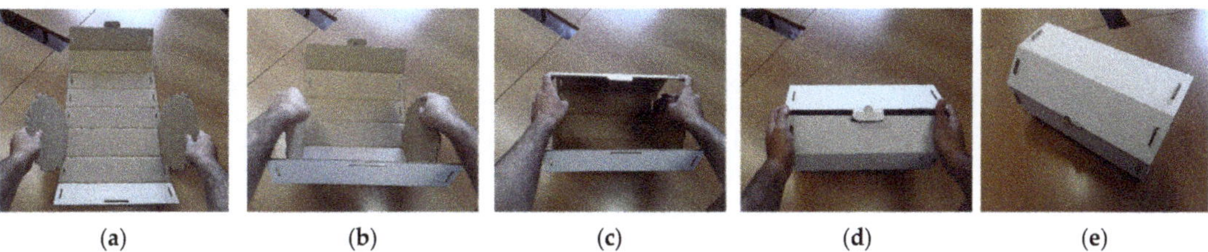

(a)   (b)   (c)   (d)   (e)

Figure 4. Some of the successive phases for packaging folding: (**a**) folding of the lateral octagonal faces, (**b**) roto-translation of the box to lock the flaps of the octagonal face with the windows of the rectangular faces, (**c**) folding by wrapping of the rectangular faces on the flaps of the octagonal faces (folding strategy alternative to (**b**)), (**d**) closing of the final flap on the longitudinal dimension of the distal rectangular face, and (**e**) the final folded box.

3.2. Quick-Release System

Four solutions were identified for the quick-release system, referred to as (a) clamps, (b) shutters, (c) cage, and (d) Y. The concepts are depicted in Figure 5, and their working principles are briefly summarized.

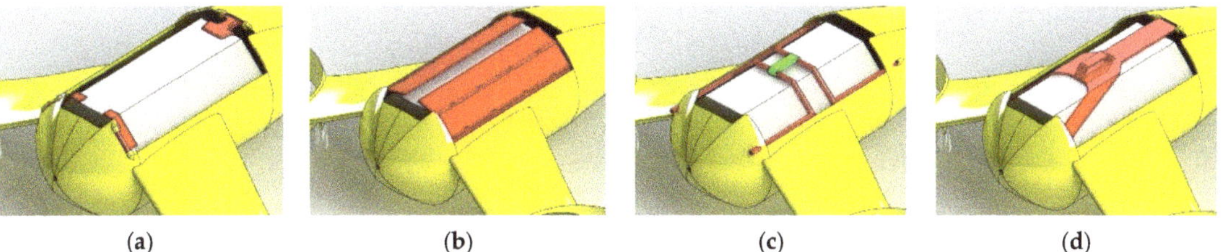

(a)   (b)   (c)   (d)

Figure 5. Identified quick-release systems: (**a**) clamp-based solution (**b**) shutter-like system, (**c**) cage-based architecture, and (**d**) Y-like structure.

3.2.1. Clamps

The first solution comprises three clamps, each rotating around a hinge and shaped to adapt to the package profile, as shown in Figure 6.

(a)   (b)

Figure 6. Detail of the clamps: one side of the package is constrained by two clamps (**a**), and the opposite side is constrained by a single clamp (**b**).

Each clamp is expected to be equipped with an additional latch (not visible in Figure 6) for safety purpose, which can mechanically force the clamps to maintain the locking configuration. Clamps actuation can be achieved with three main strategies:

(c1) The clamps are directly connected to an electric actuator that manages opening and closing of the clamps and can be remotely controlled. The system is user-independent, as locking and unlocking operations do not require user action;

(c2) Clamp-closing operations are performed automatically. Once the package is locked, the safety latch can be automatically closed and opened. To release the package, the user must wait for the remote opening of the safety latch, then manually open each clamp;

(c3) Clamp closing and safety latch locking and unlocking managed as in (c2), but every clamp is provided with a spring that is compressed when the clamps are closed. Opening of the safety latch enables the automatic release of the clamps without action by the user.

3.2.2. Shutters

The second solution involves a shutter-like closing system composed of several rectangular modular elements connected in series through sets of coaxial rotational hinges, creating two shutter units (see Figure 7).

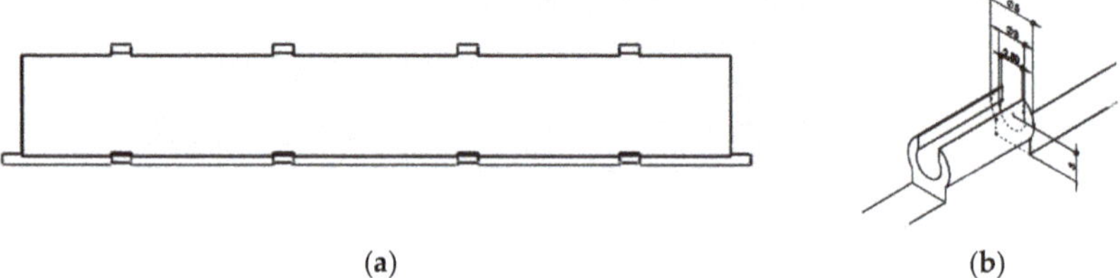

(a)   (b)

Figure 7. Details of the shutter-like system. From left to right: (**a**) the modular element and (**b**) the hinge knuckle.

The shutters move along arc-shaped rails, either automatically or manually actuated by the user through a handle (not depicted in Figure 5(b)). Once the shutters are closed, the units are locked with a hook (not shown in Figure 5b).

This solution is compliant with the presence of a safety latch, and two safety latches are included close to the rails to constrain the movement of the shutter units.

3.2.3. Cage

The third solution is composed of two symmetrical cage elements, each shaped from a tubular unit. The elements translate longitudinally along linear runners and can be locked with a hook to constrain the package when converging toward one another (Figure 8).

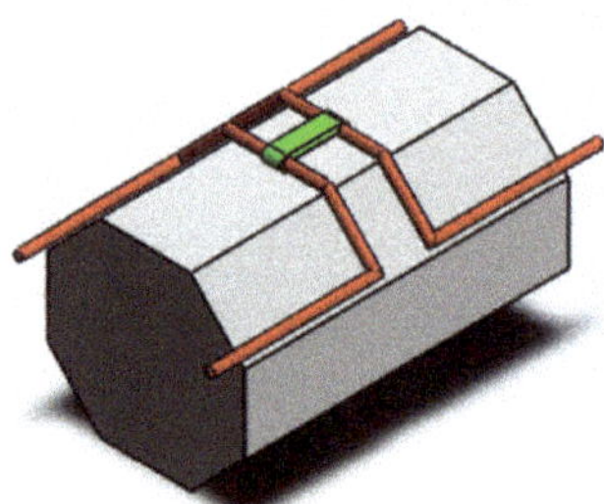

Figure 8. Closed configuration of the cage-based architecture.

Similarly to the previous cases, this solution can be integrated with safety latches (two to four) to prevent motion of the cage elements along the trails, and the central hook can be manually activated by the user or remotely actuated to reduce the actions required by the user.

3.2.4. Y

The final proposed solution comprises a monolithic Y-like structure connected to the fuselage with a hinge at the base of the Y-shaped element. Each of the distal branches of the element is equipped with a mobile latch, which enables locking of the Y module to the fuselage. The package is constrained by the lower surface of the Y element, which envelopes the box.

The mobile latches can be actuated according to many strategies. Figure 9 depicts a linkage mechanism that allows for movement of both the latches with a single degree of freedom. Figure 9a shows the kinematic chains composing the right and left sides of the system, comprising a four- and five-bar mechanism, respectively. The cranks are rigidly connected to the central rotational joint, which, once actuated, forces the distal links to slide along the system sliders. Actuation can be provided manually with a dedicated handle or automatically through electric actuation. The mechanism could be integrated in the Y element but was designed to be integrated into the drone fuselage.

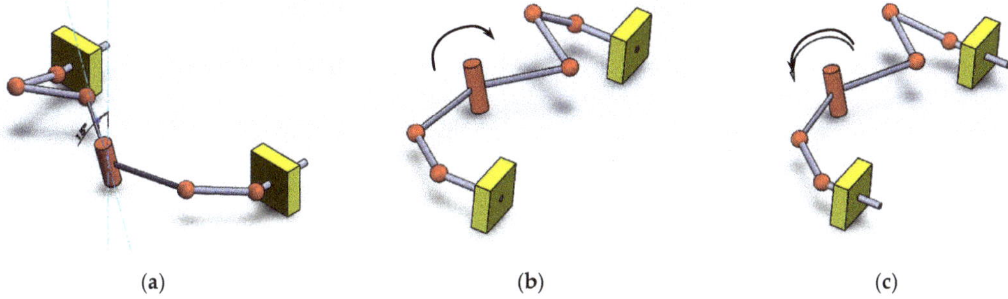

(a)   (b)   (c)

Figure 9. The proposed 1-DoF linkage mechanism in the general position (**a**), opened, (**b**) and closed configuration (**c**). Red color identifies the joints, the spheres indicate spherical joints, and the cylinder represents a rotational joint. Yellow elements represent portions of the drone fuselage.

This solution allows for integration of additional safety latches, which could be remotely controlled to prevent system movement.

3.2.5. Solution Comparison

To compare the identified solutions, the six parameters of mass, realization, accessibility, locking, protection, and resistance were evaluated, and the assigned values were normalized in percentage, as shown in Table 4.

Table 4. Normalized values of the parameters of the six identified solutions for preference ranking. Values are expressed in percentage.

Parameter	Clamps	Shutters	Cage	Y
Mass	100	0	92	81
Realization	100	0	95	95
Accessibility	100	0	0	33
Locking	18	0	55	100
Protection	0	100	11	20
Resistance	0	72	90	100

4. Discussion

Integration of packaging within the UAV main body is feasible for medium and small packages. In the present study, a payload of 5 kg was assumed for a medium-sized drone. An embedded payload system provides several advantages for the overall delivery process, for instance, repeatability and stability performance of the UAV. This solution avoids turbulence resulting from the interaction between packaging and air. Furthermore, if no external bodies interfere with the aerodynamics of the drone during the flight, aerodynamic parameters, especially drag loads, are theoretically predictable, and the final performance is independent of package dimensions.

In the case of transportation of medical items, the delivered goods require the utmost care and introduce additional constraints to an already challenging scenario, with the interaction between drones and users presenting uncertainties and generating unexpected conditions. An embedded solution would ensure improved security levels by design, as potential sources of failure, such as gimbals or ropes, are not present, and access to the package can be better managed, reducing the risk of harm resulting from interventions by unauthorized personnel.

4.1. Packaging

A study of custom packaging specific to the considered UAV allows for optimization of the payload capacity of the drone in terms of available volume.

Various materials were evaluated for the cardboard, with different grades. The starting set of possible materials was identified considering expert indications to exclude, for instance, materials that are not easily procurable, and the final grade was selected according to the results of burst and Cobb tests, as well as ECT values assessing the mechanical characteristics of the board. The selected grade for the corrugated cardboard is light-weight, with reduced dimensions and high-performance characteristics, offering high resistance to humidity, atmospheric agents, and external stresses (such as impacts and vibrations), in addition to offering thermal insulation properties and satisfactory mechanical characteristics (such as compressive strength).

The industrialization process allows for minimization of the production costs, and the identified package shape enables delivery of goods with traditional boxes, as it can envelop and protect them during transportation. Volume within the package and inner box is available for additional features; in the case of delivery of medical items, it can house small passive refrigeration units, dry ice boxes, and sensors for continuous monitoring of the good, as suggested in the literature [19]. New solutions are already under evaluation to integrate elements into the package enabling not only the detection of relevant phenomena but also active control of the desired parameters. The industrialization process also supports a careful design of the assembly procedure for the package; a set of instructions describing how to fold/unfold the package were established to be included with the box itself, if not directly printed on the box surfaces. This feature is enabled by the characteristics of the chosen material (the specific BE composition) and is in line with indications that emerge from the literature; in particular, in the case of medical delivery, user interaction with UAVs can be strongly affected by high-stress conditions, anxiety, and a sense of urgency associated with a fear that delays could result in serious or fatal consequences [17].

To the best of our knowledge, the identified packaging solution represents the first single-use packaging developed for medical delivery purposes composed of high-performance and fully recyclable material. The package is folded and closed without glue, so the end user can completely unfold the box to the original shape to easily access the delivered good, and the chosen material does not present biohazards. Furthermore, the package material is environmentally friendly, and the production process satisfies sustainability requirements, maximally exploiting the original paper sheets, and produced scraps are fully recyclable.

4.2. Quick-Release System

The considered UAV was not designed to deliver packages under conditions of station-ary flight (parcel delivery with/without parachute, mechanical arm, or a similar design) but for ground delivery. The cargo-bay position is designed to allow the user to easily approach the package and is compatible with wheelchair use. Vertical extraction of the package from the drone main body was conceived to enhance usability in critical or emergency situations. The basic rationale is that after a quick release of the package, the box can be picked up by lifting it from inside the fuselage with one or more handles. The packaging protects the delivered good, also providing structural value, given the integration of the package into the fuselage. In this sense, the quick-release system is a key element of the delivery process.

To control packaging access, quick release is expected to be performed by a combination of local physical release and a remotely controlled safety electromechanical release. All of the proposed solutions enable implementation of such a double-check strategy.

The set of described parameters presented for comparison of the quick-release solutions represents an attempt to objectively assess a mix of quantitative and qualitative aspects that characterize each concept and that could be relevant in the selection process of the most suitable solution for a specific application. Nonetheless, this approach presents important limitations: i) the list of parameters could be arbitrarily modified and integrated, and ii) the meaning and importance assigned to each parameter for a given application could vary depending on the experience and sensitivity of the designer. For instance, the concept of accessibility was assessed in the present study according to the number of actions required by the user to extract the package. In the normalization step, fewer operations were classified as the optimal condition, indicating ease of use; however, with a different mindset, a high number of operations could be preferred, indicating an index of voluntary action.

4.3. Limitations of the Study

Besides the limitations of with respect to the specific applied methods, such as the necessity of preselecting possible materials to select the final cardboard grade or the described limits related to the choice of a specific set of parameters for evaluation of the quick-release solutions, the main limitation of the current work lies in the impossibility of completely validating the effectiveness of the packaging solution for any kind of medical item. This is mainly due to two reasons: on one hand, regulations determine different sets of requirements depending on the specific materials to be delivered; on the other hand, the quick-release system represents an expected outer third level of protection of the package, which, according to regulations, needs to be considered for performance evaluation of the overall packaging system. For the same reason, a direct comparison with currently available containers would provide few indications, given the peculiarities of the proposed package. Nonetheless, a preliminary set of packaging tests is currently being planned, with the aim of assessing leak-proof and sift-proof capabilities, as well as the maximum internal pressure that can be withstood without leakage.

5. Conclusions

In the present study, we investigated payload solutions for medium and small package delivery with a medium-sized UAV, considering three main aspects: (i) an embedded solution, (ii) medical transportation compatibility, and (iii) user-oriented design.

The payload solutions were analyzed at different levels, from the package design, in terms of material, shape, and product industrialization; to package integration within the drone fuselage, focusing on possible concepts for the implementation of a quick-release system. Particular attention was given to the interaction between the user and the drone, with a focus on usability and safety aspects.

A prototype of the industrialized version of the package was realized, and a set of concepts for a quick-release system were proposed. Further analyses are currently ongoing, with the aim of integrating monitoring and control capabilities for the package to assessing the condition of the delivered good during transportation.

Author Contributions: Conceptualization, M.S., A.B., R.A., R.F. and C.A.; methodology, A.B., R.A., R.F. and C.A.; formal analysis, M.S. and C.A.; investigation, M.S. and C.A.; resources, M.S.; data curation, M.S. and C.A.; writing—original draft preparation, M.S. and C.A.; writing—review and editing, A.B., R.A. and R.F.; supervision, M.S. and C.A.; project administration, M.S. and C.A.; funding acquisition, M.S. and C.A. All authors have read and agreed to the published version of the manuscript.

Funding: This research was partially funded by Regione Lombardia, Call Hub Ricerca e Innovazione, within the project "MoSoRe@Unibs—Infrastrutture e servizi per la Mobilità Sostenibile e Resiliente", 2020–2022.

Institutional Review Board Statement: Not applicable.

Informed Consent Statement: Not applicable.

Data Availability Statement: Not applicable.

Acknowledgments: The authors thank Michele Franceschetti and Filippo Prandini for their invaluable support.

Conflicts of Interest: The authors declare no conflict of interest.

References

1. Benarbia, T.; Kyamakya, K. A Literature Review of Drone-Based Package Delivery Logistics Systems and Their Implementation Feasibility. *Sustainability* **2021**, *14*, 360. [CrossRef]
2. Amici, C.; Ceresoli, F.; Pasetti, M.; Saponi, M.; Tiboni, M.; Zanoni, S. Review of Propulsion System Design Strategies for Unmanned Aerial Vehicles. *Appl. Sci.* **2021**, *11*, 5209. [CrossRef]
3. Amici, C.; Ceresoli, F.; Saponi, M.; Pasetti, M.; Zanoni, S.; Borboni, A.; Tiboni, M.; Faglia, R. Experimental Characterization of an Electrical Propulsion Unit for Service UAVs. In *Proceedings of I4SDG Workshop 2021*; Quaglia, G., Gasparetto, A., Petuya, V., Carbone, G., Eds.; Springer International Publishing: Cham, Switzerland, 2022; pp. 307–314. [CrossRef]
4. Suzuki, K.A.O.; Filho, P.K.; Morrison, J.R. Automatic Battery Replacement System for UAVs: Analysis and Design. *J. Intell. Robot. Syst.* **2011**, *65*, 563–586. [CrossRef]
5. Ranquist, E.A.; Steiner, M.; Argrow, B. Exploring the range of weather impacts on UAS operations. In Proceedings of the 18th Conference on Aviation, Range and Aerospace Meteorology, Seattle, WA, USA, 23 January 2017.
6. Zailani, M.A.H.; Sabudin, R.Z.A.R.; Rahman, R.A.; Saiboon, I.M.; Ismail, A.; Mahdy, Z.A. Drone for medical products transportation in maternal healthcare: A systematic review and framework for future research. *Medicine* **2020**, *99*, e21967. [CrossRef]
7. Asadi, A.; Pinkley, S.N.; Mes, M. A Markov decision process approach for managing medical drone deliveries. *Expert Syst. Appl.* **2022**, *204*, 117490. [CrossRef]
8. Cheema, M.A.; Ansari, R.I.; Ashraf, N.; Hassan, S.A.; Qureshi, H.K.; Bashir, A.K.; Politis, C. Blockchain-based secure delivery of medical supplies using drones. *Comput. Netw.* **2022**, *204*, 108706. [CrossRef]
9. Khan, S.I.; Qadir, Z.; Munawar, H.S.; Nayak, S.R.; Budati, A.K.; Verma, K.; Prakash, D. UAVs path planning architecture for effective medical emergency response in future networks. *Phys. Commun.* **2021**, *47*, 101337. [CrossRef]
10. Glick, T.B.; Figliozzi, M.A.; Unnikrishnan, A. Case Study of Drone Delivery Reliability for Time-Sensitive Medical Supplies with Stochastic Demand and Meteorological Conditions. *Transp. Res. Rec. J. Transp. Res. Board* **2021**, *2676*, 242–255. [CrossRef]
11. Purahong, B.; Anuwongpinit, T.; Juhong, A.; Kanjanasurat, I.; Pintaviooj, C. Medical Drone Managing System for Automated External Defibrillator Delivery Service. *Drones* **2022**, *6*, 93. [CrossRef]
12. Bauer, J.; Moormann, D.; Strametz, R.; Groneberg, D.A. Development of unmanned aerial vehicle (UAV) networks delivering early defibrillation for out-of-hospital cardiac arrests (OHCA) in areas lacking timely access to emergency medical services (EMS) in Germany: A comparative economic study. *BMJ Open* **2021**, *11*, e043791. [CrossRef]
13. Sigari, C.; Biberthaler, P. Medical drones: Disruptive technology makes the future happen [Medizinische Drohnen: Innovative Technologie eröffnet neue Horizonte der Unfallchirurgie]. *Der Unf.* **2021**, *124*, 974–976. [CrossRef]
14. Nimilan, V.; Manohar, G.; Sudha, R.; Stanley, P. Drone-Aid: An Aerial Medical Assistance. *Int. J. Innov. Technol. Explor. Eng.* **2019**, *8*, 1288–1292. [CrossRef]
15. Claesson, A.; Bäckman, A.; Ringh, M.; Svensson, L.; Nordberg, P.; Djärv, T.; Hollenberg, J. Time to Delivery of an Automated External Defibrillator Using a Drone for Simulated Out-of-Hospital Cardiac Arrests vs Emergency Medical Services. *JAMA* **2017**, *317*, 2332–2334. [CrossRef] [PubMed]
16. Baloola, M.O.; Ibrahim, F.; Mohktar, M.S. Optimization of Medication Delivery Drone with IoT-Guidance Landing System Based on Direction and Intensity of Light. *Sensors* **2022**, *22*, 4272. [CrossRef]
17. Stephan, F.; Reinsperger, N.; Grünthal, M.; Paulicke, D.; Jahn, P. Human drone interaction in delivery of medical supplies: A scoping review of experimental studies. *PLoS ONE* **2022**, *17*, e0267664. [CrossRef]
18. Sun, J.; Shen, Y.; Rosen, J. Sensor Reduction, Estimation, and Control of an Upper-Limb Exoskeleton. *IEEE Robot. Autom. Lett.* **2021**, *6*, 1012–1019. [CrossRef]
19. Hii, M.; Courtney, P.; Royall, P. An Evaluation of the Delivery of Medicines Using Drones. *Drones* **2019**, *3*, 52. [CrossRef]
20. Quek, S.G.; Selvachandran, G.; Sham, R.; Siau, C.S.; Ramli, M.H.M.; Ahmad, N. A Fuzzy Logic Based Optimal Network System for the Delivery of Medical Goods via Drones and Land Transport in Remote Areas. In *Intelligent Systems Design and Applications*; ISDA 2021. Lecture Notes in Networks and Systems; Abraham, A., Gandhi, N., Hanne, T., Hong, T.P., Nogueira Rios, T., Ding, W., Eds.; Springer: Cham, Switzerland, 2022; pp. 1306–1312. [CrossRef]

21. Sham, R.; Siau, C.S.; Tan, S.; Kiu, D.C.; Sabhi, H.; Thew, H.Z.; Selvachandran, G.; Quek, S.G.; Ahmad, N.; Ramli, M.H.M. Drone Usage for Medicine and Vaccine Delivery during the COVID-19 Pandemic: Attitude of Health Care Workers in Rural Medical Centres. *Drones* **2022**, *6*, 109. [CrossRef]
22. Ganesan, G.S.; Mokayef, M. Multi-Purpose Medical Drone for the Use in Pandemic Situation. In Proceedings of the 2021 IEEE Workshop on Microwave Theory and Techniques in Wireless Communications, MTTW 2021, Riga, Latvia, 7–8 October 2021; pp. 188–192. [CrossRef]
23. Pavithran, R.; Lalith, V.; Naveen, C.; Sabari, S.P.; Kumar, M.A.; Hariprasad, V. A prototype of Fixed Wing UAV for delivery of Medical Supplies. *IOP Conf. Series Mater. Sci. Eng.* **2020**, *995*, 012015. [CrossRef]
24. Mohd, S.A.; Gan, K.B.; Ihsan, A.K.A.M. Development of Medical Drone for Blood Product Delivery: A Technical Assessment. *Int. J. Online Biomed. Eng. (iJOE)* **2021**, *17*, 183–196. [CrossRef]
25. Amicone, D.; Cannas, A.; Marci, A.; Tortora, G. A Smart Capsule Equipped with Artificial Intelligence for Autonomous Delivery of Medical Material through Drones. *Appl. Sci.* **2021**, *11*, 7976. [CrossRef]
26. Jacob, B.; Kaushik, A.; Velavan, P.; Sharma, M. Autonomous Drones for Medical Assistance Using Reinforcement Learning. *Stud. Comput. Intell.* **2022**, *998*, 133–156. [CrossRef]
27. Amukele, T.; Ness, P.M.; Tobian, A.A.; Boyd, J.; Street, J. Drone transportation of blood products. *Transfusion* **2016**, *57*, 582–588. [CrossRef] [PubMed]
28. Yakushiji, F.; Yakushiji, K.; Murata, M.; Hiroi, N.; Takeda, K.; Fujita, H. The Quality of Blood is not Affected by Drone Transport: An Evidential Study of the Unmanned Aerial Vehicle Conveyance of Transfusion Material in Japan. *Drones* **2020**, *4*, 4. [CrossRef]
29. Sharley, P.H.; Williams, I.; Hague, S. Blood transportation for medical retrieval services. *Air Med J.* **2003**, *22*, 24–27. [CrossRef]
30. Maity, N.M. Lightweight Thermal Insulation Systems in Medical Delivery Drones. In Proceedings of the IEEE Aerospace Conference, Big Sky, MT, USA, 6–13 March 2021; Volume 2021-March. [CrossRef]
31. Kostin, A.S.; Silin, Y.A. Development of an Insulated Container for the Implementation of the Delivery of Special Cargo Using an Unmanned Aerial System. In Proceedings of the 2022 Wave Electronics and its Application in Information and Telecommunication Systems (WECONF), St. Petersburg, Russia, 30 May 2022–3 June 2022; pp. 1–4. [CrossRef]
32. United Nations. *Agreement Concerning the International Carriage of Dangerous Goods by Road Vol. I*; UNITED NATIONS PUBLICA-TIONS: New York, NY, USA, 2021.
33. United Nations. *Agreement Concerning the International Carriage of Dangerous Goods by Road Vol. II*; UNITED NATIONS PUBLICA-TIONS: New York, NY, USA, 2021.
34. Molina, B.D.M.; Oña, M.S. The Drone Sector in Europe. In *Ethics and Civil Drones*; Springer: Cham, Switzerland, 2018; pp. 7–33. [CrossRef]
35. Chowdhury, S.; Emelogu, A.; Marufuzzaman, M.; Nurre, S.G.; Bian, L. Drones for disaster response and relief operations: A continuous approximation model. *Int. J. Prod. Econ.* **2017**, *188*, 167–184. [CrossRef]
36. Restas, A. Drone Applications for Supporting Disaster Management. *World J. Eng. Technol.* **2015**, *3*, 316–321. [CrossRef]
37. Sankey, T.; Donager, J.; McVay, J.; Sankey, J.B. UAV lidar and hyperspectral fusion for forest monitoring in the southwestern USA. *Remote Sens. Environ.* **2017**, *195*, 30–43. [CrossRef]
38. Hodgson, A.; Peel, D.; Kelly, N. Unmanned aerial vehicles for surveying marine fauna: Assessing detection probability. *Ecol. Appl.* **2017**, *27*, 1253–1267. [CrossRef]
39. Casella, E.; Collin, A.; Harris, D.; Ferse, S.; Bejarano, S.; Parravicini, V.; Hench, J.L.; Rovere, A. Mapping coral reefs using consumer-grade drones and structure from motion photogrammetry techniques. *Coral Reefs* **2017**, *36*, 269–275. [CrossRef]
40. Szantoi, Z.; Smith, S.E.; Strona, G.; Koh, L.P.; Wich, S.A. Mapping orangutan habitat and agricultural areas using Landsat OLI imagery augmented with unmanned aircraft system aerial photography. *Int. J. Remote Sens.* **2017**, *38*, 2231–2245. [CrossRef]
41. Rao, B.; Gopi, A.G.; Maione, R. The societal impact of commercial drones. *Technol. Soc.* **2016**, *45*, 83–90. [CrossRef]
42. Formicola, R.; Ragni, F.; Mor, M.; Bissolotti, L.; Amici, C. Design Approach of Medical Devices for Regulation Compatibility: A Robotic Rehabilitation Case Study. In Proceedings of the 7th International Conference on Information and Communication Technologies for Ageing Well and e-Health, Prague, Czechia, 24–26 April 2021; pp. 146–153. [CrossRef]
43. Amici, C.; Pellegrini, N.; Tiboni, M. The Robot Selection Problem for Mini-Parallel Kinematic Machines: A Task-Driven Approach to the Selection Attributes Identification. *Micromachines* **2020**, *11*, 711. [CrossRef] [PubMed]

Article

Friction-Induced Efficiency Losses and Wear Evolution in Hypoid Gears

Eugeniu Grabovic, Alessio Artoni, Marco Gabiccini, Massimo Guiggiani, Lorenza Mattei, Francesca Di Puccio and Enrico Ciulli *

Dipartimento di Ingegneria Civile e Industriale, Università di Pisa, Largo Lucio Lazzarino 1, 56122 Pisa, Italy
* Correspondence: enrico.ciulli@unipi.it; Tel.: +39-050-2218061

Abstract: A correct methodology to evaluate the friction coefficient in lubricated gear pairs is paramount for both the estimation of energy losses and the prediction of wear. In the first part of the paper, a methodology for estimating the coefficient of friction with a semi-empirical formulation is presented, and its results are also employed to analyze mechanical efficiency losses in a hypoid gearset. Hypoid gears have complex tooth surface geometries, and the entraining kinematics of the lubricant is quite involved. The second part of the paper showcases a simulated wear investigation based on the Archard model. The main focus is on the impact of the frequency adopted for updating the worn geometry of the gear and pinion teeth on the fidelity and consistency of the tribological outcomes. These are measured in terms of overall quantity of material removed and characteristics of the loaded contact pattern. More in detail, a sensitivity analysis is presented that compares the total wear of a hypoid gearset after 30 million cycles estimated using different geometry update steps. Contact pressures, which are necessary to perform the aforementioned analyses, are calculated through an accurate, state-of-the-art loaded tooth contact analysis solver.

Keywords: hypoid gears; friction; wear; elasto-hydrodynamic lubrication

Citation: Grabovic, E.; Artoni, A.; Gabiccini, M.; Guiggiani, M.; Mattei, L.; Di Puccio, F.; Ciulli, E. Friction-Induced Efficiency Losses and Wear Evolution in Hypoid Gears. *Machines* **2022**, *10*, 748. https://doi.org/10.3390/machines10090748

Academic Editors: Marco Ceccarelli, Giuseppe Carbone and Alessandro Gasparetto

Received: 30 July 2022
Accepted: 25 August 2022
Published: 29 August 2022

Publisher's Note: MDPI stays neutral with regard to jurisdictional claims in published maps and institutional affiliations.

1. Introduction

Hypoid gears are associated with high sliding speeds during meshing due to their screwing relative motion. This brings about peculiar and, in general, more severe wear patterns than other gear types. Such wear can result in a contact pattern that is drastically different from the designed one. In fact, since an optimal contact pattern is the outcome of a *micro-geometry* optimization process [1–3], a variation of just a few tens of microns can significantly change the contact properties. Hence, wear must be cautiously considered for gears operating in long-life applications.

Park et al. [4] introduced a wear model based on Archard's law interfaced with a finite element (FEM)-based contact model. The very same *loaded tooth contact analysis* (LTCA) tool, *Ansol Transmission3D* [5], is employed in this work. However, the software's capabilities have been greatly improved over the last few years, thus allowing a more accurate prediction of the contact properties and, therefore, more accurate analyses of both lubrication and wear.

The study in [6] proposed a wear model interfaced with a semi-analytical contact model developed by Kahraman and Kolivand [7]. This tool greatly reduced the computation cost to perform the LTCA simulations while trading some accuracy for efficiency. Park [8] developed a "patching" surface interpolation technique to predict wear, which allows the employment of fewer time-step discretizations of the meshing cycle. More recently, Ref. [9] investigated wear in hypoid gears and its experimental correlation with the loaded transmission error.

In the first part of this paper, a semi-empirical formulation is introduced and employed to analyze lubrication and friction-induced efficiency losses during gear meshing. Despite the simplicity of the proposed model, the estimates it provides are effective and have been validated with experimental data in [10].

Efficiency losses are an important aspect of geared transmissions. They can be classified into *load-independent* and *load-dependent* losses. The first category includes the so-called churning and windage losses (i.e., pumping of the lubricant between the mating members and splashing caused by inertial effects). The second category is related to rolling and sliding frictional losses. The rolling losses are usually negligible compared to the sliding ones, especially in hypoid gears. Our work proposes an estimation method for the friction coefficient under different lubrication regimes, which is the fundamental parameter required to accurately predict friction-induced efficiency losses.

Many friction prediction models for hypoid gears are available in the literature. Contribution [11] evaluated the lubrication performance under different possible contact paths on bevel gears, [12] carried out an elasto-hydrodynamic lubrication analysis on hypoid gears under relatively high loads, while [13] analyzed the lubrication of hypoid gears taking into account the three-dimensional surface roughness. However, despite their accurate tribological analyses, many of the cited contributions fall short in providing reliable contact analysis results. In fact, an accurate LTCA tool has a paramount importance for the subsequent estimation of the friction coefficient.

In the second part of the paper, a wear investigation is performed. The main goal is to establish how often the worn tooth flank geometry should be updated (properly accounting for the removed material) to achieve a reliable and consistent prediction of the final wear and contact pattern. To this end, we computed wear predictions after 30 million meshing cycles (counted on the pinion) using different geometry update steps. This allowed the performance of a sensitivity analysis and provide quantitative indications on the minimum number of geometry updates required to provide a reliable prediction with limited computational burden.

2. Estimation of the Lubricated Friction Coefficient

There are several factors that influence friction, such as the lubrication regime (full film, mixed, boundary), the behavior of the lubricant with varying operating conditions (temperature, pressure, shear rate) and the surrounding environment (the boundary conditions of the lubricated contact).

The friction coefficient f is evaluated from boundary to full-film lubrication conditions using a load sharing function $g(\Lambda)$, related to the portion of load supported by the full film. According to [14], f can be calculated as:

$$f = f_h g(\Lambda)^{1.2} + f_b(1 - g(\Lambda))$$ (1)

where f_h and f_b are the friction coefficients related to the hydrodynamic (full fluid) and the boundary lubrication conditions, respectively. Λ is the ratio between the film thickness h and the equivalent surface roughness of the contacting bodies:

$$\Lambda = \frac{h}{\sqrt{R_{q1}^2 + R_{q2}^2}}$$ (2)

Usually, the film thickness h is assumed to be the *central film thickness* h_c, and R_{q1} and R_{q2} are the root mean square roughnesses of the two surfaces.

Several expressions can be found for $g(\Lambda)$ in the literature. The formulation in [14] is employed in this work, which reads:

$$g(\Lambda) = 0.84\Lambda^{0.23} \text{ if } \Lambda \leq 2; \ g(\Lambda) = 1 \text{ if } \Lambda > 2$$ (3)

The coefficient of friction f_b is considered constant (with $f_b = 0.08$ being the typical value [14]), while the mean value of f_h can be evaluated as the ratio between the shear stress τ and the mean contact pressure p_m:

$$f_h = \frac{\tau}{p_m} \tag{4}$$

For elliptical nonconformal contacts, $p_m = \frac{F}{\pi a b}$, where a and b are the axis semi-widths of the contact ellipse and F is the normal load.

The estimation of τ is performed according to the Eyring constitutive model:

$$\tau = \tau_E \sinh\left(\frac{\eta \dot{\gamma}}{\tau_E}\right)^{-1}, \tag{5}$$

where τ_E is the Eyring shear stress, η is the dynamic viscosity, and the shear strain rate is approximated as $\dot{\gamma} = \frac{\Delta u}{h_c}$, with Δu being the sliding velocity between the two mating teeth at the nominal contact point.

The dynamic viscosity variation with temperature T and pressure p is described by the empirical formulas introduced by Vogel and Gold [15]:

$$\eta = \eta_0 e^{\frac{B}{T-C} + \alpha_P p}, \tag{6}$$

where the viscosity-pressure coefficient is calculated as follows:

$$\alpha_P = \frac{10^{-5}}{a_1 + a_2 T + (b_1 + b_2 T)(p 10^{-5})} \tag{7}$$

The remaining coefficients in Equations (6) and (7) need to be estimated empirically, based on the properties of the specific lubricant.

Different expressions for the evaluation of the central film thickness can be found in the literature. Four different lubrication regimes can be observed for nonconformal contacts, depending on the elastic deformation of the bodies and the variation of viscosity with pressure. The four regimes are usually indicated as isoviscous-rigid IR, piezoviscous-rigid PR, isoviscous-elastic IE and piezoviscous-elastic PE. The formulas reported in [16] were elaborated in [17] for the more general case in which the entraining velocity is not collinear with any principal direction. The final expressions employed in this work read as follows:

$$h_{c_{IR}} = 128 \frac{\eta_0^2 u^2 R_e^3}{F^2}\left(\frac{R_s}{R_e}\left(0.131 \arctan\frac{R_s}{2R_e} + 1.683\right)^2 \left(1 + \frac{2R_e}{3R_s}\right)^{-2}\right) \tag{8}$$

$$h_{c_{PR}} = 141 \frac{\eta_0^{1.25} \alpha^{0.375} u^{1.25} R_e^{1.5}}{F^{0.875}}\left(1 - e^{-0.0387\frac{R_s}{R_e}}\right) \tag{9}$$

$$h_{c_{IE}} = 11.15 \frac{\eta_0^{0.66} u^{0.66} R_e^{0.766}}{F^{0.213} E^{0.447}}\left(1 - 0.72 e^{-0.28\left(\frac{R_s}{R_e}\right)^{\frac{2}{7}}}\right) \tag{10}$$

$$h_{c_{PE}} = 3.61 \frac{\eta_0^{0.68} \alpha^{0.53} u^{0.68} R_e^{0.446}}{E^{0.087} F^{0.063}}\left(1 - 0.61 e^{-0.73\left(\frac{R_s}{R_e}\right)^{\frac{2}{7}}}\right) \tag{11}$$

Here, E is the equivalent elastic modulus, u is the entraining velocity, and R_e and R_s are the equivalent radii of curvature parallel and perpendicular to u, respectively. An accurate description of u, R_e and R_s is provided in Section 3. The proper lubrication regime is determined in a practical way by taking the highest of the values given by Equations (8)–(11).

Thermal effects are included using a reduced value for the central film thickness obtained by multiplying it by a dimensionless reduction factor Φ. Several models for Φ are available. The one described in [18] is used here:

$$\Phi = \frac{1}{1 + 0.1(1 + 8.33S^{0.83})L^{0.64}},\tag{12}$$

where $S = \Delta u / u$ and $L = \beta \frac{\eta_0 u^2}{k}$ is the dimensionless thermal loading parameter.

The coefficients that characterize the 75W90 oil, which is the lubricant used in this study, are shown in Table 1. Those experimental values have been obtained from [19].

Table 1. Coefficients of the 75W90 oil.

	Symbol	Value
	β	$2.86 \cdot 10^{-2}\ [\mathrm{K}^{-1}]$
Thermal conductivity	k	$0.14\ [\frac{\mathrm{W}}{\mathrm{mK}}]$
Eyring shear stress	τ_E	$8\ [\mathrm{MPa}]$
Vogel coefficients		
	η_0	$5.90 \cdot 10^{-5}\ [\mathrm{Pa\ s}]$
	B	$1205.5\ [^\circ\mathrm{C}]$
	C	$-125.3\ [^\circ\mathrm{C}]$
Pressure-viscosity coefficients		
	a_1	$4.929 \cdot 10^2\ [\mathrm{bar}]$
	a_2	$3.901\ [\frac{\mathrm{bar}}{^\circ\mathrm{C}}]$
	b_1	$2.479 \cdot 10^{-2}$
	b_2	$2.354 \cdot 10^{-4}\ [\frac{1}{^\circ\mathrm{C}}]$

It is worth remarking that the formulas introduced in this section provide a *first approximation* of the lubricant film thickness and coefficient of friction. In particular, very high sliding speeds are associated with large increments of temperature that may cause a significant decrease in viscosity and hence in film thickness. The problem in its full generality would be also time-dependent, but some preliminary film thickness evaluation can be made considering a stationary situation with constant speed and geometry per contact configuration of the mating teeth. It may be worth noting in passing that the film thickness would also be influenced by the presence of spin, which is neglected here for simplicity.

3. Generation and Kinematic Analysis of Hypoid Gears

The face-milled tooth flank geometries are digitally synthesized through simulation of the generation process (by envelope) during finishing (grinding). The enveloping motions are defined by the kinematic parameters (*machine settings*) of the 9-axis Gleason hypoid generator and the geometry of the grinding tool (*tool settings*). A special program developed by the authors [20] allows the obtaining of those settings from the basic macro-geometry data. A plot of the hypoid gearset under investigation is shown in Figure 1.

Through geometric modelling and simulated loaded contact analysis (LTCA), it is also possible to compute fundamental kinematic and geometric information at each contact point, such as (subscript $j = 1/2$ refers to the gear/pinion, respectively; superscript $i = 1/2$ refers to the first/second direction on each surface):

- velocities $\mathbf{u}_1$ and $\mathbf{u}_2$ of the two bodies (teeth) at the contact point;
- angular velocities $\boldsymbol{\omega}_1$ and $\boldsymbol{\omega}_2$ of the bodies;
- principal curvatures of the bodies $K_1^{(1)}$, $K_1^{(2)}$, $K_2^{(1)}$ and $K_2^{(2)}$ and their directions $\boldsymbol{\tau}_1^{(1)}$, $\boldsymbol{\tau}_1^{(2)}$, $\boldsymbol{\tau}_2^{(1)}$ and $\boldsymbol{\tau}_2^{(2)}$.

- principal relative curvatures K_x and K_y and their associated directions $\boldsymbol{\tau}_x$ and $\boldsymbol{\tau}_y$. The direction along the largest semi-width of the contact ellipse is denoted as $\boldsymbol{\tau}_y$.

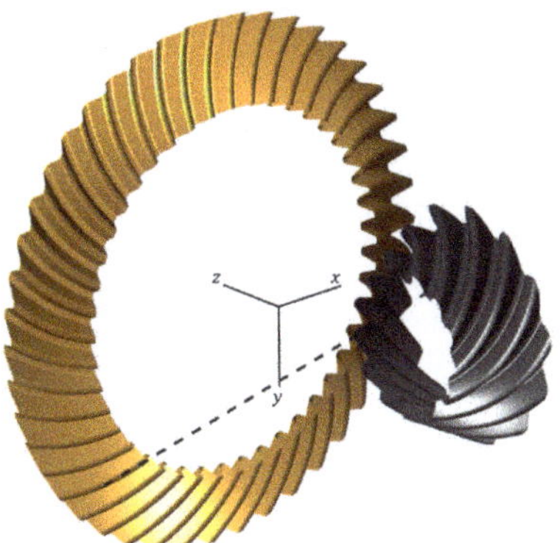

Figure 1. Hypoid gearset. The dashed line represents the instantaneous screw axis of the relative motion.

The principal normal curvature $K_j^{(i)}$ of body j along direction $\boldsymbol{\tau}_i$ is considered positive/negative if the normal plane through $\boldsymbol{\tau}_i$ cuts body j producing a section with a convex/concave boundary curve.

To employ the formulas described in Section 2, the entraining velocity **u** of the lubricant needs to be computed. In addition, the direction of such velocity does not coincide with any of the principal directions at the contact point. However, given the angular velocities of the bodies, their linear velocities at the contact point and their curvatures K_{1_x}, K_{1_y} for the gear wheel and K_{2_x}, K_{2_y} for the pinion, respectively, along the principal relative directions $\boldsymbol{\tau}_x$ and $\boldsymbol{\tau}_y$ of the equivalent contact, the components of the entraining velocity vector can be obtained as [17,21]:

$$u_x = (\omega_{2_y} - \omega_{1_y})\frac{1}{K_{1_x} + K_{2_x}} + \frac{1}{2}(u_{1_x} - u_{2_x})\frac{K_{2_x} - K_{1_x}}{K_{1_x} + K_{2_x}} \tag{13}$$

$$u_y = (\omega_{1_x} - \omega_{2_x})\frac{1}{K_{1_y} + K_{2_y}} + \frac{1}{2}(u_{1_y} - u_{2_y})\frac{K_{2_y} - K_{1_y}}{K_{1_y} + K_{2_y}} \tag{14}$$

These are the components of the entraining velocity vector in the local frame with its axes aligned with the principal directions of the equivalent contact. The norm of the entraining velocity vector $u = \sqrt{u_x^2 + u_y^2}$ is employed for the estimation of the friction coefficient formulas introduced in the previous section. It is important to remark that the validity of Equations (13) and (14) is subject to the following hypotheses:

1. body 1 (tooth of the gear wheel) is always convex along both its principal directions (which is true under the so-called *Drive* operating conditions, i.e., the gear convex tooth sides make contact with the pinion concave tooth sides);
2. the cross product $\boldsymbol{\tau}_x \times \boldsymbol{\tau}_y$ yields a unit vector always pointing inside body 1.

As a final step, the radii of curvature along the entraining and the side-leakage directions (R_e and R_s) can be obtained by [22]:

$$\begin{aligned} R_e &= (K_x \cos^2\theta + K_y \sin^2\theta)^{-1} \\ R_s &= (K_x \sin^2\theta + K_y \cos^2\theta)^{-1}, \end{aligned} \tag{15}$$

where $\theta = \arctan(u_y/u_x)$ is the angle between the entraining velocity and $\boldsymbol{\tau}_x$.

4. Numerical Application on a Hypoid Gearset

An application of the aforementioned friction coefficient model is presented in this section. The basic data of the hypoid gearset employed in our study are shown in Table 2.

Table 2. Hypoid basic parameters.

Parameter	Value
Hypoid offset	30 mm
Shaft angle	90 deg
Hand (pinion)	Left
Gear type	Formate
Pinion teeth	13
Gear teeth	44
Spiral angle (pinion)	45.5 deg
Face width (gear)	31.87 mm
Pitch diameter (gear)	192.36 mm
Cutter mean radius	63.5 mm
Nominal pressure angle	20 deg
Profile shift coefficient	0.524
Profile crowning (pinion)	60 μm
Lengthwise crowning (pinion)	100 μm
Surface roughness	0.45 μm
Operating oil temperature	100 °C

The machine-tool settings computed by a special program developed by the authors, and described in detail in [20], are given as input to Transmission3D to perform an accurate LTCA analysis. More in detail, the contact simulation is performed by discretizing the overall meshing cycle into $N_{\text{steps}} = 15$ contact configurations (time steps). The analysis results are then extracted and post-processed to carry out the calculations for the estimation of the friction coefficient. We assumed a pinion torque of 250 Nm and a pinion speed of 2000 rpm.

Figure 2 shows the load shared by a mating tooth pair over a mesh cycle (as a function of the pinion rotation angle φ_2). Figure 3 shows the entraining velocity of the lubricant. Figure 4 shows the evolution of the entraining and the side-leakage radii. Those values are strongly dependent on the micron-level flank deviations from the conjugate surfaces. As a matter of fact, micro-geometry can drastically change the evolution of the contact path, which may result in different contact zones and thus different local curvatures.

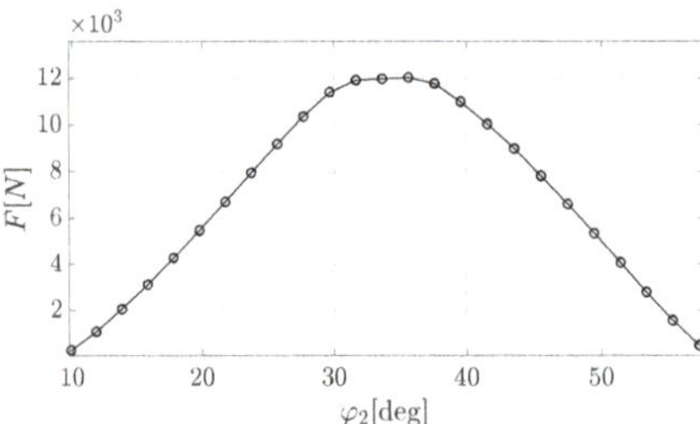

Figure 2. Load shared by the contacting teeth over a mesh cycle.

In Figure 5 we can observe the different central film thicknesses calculated for the different lubrication regimes. At each contact configuration, the lubrication regime is the piezo-viscous-elastic one, probably due to the high loading of the gearset. This results in an analogous trend of the Λ factor, which can be observed in Figure 6. It is clear that the meshing tooth surfaces always operate under mixed lubrication conditions. To increase Λ, a lower surface roughness of the tooth flanks should be aimed for. Finally, the coefficient of friction is shown in Figure 7.

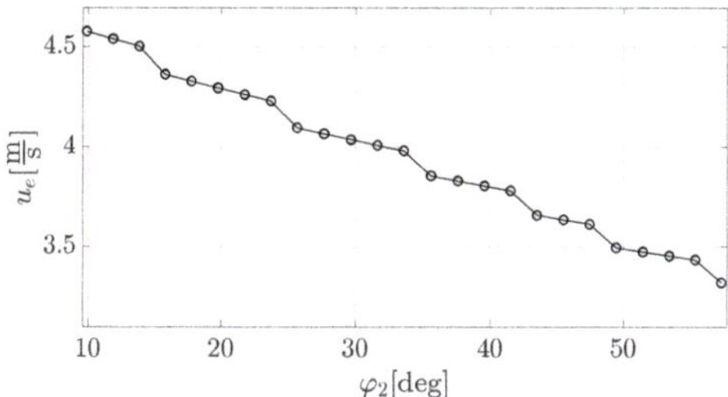

Figure 3. Entraining velocity u of the lubricant over a mesh cycle.

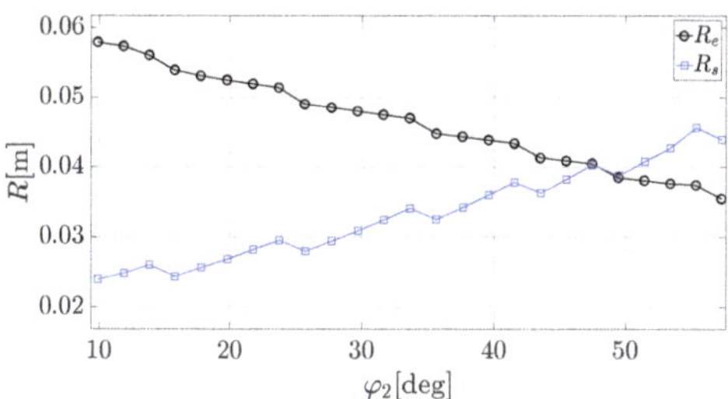

Figure 4. Entraining (R_e) and side leakage (R_s) radii over a mesh cycle.

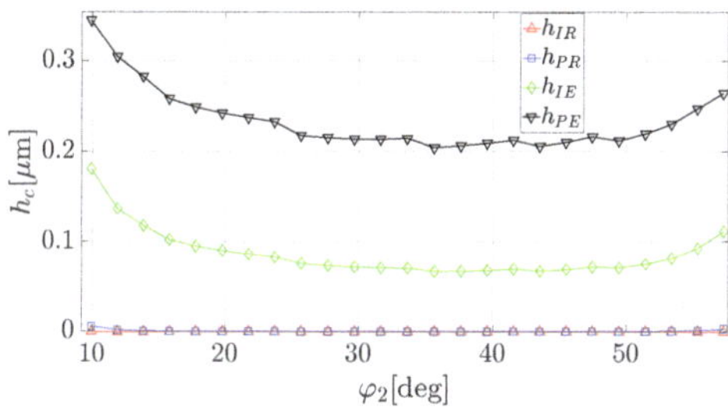

Figure 5. Central film thickness (h_c) values over a mesh cycle.

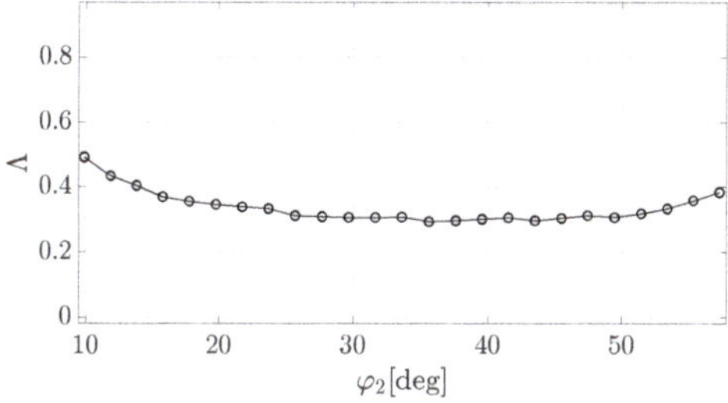

Figure 6. Lambda factor (Λ) over a mesh cycle.

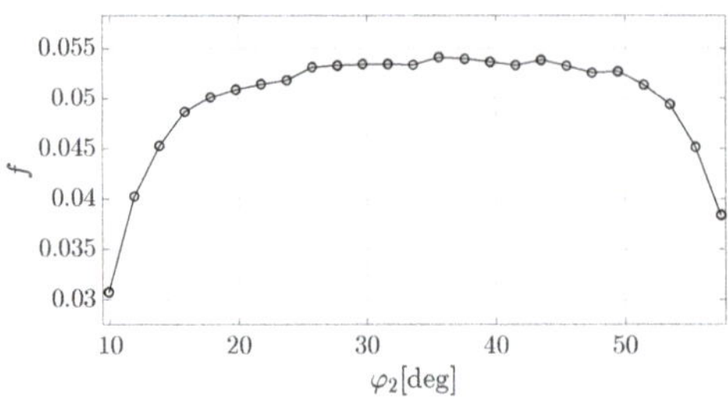

Figure 7. Coefficient of friction (f) over a mesh cycle.

The efficiency loss can be computed by numerical integration of the frictional power losses over the meshing cycle:

$$E_l = \sum_{i=1}^{N_{\text{steps}}-1} \frac{1}{2}(P_i + P_{i+1}) \frac{\Delta t}{N_{\text{steps}} - 1} \tag{16}$$

where $P_i = \sum_{k=1}^{nT} f_{ik} F_{ik} u_{ik}^{(s)}$ is the power loss at the i-th time step and f_{ik}, F_{ik}, $u_{ik}^{(s)}$ are, respectively, the friction coefficient, normal contact force and sliding velocity of the k-th tooth pair in engagement at the i-th time step.

The efficiency estimation has been performed under different torques and speeds, namely in a torque range between 50 and 300 Nm and in a speed range between 500 and 3000 rpm. The results are shown in Figure 8. According to our model, the efficiency decreases at large torque values and increases at higher speed. Those results are consistent with a mixed lubrication regime: a larger torque decreases the film thickness, resulting in a larger probability of asperity contacts. On the other hand, a higher speed increases the film thickness due to hydrodynamic effects, which mitigates the adverse effect of the boundary friction coefficient f_b (cfr. Equation (1)).

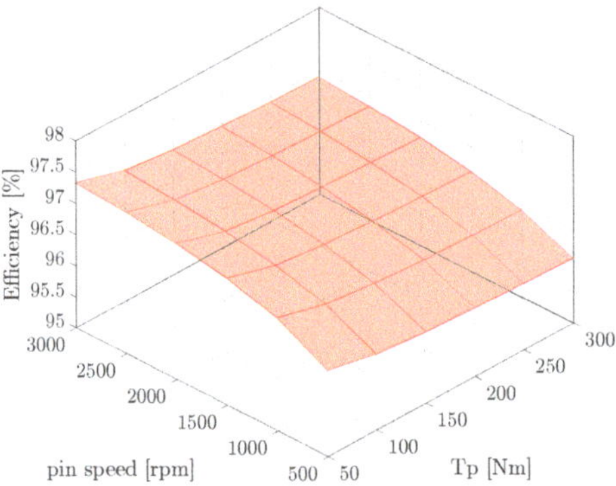

Figure 8. Efficiency map as a function of pinion torque and pinion speed.

5. Wear Model

The wear pattern is extracted directly by Transmission3D. Its wear implementation is based on the well-known Archard model:

$$\frac{\mathrm{d}w}{\mathrm{d}s} = kP \tag{17}$$

where w is the wear depth, s is the sliding distance, k is the *wear coefficient*, and P is the local pressure. The numerical value of the coefficient k needs to be determined experimentally. Kahraman et al. [23] suggested a value of 9.65×10^{-19} m^2/N, which was experimentally derived for helical gears under mixed lubrication conditions. Due to the complexity and cost of an experimental test rig, this value has been used also for the hypoid gears investigated in this work. Under comparable operating conditions in terms of loading and peripheral speeds, hypoid gears generally have more sliding than helical gears, thus we can hypothesize that the value of k is higher for the former. This might be caused by the fact that the larger sliding speeds entail an increased heating of the lubricant, which decreases the film thickness and thus induces more asperity contacts (more wear) under a mixed lubrication regime. However, obtaining an accurate k value for hypoid gears is out of the scope of the present work.

After a certain number of meshing cycles, wear can noticeably modify the tooth flanks, thus the tooth geometry needs to be updated (before simulating further wear) to re-evaluate the contact pressure distribution, which is very sensitive to micro-geometry variations. The geometric update is performed by superimposing the post-processed wear pattern (evaluated at points arranged in a grid on the tooth flanks) onto the previous tooth surfaces. At the j-th grid point:

$$\mathbf{p}_j^{(k)} = \mathbf{p}_j^{(k-1)} + \mathbf{n}_j^{(k-1)} w_j^{(k)}, \tag{18}$$

where:

- $\mathbf{p}_j^{(k)}$ contains the coordinates of the grid point after the k-th geometry update;
- $\mathbf{p}_j^{(k-1)}$ and $\mathbf{n}_j^{(k-1)}$ are the grid point coordinates and unit normal components after the $(k-1)$-th (previous) geometry update;
- $w_j^{(k)}$ is the k-th local wear depth.

The updated points $\mathbf{p}^{(k)}$ are then best-fit by a *NURBS* surface [24], which allows us to also compute the updated unit normals $\mathbf{n}^{(k)}$. The updated points and normals are then fed back to a tooth mesher for Transmission3D, and the next wear simulation step is executed.

6. Wear Analysis Application

In this section, we present the results of an investigation aimed at assessing the impact of the frequency in updating the geometry of the worn tooth surfaces on the prediction of the final accumulated wear and contact patterns. The *contact pattern* represents the envelope of all the instantaneous contact zones over the meshing cycle; the *wear pattern* is the corresponding wear distribution on the tooth surface (after a certain number of wear cycles). Each geometry update necessarily calls for a new LTCA simulation, hence frequent updates can easily lead to a significant cost in terms of CPU time for a complete wear simulation. Our goal is to evaluate the best trade-off between the number of geometry updates and the computational burden. We assume here a service life of 30 Mc (where 1 Mc = 10^6 cycles), counted as revolutions of the hypoid pinion. The initial contact patterns of pinion and gear are shown in Figure 9.

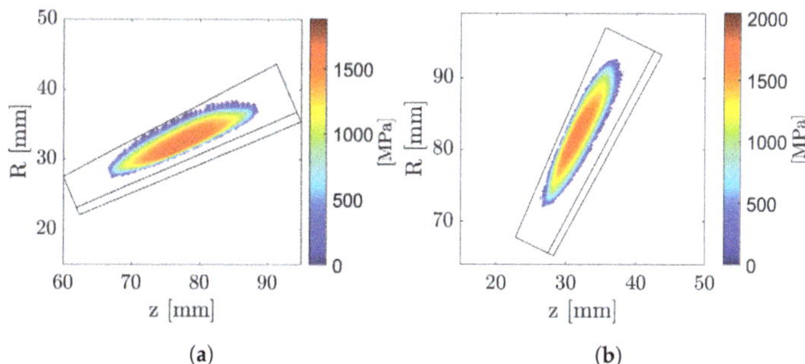

Figure 9. Initial contact patterns. (**a**) Pinion. (**b**) Gear.

As a first step, serving as a baseline result, a very coarse analysis is carried out, where the predicted wear is evaluated assuming that the initial tooth geometry does not change during the whole service life. After calculating wear distribution, a geometry update is performed only once and the resulting contact pressures are computed. The corresponding wear and pressure patterns are shown in Figure 10 for the pinion and in Figure 11 for the gear. The contact patterns exhibit remarkable differences with respect to the ones in Figure 9: a significant shift of the pressure peaks away from the tooth flank center is evident (Figures 10a and 11a), and it is due to an unrealistic wear indentation effect.

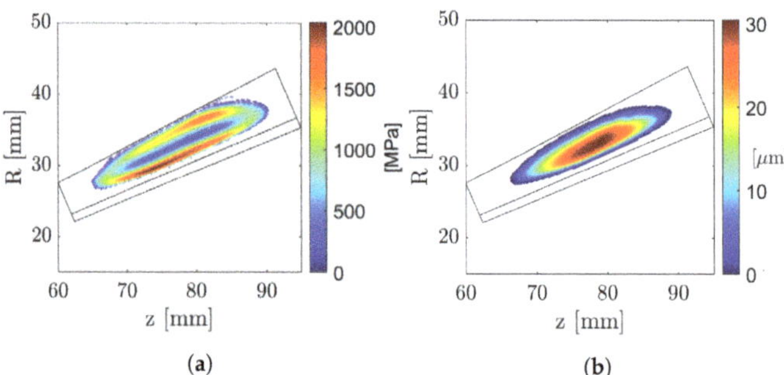

(a) (b)

Figure 10. Pinion patterns after 30 M cycles. Only one geometry update is performed. (**a**) Contact pattern. (**b**) Wear pattern.

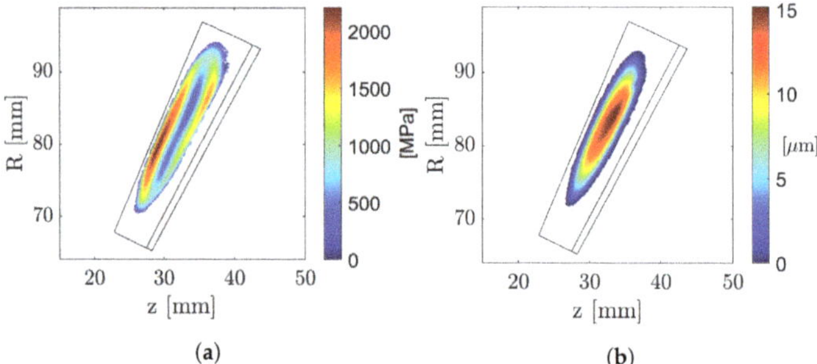

(a) (b)

Figure 11. Gear patterns after 30 Mc (pinion revolutions) cycles. Only one geometry update is performed. (**a**) Contact pattern. (**b**) Wear pattern.

In the subsequent analysis, a similar wear simulation is performed. This time, however, the tooth flank modifications due to the worn-out material, and the corresponding variation in contact pressures, are gradually updated according to a stepwise approach, where each step consists of 5 Mc. In other words, the tooth flank geometry is iteratively modified every 5 Mc as per Equation (18). This update strategy, in practice, is equivalent to updating after a given wear volume, i.e., the wear volume at the first wear cycle, multiplied by the number of wear cycles [25]. The evolution of the pinion contact and wear patterns are shown, respectively, in Figures 12 and 13. The associated gear patterns are shown in Figures 14 and 15. As expected, the overall pattern grows larger along both the profile and face directions of the active flank. However, a localization of the contact pressures near the fillet portion of the pinion (close to, but away from, the gear tip) is evident. In the same area, also the wear pattern features a localization of its peaks. Figure 16 shows the pinion instantaneous contact zones at a specific meshing configuration before and after wear. The corresponding instantaneous contact zones for the gear are shown in Figure 17. It can be noticed how wear induced a greater overlap ratio, increasing the number of mating tooth pairs (from one to three). Pressure localization is evident here as well, which appears to be caused by an abrupt change to the local curvature due to wear.

To gather more data for a sensitivity analysis, wear simulations have been performed also with geometry updates every 10 Mc, 1 Mc and 0.5 Mc. A side-by-side comparison of the final contact patterns is shown in Figure 18. For brevity, only the pinion contact patterns are shown. It can be observed that the patterns obtained with 6, 30 and 60 update steps, i.e., every 5 Mc, 1 Mc, and 0.5 Mc, seem almost identical. Even with just three updates, i.e., every 10 Mc, the final pattern is already reasonably close to our most accurate analysis.

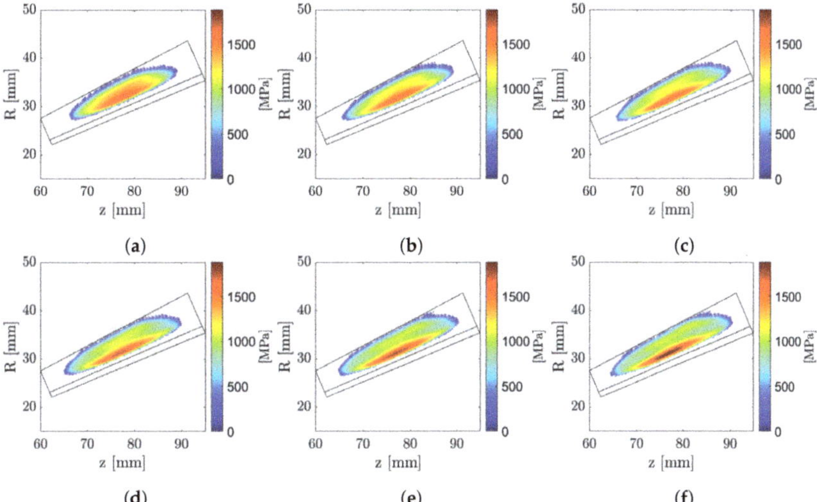

Figure 12. Evolution of the pinion contact pattern with a geometry update every 5 Mc. (**a**) 5 Mc. (**b**) 10 Mc. (**c**) 15 Mc. (**d**) 20 Mc. (**e**) 25 Mc. (**f**) 30 Mc.

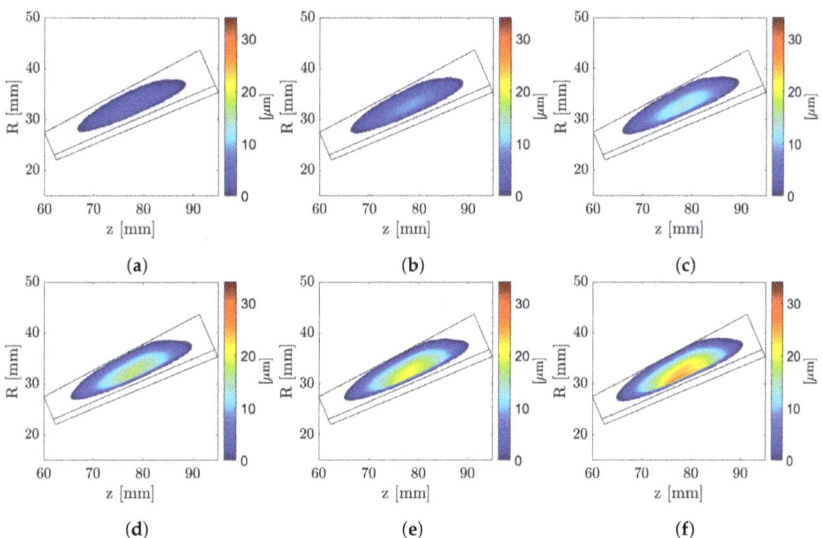

Figure 13. Evolution of the pinion (cumulative) wear pattern with a geometry update every 5 Mc. (**a**) 5 Mc. (**b**) 10 Mc. (**c**) 15 Mc. (**d**) 20 Mc. (**e**) 25 Mc. (**f**) 30 Mc.

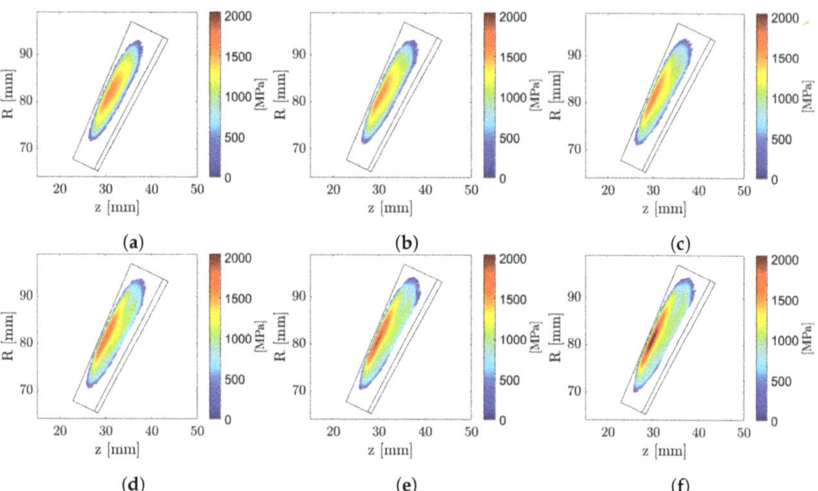

Figure 14. Evolution of the gear contact pattern with a geometry update every 5 Mc. (**a**) 5 Mc. (**b**) 10 Mc. (**c**) 15 Mc. (**d**) 20 Mc. (**e**) 25 Mc. (**f**) 30 Mc.

Figure 19 depicts a more quantitative result of this analysis by showing the maximum contact pressure (P_{max}) registered during meshing as a function of the wear cycles. An analogous representation of the maximum wear depth w_{max} registered on the pinion is shown in Figure 20. The graphs show that a geometry update performed every 5 Mc may be a good choice for balancing an accurate prediction, both in terms of contact pressure and wear depth, with affordable computations in terms of CPU time.

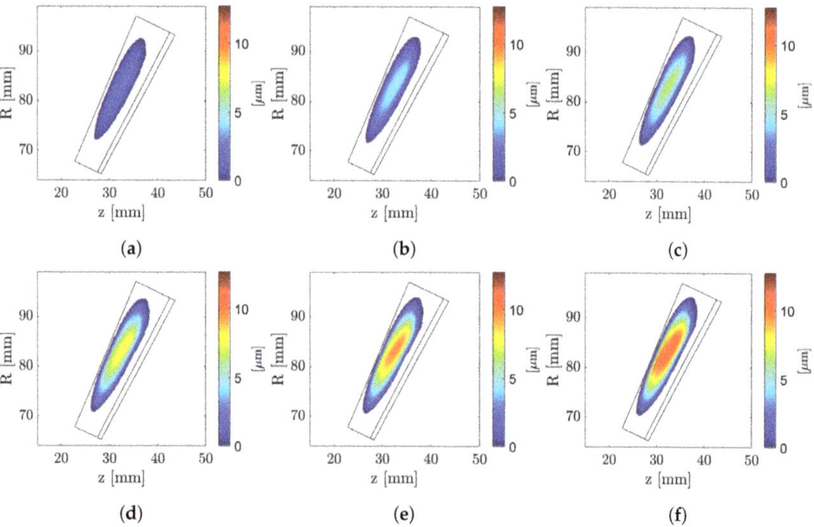

Figure 15. Evolution of the gear (cumulative) wear pattern with a geometry update every 5 Mc. (**a**) 5 Mc. (**b**) 10 Mc. (**c**) 15 Mc. (**d**) 20 Mc. (**e**) 25 Mc. (**f**) 30 Mc.

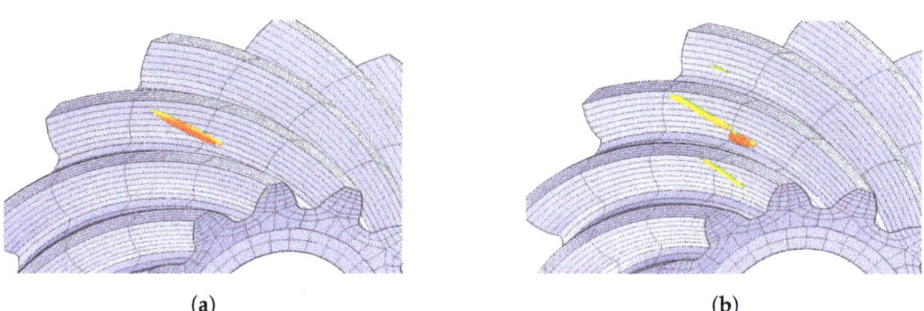

Figure 16. Instantaneous contact area(s) at the same meshing configuration on the pinion before and after wear simulation (performed updating the geometry every 5 Mc). (**a**) Before wear. (**b**) After wear.

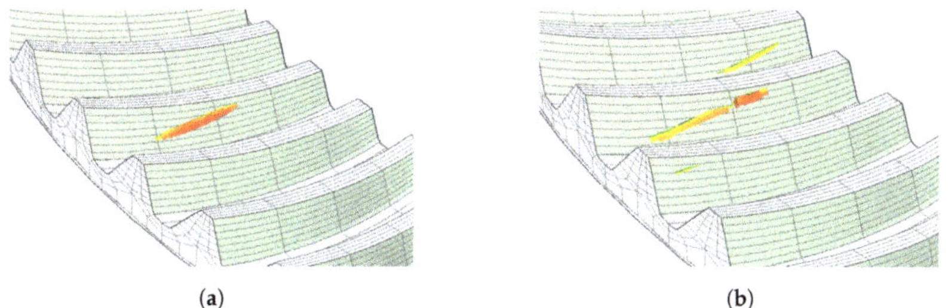

Figure 17. Instantaneous contact area(s) at the same meshing configuration on the gear before and after wear simulation (performed updating the geometry every 5 Mc). (**a**) Before wear. (**b**) After wear.

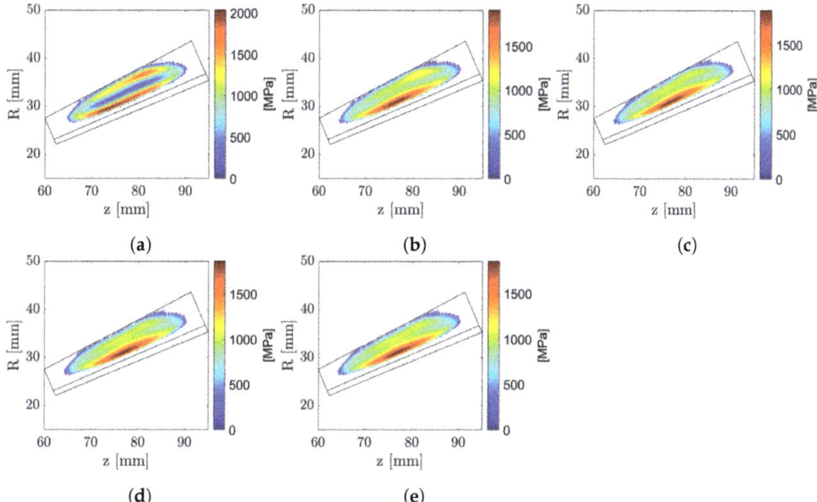

Figure 18. Comparison of the final pinion contact patterns resulting from wear simulations with different numbers of geometry updates. (**a**) Update after 30 Mc. (**b**) Update every 10 Mc. (**c**) Update every 5 Mc. (**d**) Update every 1 Mc. (**e**) Update every 0.5 Mc.

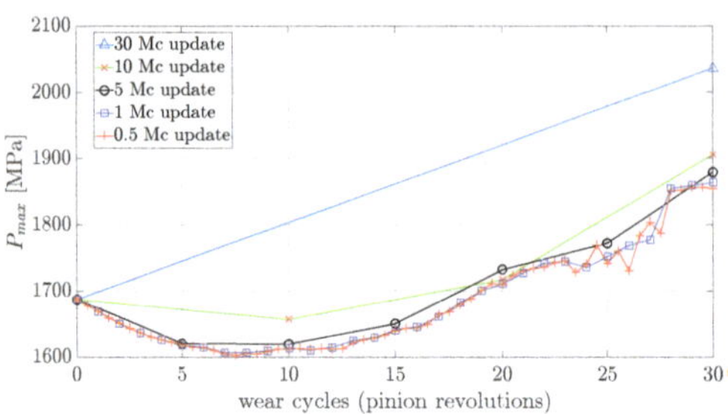

Figure 19. Maximum contact pressure from wear simulations with different geometry update frequencies.

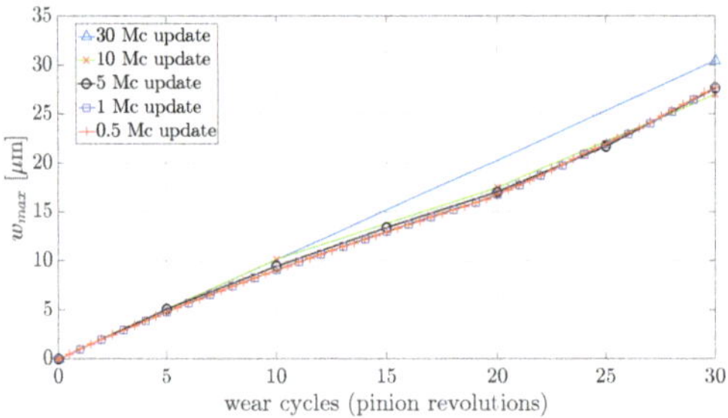

Figure 20. Maximum pinion wear depth from wear simulations with different geometry update frequencies.

7. Conclusions

In the first part of the paper, a semi-empirical formulation of the friction coefficient has been proposed to estimate friction-induced losses in lubricated hypoid gears, whose tooth surfaces have been digitally synthesized through simulation of the grinding process. The calculations leverage both special procedures developed by the authors to compute basic geometric and kinematic quantities and an *ad hoc* LTCA simulation tool for an accurate prediction of normal loads and contact pressures. Our streamlined methodology is suitable for integration into an automatic optimization pipeline where the complex interactions between the flank geometry and the coefficient of friction are profitably accounted for to develop more efficient geared transmissions.

In the second part of the paper, a wear study has been presented. The aim was to assess the impact of the frequency adopted for updating the worn geometry of the gears on the accuracy and consistency of the tribological results. Interestingly, in this preliminary study, updating the geometry every five million cycles (corresponding to a wear variation of about 5 micrometers) turned out to be sufficient to suitably capture the evolution of both the maximum contact pressure and the maximum wear depth.

Author Contributions: Conceptualization, A.A., M.G. (Marco Gabiccini), L.M., F.D.P. and E.C.; Data curation, E.G.; Methodology, E.C.; Software, E.G.; Supervision, A.A., M.G. (Marco Gabiccini), M.G. (Massimo Guiggiani), L.M., F.D.P. and E.C.; Writing—original draft, E.G.; Writing—review and editing, A.A. and M.G. (Marco Gabiccini). All authors have read and agreed to the published version of the manuscript.

Funding: This research received no external funding.

Institutional Review Board Statement: Not applicable.

Informed Consent Statement: Not applicable.

Data Availability Statement: Not applicable.

Conflicts of Interest: The authors declare no conflict of interest.

References

1. Artoni, A.; Bracci, A.; Gabiccini, M.; Guiggiani, M. Optimization of the loaded contact pattern in hypoid gears by automatic topography modification. *J. Mech. Des.* **2009**, *131*, 0110081–0110089. [CrossRef]
2. Vivet, M.; Mundo, D.; Tamarozzi, T.; DEsmet, W. An analytical model for accurate and numerically efficient tooth contact analysis under load, applied to face-milled spiral bevel gears. *Mech. Mach. Theory* **2018**, *130*, 137–156. [CrossRef]
3. Wang, Q.; Zhou, C.; Gui, L.; Fan, Z. Optimization of the loaded contact pattern of spiral bevel and hypoid gears based on a kriging model. *Mech. Mach. Theory* **2018**, *122*, 432–449. [CrossRef]
4. Park, D.; Kahraman, A. A surface wear model for hypoid gear pairs. *Wear* **2009**, *267*, 1595–1604. [CrossRef]
5. Advanced Numerical Solutions. Transmission3D User's Manual v. 2.54. 2020. Available online: http://ansol.us/Products/TX3 (accessed on 1 January 2021).
6. Park, D.; Kolivand, M.; Kahraman, A. Prediction of surface wear of hypoid gears using a semi-analytical contact model. *Mech. Mach. Theory* **2012**, *52*, 180–194. doi: [CrossRef]
7. Kolivand, M.; Kahraman, A. A load distribution model for hypoid gears using ease-off topography and shell theory. *Mech. Mach. Theory* **2009**, *44*, 1848–1865. doi: [CrossRef]
8. Park, D.; Kolivand, M.; Kahraman, A. An approximate method to predict surface wear of hypoid gears using surface interpolation. *Mech. Mach. Theory* **2014**, *71*, 64–78. doi: [CrossRef]
9. Huang, D.; Wang, Z.; Kubo, A. Hypoid gear integrated wear model and experimental verification design and test. *Int. J. Mech. Sci.* **2020**, *166*, 105228. doi: [CrossRef]
10. Grabovic, E.; Ciulli, E.; Artoni, A.; Gabiccini, M. A model for the prediction of frictional power losses in hypoid gears. In Proceedings of the 4th International Conference of IFToMM, Italy, Napoli, 7–9 September 2022; *in press*.
11. Cao, W.; Pu, W.; Wang, J.; Xiao, K. Effect of contact path on the mixed lubrication performance, friction and contact fatigue in spiral bevel gears. *Tribol. Int.* **2018**, *123*, 359–371. [CrossRef]
12. Mohammadpour, M.; Theodossiades, S.; Rahnejat, H. Elastohydrodynamic lubrication of hypoid gear pairs at high loads. *Proc. Inst. Mech. Eng. Part J J. Eng. Tribol.* **2012**, *226*, 183–198. [CrossRef]
13. Pu, W.; Wang, J.; Yang, R.; Zhu, D. Mixed Elastohydrodynamic Lubrication with Three-Dimensional Machined Roughness in Spiral Bevel and Hypoid Gears. *J. Tribol.* **2015**, *137*. [CrossRef]
14. Castro, J.; Seabra, J. Coefficient of friction in mixed film lubrication: Gears versus twin-discs. *Proc. Inst. Mech. Eng. Part J J. Eng. Tribol.* **2007**, *221*, 399–411. [CrossRef]
15. Gold, P.W.; Schmidt, A.; Dicke, H.; Loos, J.; Assmann, C. Viscosity–pressure–temperature behaviour of mineral and synthetic oils. *J. Synth. Lubr.* **2001**, *18*, 51–79. [CrossRef]
16. Hamrock, B. *Fundamentals of Fluid Film Lubrication*; McGraw-Hill: New York, NY, USA, 1994; pp. 499–505.
17. Bassani, R.; Ciulli, E.; Guiggiani, M.; Piccigallo, B.; Andrei, G. Elastohydrodynamic lubrication in face gears drives. In Proceedings of the International Tribology Conference, Nagasaki, Japan, 29 October–2 November 2000.
18. Castro, J.; Seabra, J. Scuffing and lubricant film breakdown in FZG gears Part I. Analytical and experimental approach. *Wear* **1998**, *215*, 104–113. [CrossRef]
19. Hoppert, M. Analytische und Experimentelle Untersuchungen zum Wirkungsgradverhalten von Achsgetrieben. Ph.D. Thesis, Fakultät für Maschinenbau der Technischen Universität Ilmenau, Ilmenau, Germany, 2015.
20. Grabovic, E.; Artoni, A.; Gabiccini, M. Holistic Optimal Design Of Face-Milled Hypoid Gearsets. In Proceedings of the IDETC-CIE 2022, St. Louis, MO, USA, 14–17 August 2022; number IDETC2022-89598.
21. Ciulli, E. Formulas for Entraining Velocity in Lubricated Line Contacts. *J. Tribol.* **2002**, *124*, 856–858. [CrossRef]
22. Gohar, R. *Elastohydrodynamics*; E. Horwood Halsted Press: Chichester, UK; New York, NY, USA, 1988.
23. Bajpai, P.; Kahraman, A.; Anderson, N.E. A Surface Wear Prediction Methodology for Parallel-Axis Gear Pairs. *J. Tribol.* **2004**, *126*, 597–605. doi: [CrossRef]
24. Piegl, L.; Tiller, W. *The NURBS Book*, 2nd ed.; Springer: New York, NY, USA, 1996.
25. Mattei, L.; Di Puccio, F. Influence of the wear partition factor on wear evolution modelling of sliding surfaces. *Int. J. Mech. Sci.* **2015**, *99*, 72–88. [CrossRef]

Article

Optimal Design for Vibration Mitigation of a Planar Parallel Mechanism for a Fast Automatic Machine

Federico Zaccaria [1], Edoardo Quarta [2], Simone Badini [2] and Marco Carricato [1,*]

[1] DIN—Department of Industrial Engineering, University of Bologna, 40126 Bologna, Italy
[2] IMA Group, 40064 Ozzano nell'Emilia, Italy
* Correspondence: marco.carricato@unibo.it

Abstract: This work studies a planar parallel mechanism installed on a fast-operating automatic machine. In particular, the mechanism design is optimized to mitigate experimentally-observed vibrations. The latter are a frequent issue in mechanisms operating at high speeds, since they may lead to low-quality products and, ultimately, to permanent damage to the goods that are processed. In order to identify the vibration cause, several possible factors are explored, such as resonance phenomena, elastic deformations of the components, and joint deformations under operation loads. Then, two design optimization are performed, which result in a significant improvement in the vibrational behaviour, with oscillations being strongly reduced in comparison to the initial design.

Keywords: automatic machines; planar parallel mechanism; five bar linkage; design optimization; vibration reduction

Citation: Zaccaria, F.; Quarta, E.; Badini, S.; Carricato, M. Optimal Design for Vibration Mitigation of a Planar Parallel Mechanism for a Fast Automatic Machine. *Machines* **2022**, *10*, 770. https://doi.org/10.3390/machines10090770

Academic Editors: Marco Ceccarelli, Giuseppe Carbone and Alessandro Gasparetto

Received: 29 July 2022
Accepted: 2 September 2022
Published: 5 September 2022

1. Introduction

Parallel mechanisms were proposed in the literature for the most disparate tasks [1], since they present significant performances in terms of accuracy and rigidity, as well as the ability to operate at high speed. For fast pick-and-place operations, the problem of vibration mitigation is of paramount importance.

In this work, we analyze and optimize a planar parallel mechanism employed in a fast automatic machine. Our study is driven by the fact that the mechanism experiences vibrations of unacceptable magnitude, and we seek to mitigate this phenomenon by optimizing its design. First, we investigate the possible sources of vibrations. By using a multibody simulation software, we develop FEM-based modal and elastodynamic analyses to investigate the possible occurrence of resonance phenomena and elastic deformation of the mechanism components. We also investigate possible vibrations generated by the joint compliance. On the basis of these analyses, we mitigate vibrations by optimizing the mechanism design based on (i) reduction of the moving masses, and (ii) change of the link geometry.

The paper is structured as follows. Section 2 describes the main components of the studied mechanism and its working principle. Section 3 investigates the possible sources of vibrations through the dynamic simulation of the mechanism: in particular, modal, elastodynamic, and joint-compliance analyses are conducted. Then, Section 4 proposes two design optimizations aiming at reducing the mechanism vibrations. Finally, conclusions are drawn in Section 5.

2. Mechanism Description

In this paper, we focus on the mechanism illustrated in Figure 1a, which serves as a connection between two stages of a more complex automatic machine. Paper tags are taken from the exit location of the previous machine stage by a custom gripper, and then delivered to the entry station of the next stage. Thus, the aim of the mechanism is to execute a pick-and-place task at high rate (specifically, 1000 cycles/min). In particular, since the components to be delivered are light and thin (i.e., paper tags), high accuracy is required to safely perform the pick-and-place task.

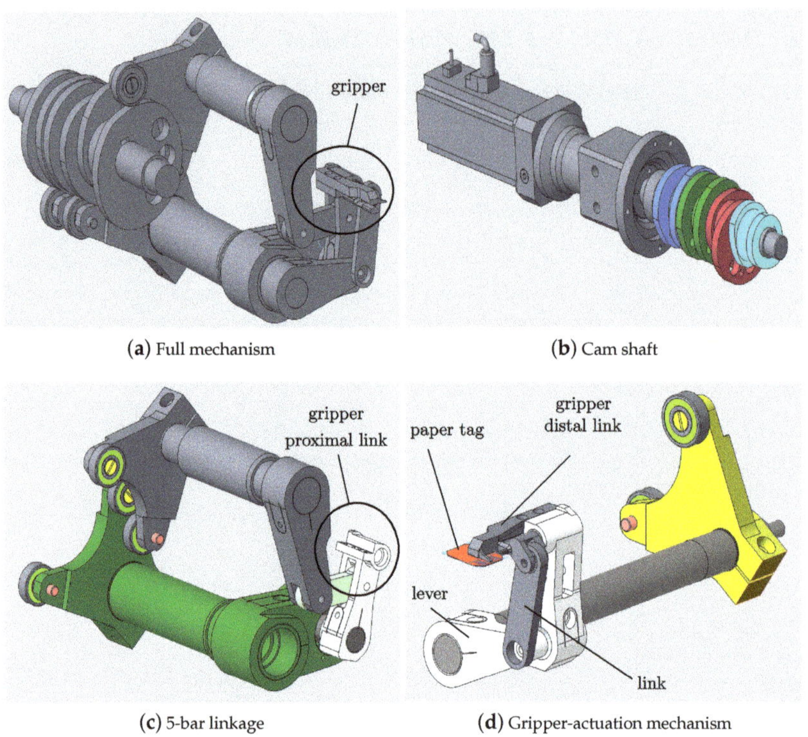

(a) Full mechanism

(b) Cam shaft

(c) 5-bar linkage

(d) Gripper-actuation mechanism

Figure 1. An overview of the complete mechanism (**a**), the cam shaft (**b**), the 5-bar linkage used for the gripper movement (**c**), and the gripper actuation mechanism (**d**).

The device in Figure 1a comprises two submechanisms: one, with two degree of freedom (*DoF*), displaces the gripper over a plane, whereas the other actuates the gripper opening/closing motion. Despite the possibility of employing three different motors to position and actuate the gripper, a single motor rotating at constant speed is preferred coupled to a set of cams. This is mainly done because the two extreme poses of the gripper are assigned, and also to avoid motor-synchronization problems at high speed.

In order to give a detailed description of the mechanism operation, we can consider three main subgroups: a motorized shaft equipped with cams, a five-bar 2-*DoF* linkage, and a gripper-actuation mechanism. The aim of the cam-shaft group (Figure 1b) is to convert the continuous rotation of the electric drive to the prescribed alternate motions of a set of rocking levers. While the first set of cams is devoted to actuate the five-bar linkage, a second cam set is employed for the gripper actuation. Each cam group is made by a principal and a conjugate profile, in order to avoid the use of call-back springs. The five-bar mechanism is used to displace the gripper between two assigned locations (Figure 1c). Two levers receive the alternate motion from the first cam group and, by means of two shafts, actuate the five-bar input links. The five-bar is composed of four mobile aluminum members connected by revolute joints, and the distal end of a member serves as a proximal link of the gripper (see Figure 1c) Finally, the gripper-actuation subgroup (Figure 1d) receives actuation by a cam through another lever and a shaft. The latter rotation actuates a leverage that displaces the distal link of the gripper; the movement of the lever combined with that of the proximal link enables the closing/opening motion that allows paper tags to be grasped.

Given the required high production rate, and the light weight of the products to be handled (a few grams), the accuracy of the pick-and-place task must be relevant. The mechanism is supposed to operate among the two poses depicted in Figure 2. For each rotation of the main motor shaft, the working cycle of the mechanism can be summarized as follows:

1. The gripper is positioned at the grasp position (Figure 2a) and it is open, waiting for paper tags to be received;
2. Once a paper tag is available, the gripper closes and the product is grasped thanks to the movement of the gripper-actuation mechanism (Figure 2b);
3. The gripper moves to the deliver position (Figure 2c) thanks to the movement of the five-bar linkage;
4. At the deliver position, the gripper opens and the product is released (Figure 2d);
5. The gripper moves back to the grasp position and the cycle restarts.

The mechanism operation was monitored at its nominal production rate of 1000 cycles/min, and oscillations were experimentally observed when the gripper was at the grasp position (Figure 2a), where on the contrary it was supposed to remain still. Oscillations were measured to reach an amplitude of 1 mm, which is unacceptable for the quality standards of the products that are to be delivered. In the next section, we investigate the possible causes of such vibrations.

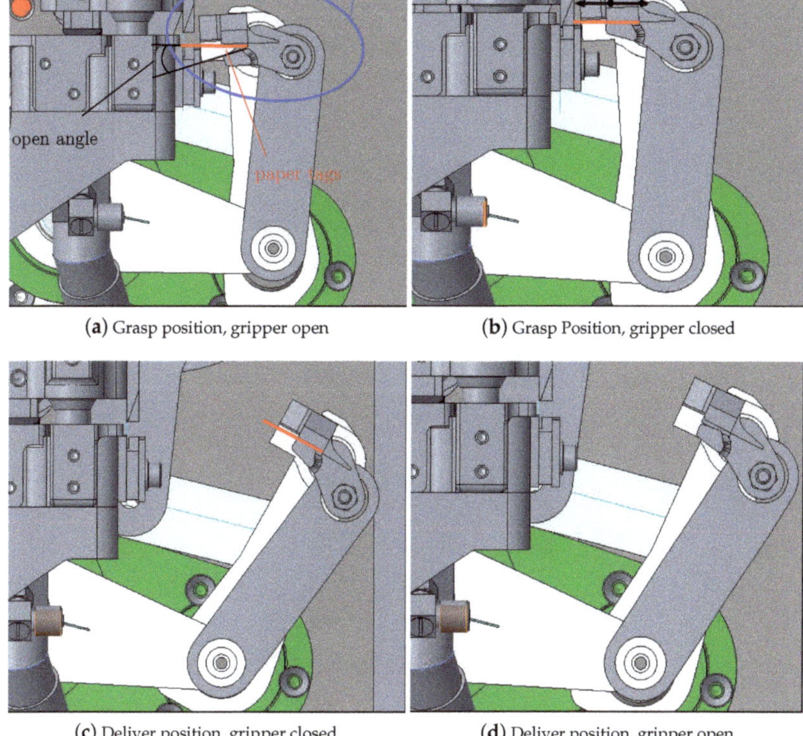

(**a**) Grasp position, gripper open

(**b**) Grasp Position, gripper closed

(**c**) Deliver position, gripper closed

(**d**) Deliver position, gripper open

Figure 2. (**a**) Gripper position where the paper tags are received, (**b**) experimentally-identified direction of oscillation, (**c**) deliver position, (**d**) deliver position with paper tag delivered.

3. Dynamic Analysis

The main aim of this section is to identify the possible causes of the vibratory phenomena observed in the real mechanism. We start from a rigid-body dynamic simulation of the five-bar mechanism, and then we carry out a modal analysis to explore possible resonance phenomena. Then, the flexibility of the system is taken into account by performing an elastodynamic simulation of the parallel linkage. Finally, we investigate the compliance of the base joints and possible solutions.

3.1. Rigid-Body Dynamic Analysis

The first step toward the identification of the vibration cause is the computation of the ideal behavior of the system. This is performed by developing a rigid-body dynamic analysis where, for a given input of the motor, we identify the position, velocity, and accelerations of each component of the system.

At this stage, we focus on the five-bar mechanism illustrated in Figure 3, and we consider the angles θ_1, θ_4 provided by the cams as assigned inputs. By assuming all components as rigid, the closure-loop equations of the five-bar linkage can be solved to obtain the theoretical location x, y of the gripper reference point E, and the intermediate angles θ_2, θ_3. Then, the velocities and acceleration of each component of the five-bar may be determined by the solution of the velocity/acceleration linear systems obtained by successive derivation of the closure-loop equations for given velocities $\dot{\theta}_1, \dot{\theta}_4$ and accelerations $\ddot{\theta}_1, \ddot{\theta}_4$. By knowing the mass distribution of each component, the inverse dynamic problem can then be solved to obtain the joint reactions that stress the system, as well as the actuation actions. By employing the Newton–Euler approach [2], unknown reaction forces are introduced at each joint of the mechanism. Then, the dynamic equilibrium of the forces and couples is established for each mechanism member. The corresponding equations can be written as a linear system of the form:

$$\mathbf{AF} = \mathbf{B} \tag{1}$$

with $\mathbf{F}$ being the vector of unknown joint reaction forces and actuation actions, $\mathbf{B}$ a known term including inertial and Coriolis effects, and $\mathbf{A}$ a matrix that collects the coefficients that multiply the unknown force array $\mathbf{F}$. These coefficients, which depend on the mechanism configurations, can be obtained by the solution of the kinematic problem.

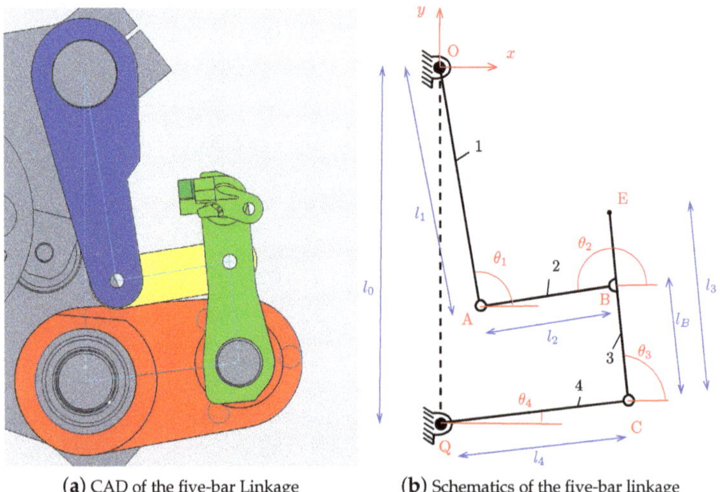

(**a**) CAD of the five-bar Linkage   (**b**) Schematics of the five-bar linkage

Figure 3. (**a**) CAD of the five-bar mechanism, (**b**) schematics of the five-bar mechanism.

The inverse dynamic problem is solved by a multibody simulation software (ANSYS), where all the components are modeled as rigid, and friction is disregarded. Revolute joints model the bearings that support the rotating shafts and the hinges between components, whereas ideal contact constraints are used to simulate the cam-lever motion transmission. As a result of the simulation, we obtain the position, velocity, and acceleration of each body, as well as joint reactions. As an example, the magnitude of the base reaction $\mathbf{F}_0$ for a single cycle, which will play a crucial role in the mechanism optimization, is depicted in Figure 4.

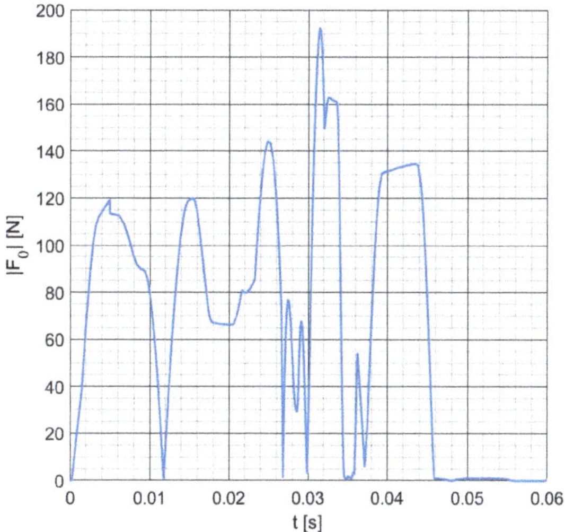

Figure 4. Magnitude of the base force $\mathbf{F}_0$ for a single operating cycle.

3.2. Modal Analysis

Modal analysis is the process of determining the dynamic properties of mechanical systems in the form of natural frequencies and/or mode shapes [3]. When a system is excited or it is working close to one of its natural frequencies, resonance occurs and the system may display large vibrations.

In general, two different approaches are employed for modal analysis: numerical simulations and experimental analysis [3]. When a mechanism is subjected to a known external excitation, frequency response functions can be experimentally measured to obtain the natural frequencies and mode shape characteristics of the system. However, this approach generally requires complex measurements that may not be easily carried out in a complex machine. On the opposite, numerical simulations aim at predicting the dynamic behavior of the system by the use of a mathematical model [4]. In this context, the finite-element method is commonly used to derive the model of complex systems such as parallel mechanisms, leading to an eigenproblem in the form:

$$\left(\mathbf{M} - \omega_j^2 \mathbf{K}\right)\boldsymbol{\phi}_j = \mathbf{0} \tag{2}$$

where $\mathbf{M}$ is the mass matrix, $\mathbf{K}$ is the stiffness matrix, ω_j, and $\boldsymbol{\phi}_j$ are the natural frequency and the vibration mode associated to the j-th resonance, respectively.

In order to investigate the possible occurrence of resonance, we considered the mechanism at the grasp configuration (Figure 2a), where oscillations were measured. The full mechanism is discretized according to the finite-element approach. The coupling between the mechanical components is modeled in the same fashion as in rigid-body simulation: revolute joints represent bearings and rotative connections between links, and contact joints represent the motion transmission by cams. Damping is assumed to be negligible.

The material properties are those of aluminum components, and solid elements are used to discretize the geometry. The number of elements is gradually increased until convergence is achieved, and the final discretization is represented in Figure 5. The finite-element discretization and the modal analysis are performed in a multibody simulation software (ANSYS), and the first six natural frequencies associated with vibration modes are reported in Table 1. By inspection of mode shapes, only the first two correspond with the vibratory phenomena that were experimentally observed, and these two modes have frequencies considerably distant from the frequencies that excite the system. Therefore, we can assume that the oscillations of the gripper are not caused by resonance phenomena.

Figure 5. Discretization of the mechanism employed for the modal analysis.

Table 1. Natural frequencies associated with the first six resonance modes.

Mode Number	Frequency [Hz]
1	463.52
2	596.44
3	695.68
4	866.64
5	1182.60
6	1269.10

3.3. Elastodynamic Simulation

A typical cause of oscillations in mechanisms working at high frequency is the intrinsic elasticity of their components. In order to reduce the high inertial forces and actuation torques required to operate at high speed, mechanisms for fast-operating machines are usually made as lightweight as possible. However, light members generally bring reduced stiffness and elastic deformations may occur. The performance of the mechanism, in terms of position accuracy, may consequently be reduced by elastic oscillations.

While rigid simulations play a dominant role in the design and synthesis of standard mechanisms, elastodynamic simulations are usually performed when the operating speed is relevant, and the influence of the component elasticity is not negligible [5]. Continuous elastic systems and lumped parameter models aim at finding approximate solutions in reduced computational time [6]. On the other hand, the finite-element method [7,8] provides a more general and accurate modeling technique for complex mechanisms, and the resulting dynamic model is a set of differential equations that can be formulated as:

$$\mathbf{M\ddot{q} + C\dot{q} + Kq = f} \tag{3}$$

where $\mathbf{q}$ is the vector of the nodal coordinate of the discretized system, $\mathbf{C}$ is the compliance matrix, and $\mathbf{f}$ is the vector of external forces. Thus, by the numerical integration of Equation (3) over a defined temporal interval, the position, velocity, and acceleration of each member of the mechanism can be evaluated with the inclusion of elastic effects.

In this work, we used the finite-element approach to investigate the effect of elasticity on the vibrations of the mechanism. Despite the possibility of considering the elasticity of all members of the mechanism, in order to reduce the computational cost, we decided to evaluate the influence of the five-bar linkage deformations only, and to assume the other components as rigid. This decision is mainly driven by the considerable computational time needed to achieve accurate solutions if the full mechanism is discretized. Solid elements are employed to discretize the five-bar mechanism, and the resulting discretization is displayed in Figure 6a. As for the modal analysis, the number of elements is gradually increased to achieve convergence of the numerical results. Because of the high speed of the mechanism, the material damping is assumed to be negligible. The mechanism is simulated for a single cycle at the operating speed of 1000 cycles/min. At the end of the simulation, we compare the displacement of the gripper between the theoretical position given by a rigid simulation, and the prediction of the elastic model. As shown in Figure 6b, slight oscillations are predicted by the flexible model, with an amplitude on the order of 0.02 mm. We can consequently assume that the cause of the vibrations is not the elasticity of the system, since the predicted amplitude is significantly different from the measured displacement of 1 mm.

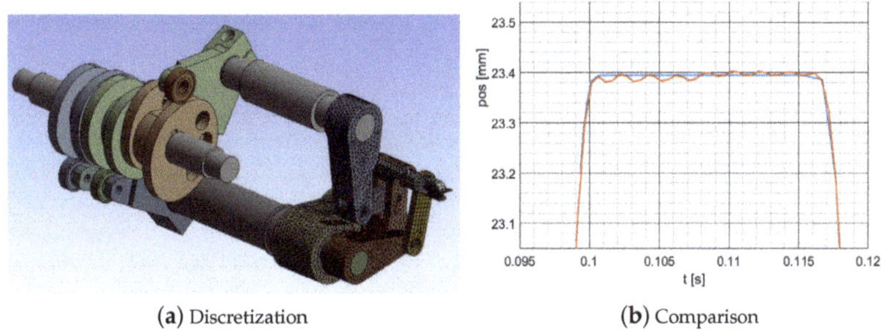

(**a**) Discretization   (**b**) Comparison

Figure 6. (a) FEM discretization of the five-bar linkage employed for the elastodinamic simulations, (b) comparison between rigid and flexible simulations at the grasp position.

3.4. Joint Stiffness and Possible Solutions

Joint clearance and deformation under external load is a known cause of oscillations, especially for high-speed mechanisms [9].

The influence of joint clearance and deformation under external loads can be numerically evaluated by establishing the kinetostatic model of each joint affected by clearance/deformation, and then evaluating the corresponding effect on the mechanism end-effector [10]. However, in this work, we are interested in understanding if joint compliance is the source of the experimentally-measured vibrations, and then mitigating such oscillations. As shown in Ref. [11], vibrations induced by such phenomena are mainly governed by the entity of the joint clearance, the magnitude of forces that act on the joint, and the joint stiffness.

In comparison to complex mathematical models, a more practical way to evaluate if joint deformation may influence the mechanical behavior is to stress the mechanism in static configuration with loads corresponding to the operating efforts, and experimentally measure if joint displacements are relevant.

To this purpose, the five-bar linkage was statically positioned in the grasp configuration, and by means of a dynamometer we applied to joint A (see Figure 3b) an external load that corresponds to the peak load simulated during the rigid-body analysis described in Section 3.1. When this load was statically applied, we measured joint displacements of 0.020 mm and 0.050 mm at points 1 and 2 of Figure 7, respectively. These deformations may significantly influence the behavior of the system at the nominal speed of 1000 cycles/min. A possible solution aims at increasing the joint stiffness, which requires the re-design of the base joints. However, due to the high complexity and interconnection of the whole machine, we preferred to aim at a force magnitude reduction by introducing modifications to the five-bar mechanism only.

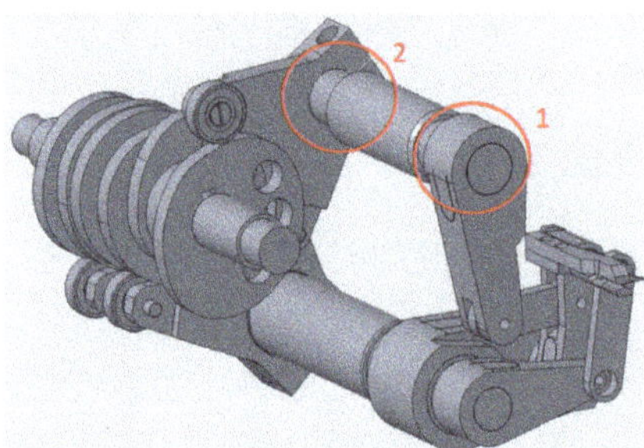

Figure 7. Locations where joint deformations are experimentally identified.

4. Optimization

In this Section, we discuss two possible strategies that aim at reducing the gripper oscillations. Since we consider the base joint deformation under operative loads as the cause of vibrations, we seek at reducing the base reaction force $\mathbf{F}_0$. Additional changes in the overall mechanism may be considered, but we limit ourselves to modifications of the five-bar only, to reduce the cost of the intervention. First, we discuss an iterative optimization process that reduces the base reaction magnitude by reducing the overall moving mass of the parallel linkage. Then, a second optimization is proposed where the lengths of the five-bar links are varied to reduce the base joint efforts.

4.1. Mass Reduction

In parallel mechanisms working at high speed, the main source of stress on components is given by their own inertial effects. Thus, it is legitimate to assume that a reduction of the overall moving masses may reduce the base joints reaction and, consequently, reduce the gripper vibrations.

Topology optimization can be a solution to reduce the mass of each component by reallocating the material only where needed [12]. In topology optimization, a mathematical method is employed to optimize the distribution of the material in a finite domain, while satisfying the given constraints. Usually, based on the constrained minimization of a cost function, the main steps of topology optimization requires the identification of design variables, the cost function, and the constraints to be satisfied. However, the result of topology optimization frequently conducts to components with highly complex geometries, which can be difficult to realize with traditional tools or which requires long production times.

A more practical solution is to directly modify the existing components, in order to save production time and to keep modifications to a minimum. In this way, simple mechanical modifications can be carried out to remove material on the components (e.g., drilling, holes, chambers, as shown in Figure 8). However, these modifications influence the dynamic properties of the system, and verifications are required to check if the mechanism performance is not deteriorated. To do that, an iterative process can be set up, as follows:

1. Each five-bar linkage component is manually modified with simple geometry modifications;
2. The modified mechanism is rigidly simulated by means of a multibody simulation software (ANSYS) in order to calculate the reaction force that acts on each joint;
3. Then, each link is independently simulated by taking into account its elasticity to estimate the corresponding deformations. Despite the possibility of simulating the overall system in a single time, we preferred to independently verify each component to reduce the simulation time;
4. If deformations are negligible, the process continues;
5. Modifications are repeated until the base force reaches a sufficiently low value.

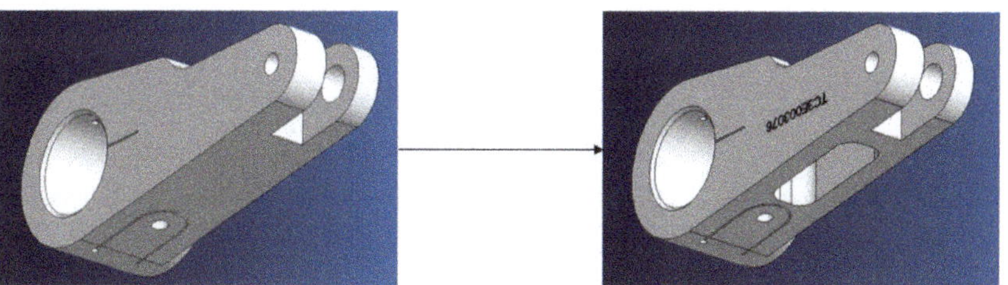

Figure 8. Example of component modifications: a milling operation is performed at the center of the components to reduce its overall mass.

We performed the previously described iterative process on the five-bar linkage. Starting from an initial mass of the mechanism of 0.460 kg, a 19% reduction is obtained with a corresponding final mass of 0.373 kg. Then, a rigid dynamic simulation is performed to quantify the joint load reduction, whose magnitude is displayed in Figure 9 and compared with the original values. The peak magnitude of $\mathbf{F}_0$ is reduced by roughly 30%, passing from 230 N to 160 N.

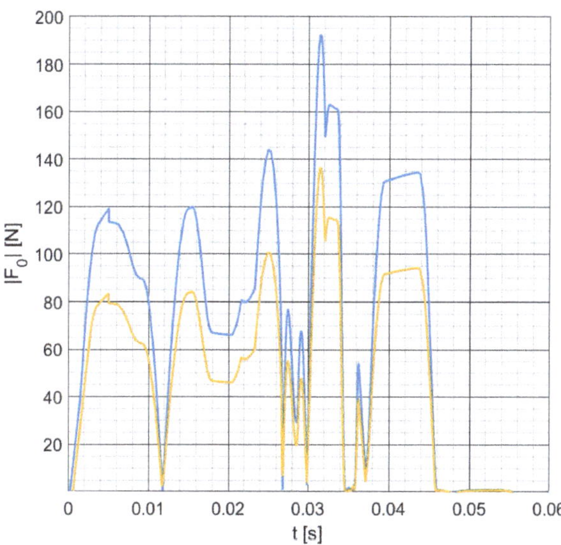

Figure 9. Magnitude of the base reaction $\mathbf{F}_0$: comparison between the initial design (blue) and the reduced-mass design (yellow).

Before manufacturing the mechanism, several tests were conducted to verify that other issues were not introduced. Firstly, a modal analysis was conducted and we verified that no resonance was excited. Then, a flexible dynamic simulation was performed, and oscillations were predicted with a magnitude of 0.02 mm, as in Section 3.3. Finally, the real mechanism was manufactured and tested at nominal working conditions of 1000 cycle/min. Oscillations were observed with a magnitude of 0.3 mm, thus significantly smaller that the original ones.

4.2. Dimension Optimization

Oscillations were drastically reduced by designing a lighter mechanism. However, the presence of residual vibrations can still be an issue, which may cause early wear. Therefore, we tried to further reduce the vibrations of the gripper at its grasp configuration.

In general, all lengths of the five-bar components can be varied to minimize the base reaction forces. This may lead to a constrained optimization problem where we seek to minimize the base force reactions while preserving the original trajectory of the gripper. Optimality conditions may be formally derived, and the optimization problem may be solved by numerical schemes. However, the modification of links 3 and 4 would also require the re-design of the gripper actuation mechanism, with further cost and production time.

Therefore, we proceeded in a different direction, by considering links 3 and 4 as assigned, and consider as design variables the lengths l_1, l_2, and the placement of the joint B over link 3 defined by the distance l_B (see Figure 3b). Due to mechanical limitations of the available space, l_1, l_2, l_B can be chosen with predefined boundaries. Since the number of design variables is limited to 3, we decided to simply sample the design space at uniform steps and to select the triplet l_1, l_2, l_B that ensures the minimum magnitude of the base joint reaction $\mathbf{F}_0$, among the design set. In particular, the optimization process is carried out as follows.

1. Considering the initial design, we extracted the joint values $\theta_3, \theta_4, \dot{\theta}_3, \dot{\theta}_4, \ddot{\theta}_3, \ddot{\theta}_4$ for a single cycle;
2. Then, we studied the five-bar mechanism by considering joints Q and C in Figure 3b as motorized. Since the motion of the kinematic chain QCE is not varied in comparison to the original design, the x, y trajectory of the gripper is preserved;
3. By the solution of the inverse dynamic problem with new inputs, we can recover the new joint values, and the joint reactions.

The aforementioned design optimization is performed by means of a custom Matlab code, where the inverse dynamic problem is solved for each sample of the design variables. The design space is defined as $l_1 \in [61, 85]$ mm, $l_2 \in [48, 60]$ mm, $l_B \in [41, 65]$ mm, and the lengths are explored with a 1 mm sampling. At the end of the design optimization, the optimal triplet is identified as $l_1 = 71$ mm , $l_2 = 48$ mm, $l_B = 65$ mm. In particular, the peak value of the $\mathbf{F}_0$ magnitude is reduced to 118 N (Figure 10), which corresponds to a reduction of 49% with respect to the 230 N of the initial mechanism.

Before prototyping the mechanism, several tests were conducted to exclude the presence of other issues. Modal analysis resulted in no natural frequencies excited, and a flexible dynamic simulation predicted negligible vibrations due to the elasticity of the components. Finally, the mechanism was manufactured and monitored at its nominal speed of 1000 cycle/min: no vibration was observed at all.

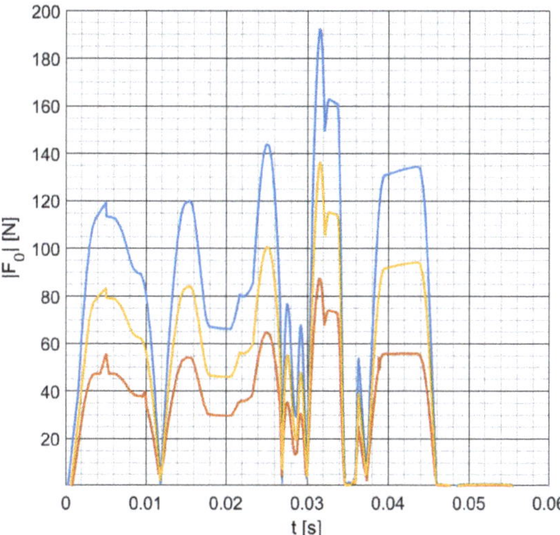

Figure 10. Magnitude of the base reaction $\mathbf{F}_0$: comparison between the initial design (blue), the reduced-mass design (yellow), and the length-optimized design (red).

5. Conclusions

In this work, we investigated the vibration phenomena of a planar parallel mechanism for application in a fast-operating industrial automatic machine. We firstly investigated the possible sources of vibrations, by excluding the presence of resonance phenomena, and oscillations inducted by the intrinsic elasticity of components. We identified as the cause of vibrations the deformation that occurs at the base joints during the operation at the nominal speed. Then, in order to reduce vibrations, two optimization approaches were conducted to reduce the loads that act on the base joints. By reducing the overall mass of the mechanism, the vibrations were strongly reduced. Then, a second optimization was carried out by modifying links lengths, achieving a further reduction of vibrations up to negligible values.

Author Contributions: Conceptualization, E.Q. and S.B.; methodology, E.Q. and S.B.; software, E.Q. and S.B.; validation, E.Q. and S.B.; formal analysis, E.Q. and S.B.; investigation, E.Q. and S.B.; resources, E.Q. and S.B.; data curation, E.Q. and S.B.; writing—original draft preparation, F.Z.; writing—review and editing, F.Z. and M.C.; visualization, F.Z. and M.C.; supervision, F.Z. and M.C.; project administration, S.B. and M.C.; funding acquisition, S.B. and M.C. All authors have read and agreed to the published version of the manuscript.

Funding: This research received no external funding.

Acknowledgments: The authors would like to thank the IMA Group for providing the equipment and the use-case this work is built upon.

Conflicts of Interest: The authors declare no conflict of interest.

References

1. Merlet, J.P. *Parallel Robots*; Springer Science & Business Media: Berlin/Heidelberg, Germany, 2005; Volume 128.
2. Soto, I.; Campa, R. On dynamic modelling of parallel manipulators: The Five-Bar mechanism as a case study. *Int. Rev. Model. Simulations (IREMOS)* **2014**, *7*, 531–541. [CrossRef]
3. Fu, Z.F.; He, J. *Modal Analysis*; Elsevier: Amsterdam, The Netherlands, 2001.
4. Chen, X.L.; Li, W.B.; Deng, Y.; Li, Y.F. Analysis of stress and natural frequencies of high-speed spatial parallel mechanism. *J. Cent. South Univ.* **2013**, *20*, 2676–2684. [CrossRef]
5. Wasfy, T.M.; Noor, A.K. Computational strategies for flexible multibody systems. *Appl. Mech. Rev.* **2003**, *56*, 553–613. [CrossRef]

6. Song, J.O.; Haug, E.J. Dynamic analysis of planar flexible mechanisms. *Comput. Methods Appl. Mech. Eng.* **1980**, *24*, 359–381. [CrossRef]
7. Nath, P.; Ghosh, A. Kineto-elastodynamic analysis of mechanisms by finite element method. *Mech. Mach. Theory* **1980**, *15*, 179–197. [CrossRef]
8. Zhang, X.; Liu, H.; Shen, Y. Finite dynamic element analysis for high-speed flexible linkage mechanisms. *Comput. Struct.* **1996**, *60*, 787–796. [CrossRef]
9. Shiau, T.N.; Tsai, Y.J.; Tsai, M.S. Nonlinear dynamic analysis of a parallel mechanism with consideration of joint effects. *Mech. Mach. Theory* **2008**, *43*, 491–505. [CrossRef]
10. Parenti-Castelli, V.; Venanzi, S. Clearance influence analysis on mechanisms. *Mech. Mach. Theory* **2005**, *40*, 1316–1329. [CrossRef]
11. Li-Xin, X.; Yong-Gang, L. Investigation of joint clearance effects on the dynamic performance of a planar 2-DOF pick-and-place parallel manipulator. *Robot. Comput. Integr. Manuf.* **2014**, *30*, 62–73. [CrossRef]
12. Bendsoe, M.P.; Sigmund, O. *Topology Optimization: Theory, Methods, and Applications*; Springer Science & Business Media: Berlin/Heidelberg, Germany, 2003.

Article

Multiobjective Design Optimization of Lightweight Gears

Francesco Cosco [1], Rocco Adduci [1], Leonardo Muzzi [1], Ali Rezayat [2] and Domenico Mundo [1],*

[1] Department of Mechanical, Energy and Management Engineering, University of Calabria, DIMEG, cubo 45C, 87036 Rende, Italy
[2] Siemens Industry Software NV, Interleuvenlaan 68, B-3001 Leuven, Belgium
* Correspondence: domenico.mundo@unical.it

Abstract: Lightweight gears have the potential to substantially contribute to the green economy demands. However, gear lightweighting is a challenging problem where various factors, such as the definition of the optimization problem and the parameterization of the design space, must be handled to achieve design targets and meet performance criteria. Recent advances in FE-based contact analysis have demonstrated that using hybrid FE–analytical gear contact models can offer a good compromise between computational costs and predictive accuracy. This paper exploits these enabling methodologies in a fully automated process, efficiently and reliably achieving an optimal lightweight gear design. The proposed methodology is demonstrated by prototyping a software architecture that combines commercial solutions and ad hoc procedures. The feasibility and validity of the proposed methodology are assessed, considering the multiobjective optimization of a transmission consisting of a pair of helical gears.

Keywords: lightweight gears; finite-element analysis; multibody simulation; design space exploration; transmission error

Citation: Cosco, F.; Adduci, R.; Muzzi, L.; Rezayat, A.; Mundo, D. Multiobjective Design Optimization of Lightweight Gears. *Machines* **2022**, *10*, 779. https://doi.org/10.3390/machines10090779

Academic Editor: Ning Sun

Received: 1 August 2022
Accepted: 5 September 2022
Published: 7 September 2022

Publisher's Note: MDPI stays neutral with regard to jurisdictional claims in published maps and institutional affiliations.

1. Introduction

Gears are crucial components for a wide range of applications: from recreational equipment to transportation, and from energy production (aka wind turbine gearboxes) to industrial machinery. Gear design has been perfected over millennia, as it encompasses a fundamental theory that enables efficient mechanical power transmission. Nevertheless, some margin of technological improvement can still be pursued. Gears design could be refined even further by exploiting the more recent advances in material science and modern designs, and manufacturing processes.

Over the last few decades, mass reduction was pursued as one of the main drivers of performance enhancement in the aerospace sector, from which lightweight design methodologies originated. Later, the usage of lightweight designs, including lightweight gears, expanded to the automotive sector, playing a crucial role in satisfying increasingly stricter regulations on combustion engine emission and fuel efficiency. The recent literature shows that weight reduction greatly benefits system efficiency in vehicles [1] and aircraft. In the latter, gearboxes can account for up to 15% [2] of the total mass saving, significantly lowering fuel consumption [3].

The improving computational performance of modern computers is opening new horizons concerning the adoption of physics-inspired models in optimization routines. In this context, model-based optimization strategies employ high-fidelity models to capture the relevant physics and obtain the proper model parameterization, which allows for netting a direct link between design parameters and model variables. As a result, the more expensive optimization processes based on prototypes and physical testing can be substituted by reliable model-based processes.

More specifically, the design of lightweight gears relies on the possibility of effectively modeling their dynamic behavior in the context of lumped or detailed system-level simulations. State-of-the-art solutions to decrease gear mass rely either on a geometrical approach where the material is removed from the gear blank [4,5] or on a multimaterial approach that combines lightweight materials with high-performance steel [6]. The geometrical approach typically exploits thin-rim geometries and the subtraction of features (e.g., holes or slots) from the gear body. In this regard, attention must be paid to prevent the deterioration of noise and vibration (N and V) performance [4] and even the impairment of the structural integrity of the geared transmission. Nonetheless, the analysis-driven design of lightweight gears requires advanced simulation methods to properly consider the body geometry and its impact on the gear flexibility [5]. The multimaterial approach exploits material-level properties to improve component performance, and has demonstrated potential for N and V improvement in lightweight gears [6]. However, several technological gaps must still be covered before achieving a maturity level that enables its industrial applicability.

Recent trends suggest that transmission lightweighting may pave the way to system-level performance improvement. The goal of concurrently reducing the gears' weight and the vibrations in transmission was pursued by Yang et al. in [7], and by Ramadani et al. in [8]. In [9], the effects of lightweight gear blank on static and dynamic behavior for electric-drive systems in electric vehicles were studied using a hybrid finite-element–analytical method in conjunction with a rigid–flexible coupled dynamic model that took into account the flexibility of the shaft, bearings, and housing.

Recent advances in FE-based contact analysis demonstrated that hybrid FE–analytical gear contact models [10–12] could better compromise computational costs and predictive accuracy. Their usage enables optimally tuning all relevant lightweighting parameters without sacrificing dynamic performance.

Built on the experience that matured in [10], this paper proposes a complete workflow required for accomplishing a multiobjective design space exploration of lightweight transmissions. We exploit the geometrical approach to derive a lightweight helical gear design. The joint objectives of optimizing the weight, and N and V performance of geared transmissions are pursued by exploiting a hybrid FE–analytical gear contact model in conjunction with multiobjective design optimization software resulting in a fully automatized optimization toolchain. Moreover, the obtained framework could easily be expanded to further generalize the optimization problem by including more complex features or using different materials.

The remainder of the paper is structured as follows: an overview of the proposed multiobjective optimization strategy is illustrated in Section 2, while Section 3 describes the hybrid FE–analytical method employed to analyze the gears meshing. In Section 4, the basic steps of the optimization workflow are illustrated. Section 5 describes an optimization case where a pair of helical gears, one of which with a lightweight design, was optimized, and presents the obtained results. Section 6 closes the paper by discussing the achieved results and proposing further advances.

2. Multiobjective Optimization Strategy

The optimal design of lightweight gears can be formulated as a multiobjective optimization problem (MOP). Such a problem involves the joint minimization of multiple, usually conflicting, objective functions while varying a set of decision variables. Mathematically, a MOP is described as:

$$\min_{x} \mathbf{F}(x), \qquad \text{subject to } x \in \mathbf{X} \subseteq \mathbb{R}^n \tag{1}$$

which describes the joint minimization of all the components of objective function $\mathbf{F}$: $\mathbf{F}(x) = \{f_1(x), f_2(x), \ldots, f_n(x)\}$, as a function of decision vector x. The effect of enforcing any constraint results in a reduction in the feasible decision space, $\mathbf{X}$. The image of feasible decision set $\mathbf{X}$ is defined as feasible objective space Z:

$$Z = \mathbf{F}(x), \ \forall x \in \mathbf{X} \tag{2}$$

Optimal solutions to the MOP usually comprise a subset of feasible solutions that are not dominated by any other feasible alternative. Geometrically, they are confined at the edges of the feasible objective space, constituting the well-known Pareto front. The inverse image of the Pareto front is called the Pareto optimal set. As depicted in Figure 1, such a representation offers valuable insight to designers, as it allows for them to immediately distinguish the optimal solutions and select the more suitable one depending on the considered criterion for prioritizing between the different objectives.

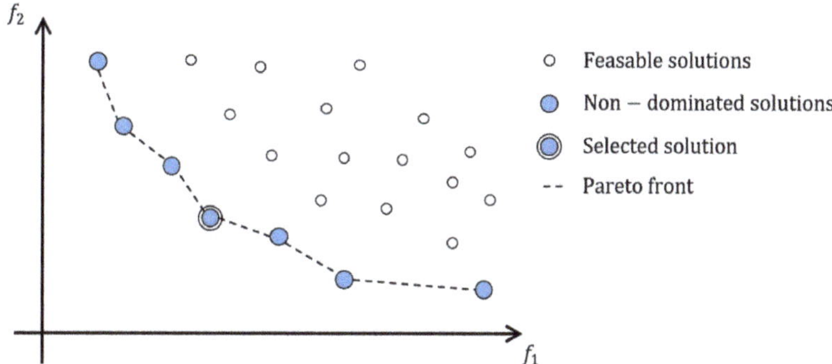

Figure 1. Example of a Pareto front.

In engineering optimization, the goodness of an algorithm is measured by looking at two metrics: efficiency and robustness. Efficiency refers to the ability of an algorithm to achieve the optimal solution within a prescribed confidence interval with a minimal number of iterations. Robustness instead concerns the ability of the algorithm to achieve the same objective with a similar number of iterations, irrespective of the starting design. In this contribution, both efficiency and robustness were attempted thanks to the usage of the Siemens Simcenter software ecosystem (Release 2020.2, Siemens PLM, Leuven, Belgium). In particular, we relied on the Siemens Simcenter HEEDS I MDO package [13] (Release 2020.2.1, Siemens PLM, Leuven, Belgium), considered for its design space exploration and optimization capabilities, in conjunction with Siemens Simcenter 3D, which is a fully integrated CAE solution, of which we used two modules: the Simcenter 3D Transmission Builder [14] and the Simcenter 3D Motion solver [15], enabling an efficient simulation of multibody models.

HEEDS I MDO Optimization Software

As depicted in Figure 2, the HEEDS user interface allows for a very natural definition of complex optimization scenarios. It provides a plethora of native plugins, each enabling input/output interfaces for specific third-party software modules such as Simcenter 3D, MATLAB, and Abaqus.

In the context of this work, the "best" solution candidates were obtained by relying on the SHERPA proprietary optimization algorithm. SHERPA stands for Simultaneous Hybrid Exploration that is Robust, Progressive and Adaptive; it is an optimization strategy designed to adapt dynamically to the specific features of the considered optimization problem. By exploiting a set of dedicated heuristics, SHERPA simultaneously analyzes several optimization strategies, and selects the more efficient and robust one. Confidentiality restrictions limit the amount of disclosable information regarding the solver's background methodologies; therefore, they are omitted in the remainder of this contribution.

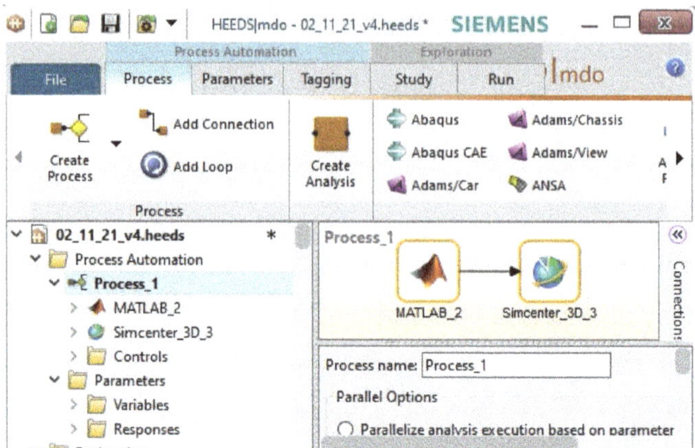

Figure 2. HEEDS graphical user interface.

3. Modeling Strategy for Gear Meshing Analysis

Meshing phenomena in drivetrain systems are typically governed by the contact mechanism occurring between the interacting bodies. A correct description of such phenomena in the modeling environment is vital to accurately describe the transmitted mechanical energy and obtain realistic results.

In the design optimization context, optimal candidates are chosen by exploring the feasible design space. The fitness of each design configuration is evaluated by measuring the performance of different simulations. In this regard, the feasibility of the adopted modeling solution requires relying on time-efficient models to ensure that the evaluation of the objective functions is accomplished in a reasonable time. Different modeling strategies are available in the literature to cope with different analysis objectives. For geared drivetrain systems, the most used solutions can be grouped as follows:

i. **Lumped-parameter (LP)** modeling [16–20] aims to (semi)analytically lump mechanical and contact properties of a system through the most relevant, yet few, parameters (e.g., meshing stiffness, damping, and inertias) and states (e.g., angular positions and velocities). It allows for efficiently representing the overall load distribution and approximating the system-level statics/dynamics with simple and quick-to-solve equations. However, it is limited to simple gear topologies and geometry-dependent parameter sets that are often difficult to obtain [21,22]. Therefore, although they are very computationally efficient, LP models require the careful parameterization of the design space (not always possible) to be used in optimization problems.

ii. **Finite-element (FE)** modeling a gear train [12,23,24] relies on using finite elements to achieve a geometrical domain discretization and approximate the behavior of the continuous bodies through numerical integrations. It is a general modeling approach since it does not rely on any gear topologies assumption and is a first-principle-based strategy. It can account for micro- and macrogeometry deviations from the nominal operating conditions, teeth coupling, and the contacting bodies' dynamic behavior. Nevertheless, it usually requires the fine discretization of the contacting bodies in the contact zones due to the high-stress gradients involved, and computationally expensive contact detection, which renders the FE method computationally prohibitive to use in design optimization problems.

iii. **Multibody (MB)** modeling [25–28] is used to analyze the dynamics of systems composed of several components interconnected in space with different specifications. It enables the representation of the different bodies as rigid or deformable (flexible) components linked together through permanent (e.g., bushings) or variable (e.g., meshing gears) connection elements. It is modular, efficient (depending on the formulation and system topology), and allows for including other modeling strategies, i.e., LP and FE. Moreover, it accounts for large and complex body motions in space, macro misalignments, and microgeometry modifications [29] with respect to nominal operating and geometrical specifications. Nevertheless, it is challenging to accurately and efficiently formulate the body flexibility effects in a MB environment jointly with contact phenomena. In particular, most MB formulations are based on small (body) deformations that represent a limitation under high loading conditions.

This work relies on an advanced contact formulation already integrated within a MB tool of Siemens Simcenter Motion [30]. In particular, the floating frame of reference (FFR) formulation [31] was adopted to describe the motion of the gears in space and their interactions, resulting in the following (at most) quadratic equations of motion (EOMs):

$$M(q)\ddot{q} + Kq + G^T\lambda + f_v + f_{ext} = 0 \qquad\qquad (3)$$
$$\phi(q) = 0$$

where q and $\ddot{q} \in \mathbb{R}^{n_q}$ represent the generalized system coordinates and accelerations, respectively; $M(q)$ is the configuration-dependent mass matrix of the system; K is the constant stiffness matrix; f_v is the velocity-dependent force vector that accounts for the gyroscopic effects; f_{ext} is the generalized external force vector; $\phi(q)$ represents the set of constraint equations, and $G = \partial\phi/\partial q$ is the constraint Jacobian.

3.1. Advanced Contact Strategy in Multibody Simulations

If we focus on the dynamics of two generic meshing gears, the interaction phenomena governing the system dynamics are in the EOMs through generalized external forces f_{ext}, as depicted in Figure 3a, which can be written as:

$$f_{ext} = f_{12}(F_{12}) + f_{21}(F_{21}) + f_{in}(T_{in}) + f_{res}(T_{res}) \qquad\qquad (4)$$

As a result of the considered boundary conditions, T_{in} is the input torque applied to the driving gear, whereas T_{res} is the resistant torque applied to the driven gear. The remaining terms required to solve the EOMs are the action (F_{12}) and reaction (F_{21}) contact forces. The $f_\bullet(\cdot)$ operator instead represents the projections of forces and torques onto the generalized coordinates space of the assembled MB system.

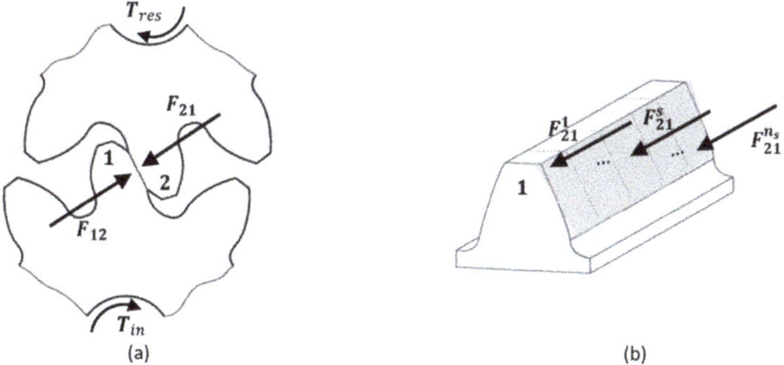

Figure 3. Contact problem: (**a**) external forces acting on the meshing gears; (**b**) slicing approach.

In such a context, the gear contact is solved using three steps:

1. Detecting the contact locations of the meshing bodies.
2. Formulating the amount of contact deformation (compliance) or inversely the contact stiffness.

3. Formulating the contact problem to be solved.

To deal with Points 1 and 2, the tracked tooth or teeth in contact are divided into n_s 2D slices (see Figure 3b), such that an analytical contact detection is performed. The slicing approach enables a very efficient evaluation of the contact location and thus of the localized geometrical properties of the meshing gears. For each slice, the software exploits the involute gear geometries, accounting for gear misalignments in all DOFs. Moreover, microgeometry modifications of the tooth profile typically used to optimize the gear transmission error are considered at this stage.

Moreover, for each s-th slice, it is assumed that the total deformation pattern results from the superposition of three effects (see Figure 4):

1. The residual static deformation based on the Andersson and Vedmar idea is described in [32]. It is computed through an FE-based approach [21,33] where a static unit nodal load, normal to the tooth surface, is applied in a preprocessing step; subsequently, the deformation pattern is subtracted from the previous one considering the same loading condition while clamping the tooth in its middle plane. In this way, a locally incorrect solution is overcome [12].
2. The nonlinear local contact deformation is approximated as two contacting cylinders in line contact according to Hertz theory, and theWeber and Banaschek formula [34].
3. The dynamic deformation pattern is taken into account using the FE-based component mode synthesis approach [35–37] and creating a set of mass-orthonormalized eigenmodes.

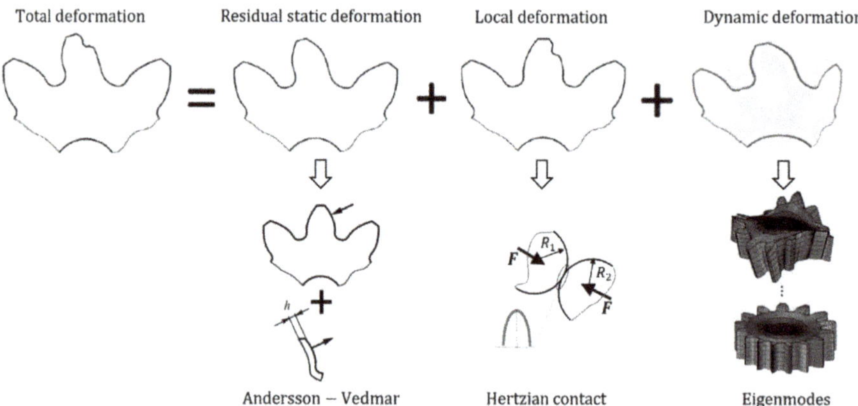

Figure 4. Computation of the required amount of deflection.

Once the compliance model had been defined for the meshing gears, the following nonlinear contact problem was constructed for each s-th slice to link the kinematic and dynamic displacements with the related contact forces F_{21}^s:

$$\delta_{21}^s - \alpha_{21}^s \left(\overline{F_{21}^s \cdot n_{21}^s}, \ geom_{1,2}, E_{1,2}, \nu_{1,2} \right)$$
$$+ [C_1(geom_1, E_1, \nu_1) + C_2(geom_1, E_1, \nu_1)] \left[\overline{F_{21} \cdot n_{21}^s} \right] \tag{5a}$$
$$= g\left(\overline{F_{21} \cdot n_{21}^s}, \ geom_{1,2}, E_{1,2}, \nu_{1,2} \right) \geq 0$$

$$\overline{F_{21} \cdot n_{21}^s} \geq 0 \tag{5b}$$

$$g\left(\overline{F_{21} \cdot n_{21}^s}, \ geom_{1,2}, E_{1,2}, \nu_{1,2} \right)^T \left[\overline{F_{21} \cdot n_{21}^s} \right] = 0 \tag{5c}$$

where δ is the corrected penetration for dynamic effects, misalignment, and microgeometry; C_i is the residual compliance maps based on the Andersson and Vedmar approach [32];

$geom_i$, E_i, v_i are the geometrical properties, the Young modulus, and Poisson's ratio of the i-th gear, respectively; n is the unit vector representing the normal to the contact surface at the contact location; α is the nonlinear Hertzian penetration described by the following Weber and Banaschek formula [34]:

$$\alpha = \frac{F}{\pi l}\left(\frac{1-v_1^2}{E_1} + \frac{1-v_2^2}{E_2}\right)\left[\ln\left(\frac{4h_1 h_2}{a^2}\right) - \frac{1}{2}\left(\frac{v_1}{1-v_1} + \frac{v_2}{1-v_2}\right)\right]. \tag{6}$$

where a is the half contact width that can be computed as

$$a = \left[\frac{4F}{\pi l}R_{eq}\left(\frac{1-v_1^2}{E_1} + \frac{1-v_2^2}{E_2}\right)\right]^{\frac{1}{2}}; \tag{7}$$

h_i is the half tooth thickness computed at the pitch point of the i-th gear, and R_{eq} is the equivalent radius of curvature of the contacting cylinders:

$$\frac{1}{R_{eq}} = \frac{1}{R_1} + \frac{1}{R_2}. \tag{8}$$

Lastly, the overall MB problem was fully defined, and the assembled EOMs of Equation (3) were solved with respect to the generalized coordinates and accelerations by means of a nonlinear iterative solver [38].

3.2. Performance Metric: Transmission Error

Although nominal gear profiles are conjugate by design, tooth and body flexibility, and other disturbing factors such as misalignment, deterioration, and micromodifications produce a slight deviation from the ideal kinematic conditions. Analysts use the transmission error (TE) to quantify the degree of offset between the ideal (or conjugate) and actual (or real) behavior of the driven gear. It is defined as the difference between the ideal kinematic motion of a gear pair and its actual realization:

$$TE = \frac{1}{\tau}\Delta\theta_2 - \Delta\theta_1, \tag{9}$$

where $\Delta\theta_1$ and $\Delta\theta_2$ represent the angular motion of the driving and driven gears, respectively, and τ is the transmission ratio of the gear pair. For practical reasons, the TE is often reported in the equivalent linear form:

$$TE = r_{b2}\Delta\theta_2 - r_{b1}\Delta\theta_1, \tag{10}$$

where r_{b1} and r_{b2} represent the base radii of the corresponding gears.

As demonstrated by Palermo et al. in [39], an interesting metric to assess the N and V performance of a geared transmission is the pick-to-pick (PtP) value of the TE, which inherently indicates the severity of parametric excitation for different levels of load and velocity occurring during the meshing cycle. The recent literature on the multiobjective optimization of macrogeometry [40,41], and combined macro- and microgeometry [42] uses the TE-related metric as an indicator of the N and V performance of geared transmissions.

4. Adopted Optimization Strategy

In complex problems such as the optimization of geared transmissions, the main challenges stand on setting up a robust framework that often requires to set up fast and reliable interconnections among several tools with different communication protocols. Moreover, due to the high nonlinear nature of contact problems in gear transmissions, efficient and accurate simulation models are required to evaluate the system performance at each design configuration.

4.1. Optimization Workflow

In this contribution, most limitations concerning the communication and solution of the optimization problem are overcome by exploiting the versatile nature of HEEDS. In addition, the optimization metrics per design are evaluated through Simcencer Motion, as described in the previous sections.

On the one hand, HEEDS receives input parameters (i.e., design variables) that are then manipulated to be compatible with the program where the selected optimization metrics are evaluated, e.g., performing a dynamic simulation in Simcenter Motion.

On the other hand, the Simcenter 3D Transmission Builder takes as input the description of the gear train in terms of its basic geometrical and topological properties, expressed according to industry standards as described in [14]. Then, a multibody model of the assembled transmission is procedurally generated within the Simcenter 3D motion environment and an advanced gear contact method [14] is considered, as described in Section 3.1, which allows for capturing relevant static and dynamic phenomena that influence the system-level behavior of the considered gear pair.

The influence of the gear crown and gear-body design is taken into account without compromising the computational efficiency of the dedicated gear-contact modules embedded in the motion solver. The results of the multibody simulations are used to compute the performance metrics required to evaluate the objective functions. Once the simulation is completed, HEEDS extracts the results and converts them back into the format that the input values previously had. The analysis of the characteristics or quantities that are optimized is entrusted to the optimization algorithm, which produces new input parameters by repeating the analysis until the stopping criteria are reached.

As depicted in Figure 5, the proposed methodology for achieving a multiobjective optimal design of a lightweight transmission is initiated by generating a baseline solution. This is achieved by means of the Simcenter 3D Transmission Builder module, which can generate the CAE models for each component of the transmission system and assemble them into the multibody model required for the quasistatic or dynamic simulations.

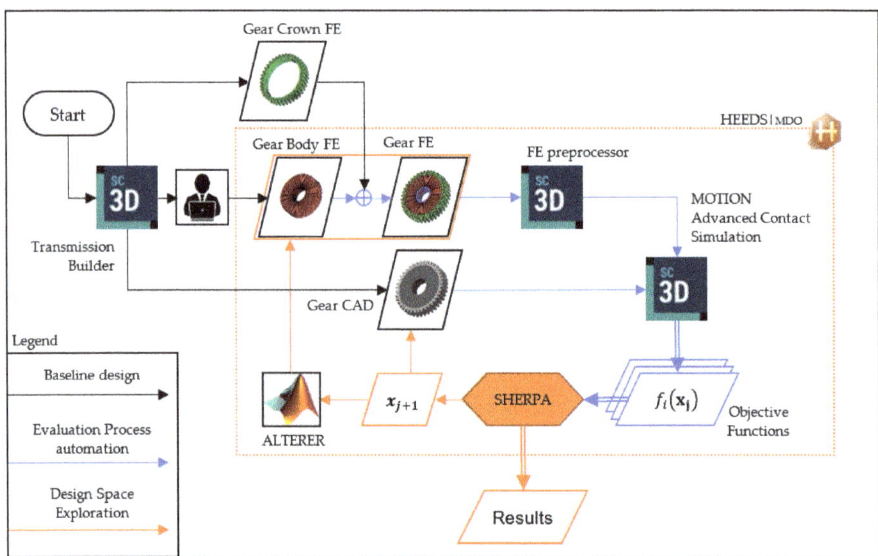

Figure 5. Multiobjective lightweight design: flowchart.

4.2. Alterer

In order to enable the required lightweight design parameterization and the use of the advanced contact solution in the presence of a lightweight gear body geometry, the current version of the Simcenter software requires manual intervention from highly qualified users for complementing the automatically generated FE model of the gear crown with the preferred lightweight geometry. The manually assembled FE model of the lightweight gear is given as input to the FE preprocessor, which computes the necessary information to enable the simulation of the gear contact problem.

The HEEDS software can be used to completely automate the above-described process except for the gear-body FE definition, which is still linked to manual user interaction. To overcome this issue, a set of MATLAB routines were developed forming a software package named ALTERER (Automated LighTweight gEaR body 9orphing). The latter defines the FE mesh of the lightweight gear body and automatically assembles the FE body part to the existing gear crown, starting from a set of design parameters. As a result, the program generates the assembled gear body FE input files already complying with the format expected by the FE Preprocessor.

Thanks to the obtained functionality, the HEEDS software can take the lead in orchestrating the ordered execution of all necessary modules, obtaining a fully automated evaluation process without any manual human intervention. HEEDS relies on its patented Sherpa proprietary algorithms and strategy in order to optimally and constantly tune the next decision set to be explored. As a result, the lightweight gear design space exploration process is fully automated. Moreover, HEEDS allows for aborting prematurely, pausing and restarting the optimization, or even extending it by launching a set of additional evaluations in case the analyst deems it necessary.

5. Application Case

In this section, the previously described optimization workflow and tools are applied to an industrially relevant use case: the optimization of the gear body of a helical gear pair.

5.1. Use-Case Description

In order to illustrate the application of the proposed methodology, we considered a multiobjective design of the gear pair starting from a reference or initial design. The baseline gear pair was formed by two identical gears, a driver and a driven one, and the geometrical properties are summarized in Table 1.

Table 1. Gears design specifications.

Parameter Name	Baseline Gears
Teeth number	40
Normal module (m)	2.5 mm
Normal pressure angle	20 deg
Helical angle	10 deg
Tooth width	20 mm
Addendum	1
Dedendum	1.25
Working center distance	101.6 mm
Rim width	-
Web thickness	20 mm

As depicted in Figure 6, the optimal design was limited to the driven gear while keeping the other fixed at its baseline geometry.

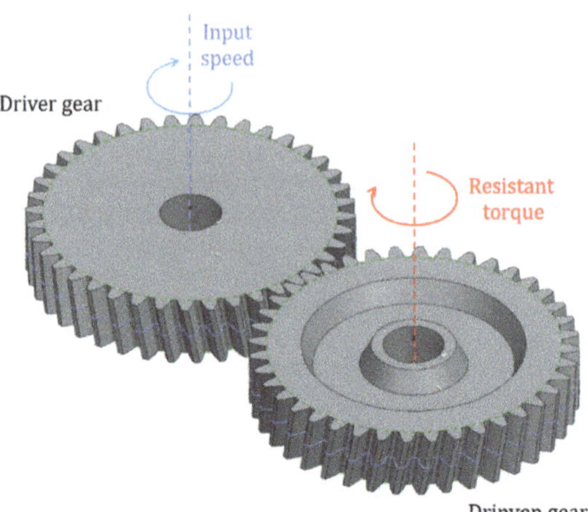

Figure 6. CAD model of the considered lightweight gear transmission: the multiobjective design optimization was formulated by removing material from the body of the driven gear while maintaining the driving one in its full configuration.

As illustrated in Figure 7, the driven gear was parameterized enabling the mass reduction considering only two geometrical decision variables: the Rim length and the Web thickness. For a given combination of the design variables, HEEDS orchestrates the evaluation of the objective functions by executing the evaluation pipeline described in Section 4.

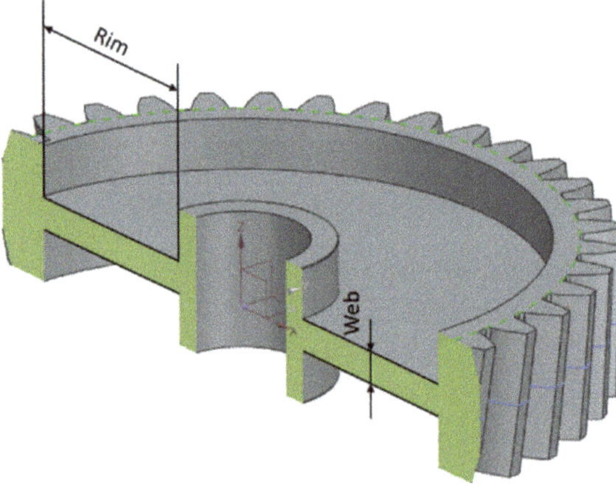

Figure 7. Lightweight gear's geometry, parametrically controlled by varying two decision variables: rim length and web thickness.

Therefore, the automated design-exploration process first obtains an updated CAD model and then the corresponding FE model of the driven gear, as depicted in Figure 8. As described above, the HEEDS | MDO proprietary multiobjective Pareto search algorithm simultaneously uses multiple search strategies to more effectively explore the Pareto front. Its superior efficiency to that of other technique available in the literature was also documented in [43]. Table 2 summarizes the interval ranges used for the reported study.

Figure 8. FE model of the lightweight driven gear.

Table 2. Design variables ranges.

Decision Variable	Min	Max	Baseline
Web thickness (mm)	5	20	20
Rim length (mm)	36.25	43.75	43.75

Additionally, the optimization problem was formulated to minimize the total mass of the driven gear while reducing the peek-to-peek value of the TE under dynamic conditions and for different levels of the transmitted torque.

To avoid convergence problems, the driver gear was enforced to reach the desired speed value of 10 rpm by following a ramp of one-second duration (as illustrated in Figure 9) while applying a specific resistant torque to the driven gear. The ramp also allows for simulating a more realistic scenario representing a motor from which the torque is taken and gradually transmitted to the gear through a clutch or simply the time required to overcome the system's inertia.

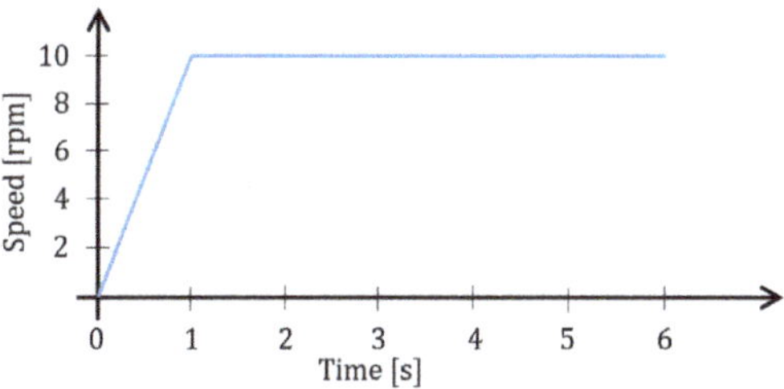

Figure 9. Angular velocity profile assigned to driver gear.

5.2. Optimization Results

Figure 10 shows the results obtained using the methodology under investigation for three levels of the driving torque: 50, 100, and 200 nm. As expected, increasing the torque had a direct and proportional impact on the PtP value of the TE. For each analyzed level of torque, we could observe two clusters of solutions. The first cluster contained all those solutions starting from the baseline and moving in the direction of reducing the mass while mildly compromising PtP performance: in this cluster, the two conflicting design goals were reflected in a typical hyperbolic trend. Interestingly, a second cluster of solutions appeared when looking at the leftmost side of the chart. Below 0.9 kg of mass, different hyperbolic branches dominated the Pareto front, featuring the possibility of substantially improving PtP performance while also reducing the mass of the gear.

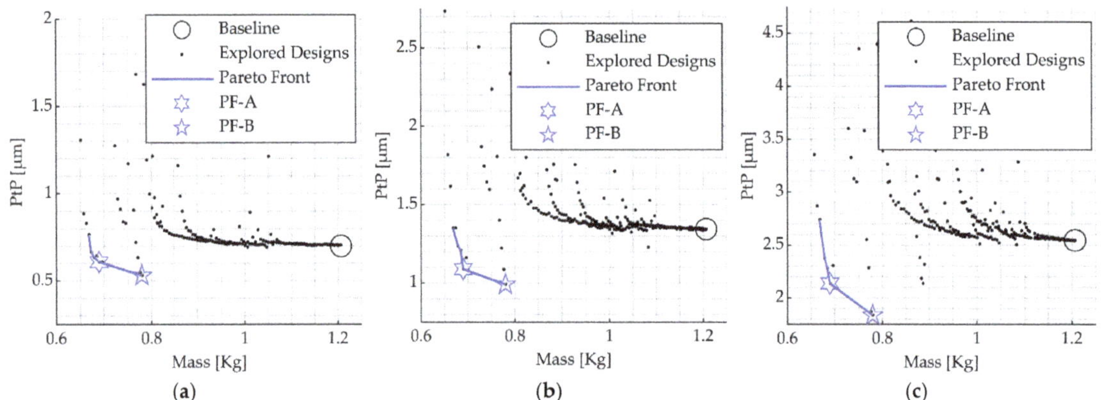

Figure 10. PtP vs. mass trend for different values of torque: (**a**) 50 nm; (**b**) 100 nm; (**c**) 200 nm.

The results of MOP analysis are summarized in Table 3. From each Pareto front, two extreme solutions were extracted for further analysis: solutions PF-A, which were selected as the best compromise along the Pareto Front, and solutions PF-B were selected as the best PtP solution; thus, they are at the bottom corner of each Pareto front.

Table 3. Optimal design Results.

	Design Parameters		Design Goals							
	Web Thickness (mm)	Rim Length (mm)	Mass		PtP					
			(kg)	(%)	50 nm		100 nm		200 nm	
					(μm)	(%)	(μm)	(%)	(μm)	(%)
Baseline	20	43.75	1.205		0.75		1.35		2.54	
PF-A	6.04	43.75	0.691	−42.7%	0.61.	−19.0%	1.09	−19.1%	2.14	−15.7%
PF-B	5.60	40	0.780	−35.3%	0.53	−29.3%	0.99	−26.3%	1.84	−27.6%

The computational time required for the optimization of the gear pair ranged between 54.75 h for the 100 nm case and 69.95 h for the 200 nm case, with an average computational time per evaluated design of 10.84 and 13.94 min, respectively. The computational overhead of executing HEEDS proved to be minimal compared to the complexity of the simulation software used for the evaluation of the objective function. All examples were computed using a commodity laptop featuring an Intel Core i7-h and 12 GB RAM.

6. Conclusions

Fueled by recent advances in the field of FE-based contact analysis, this paper extended the usage of modern design exploration software to the field of lightweight gear transmissions. The envisioned methodology was built on top of existing state-of-the-art commercial solutions. HEEDS and its patented SHERPA multiobjective search algorithms were used to efficiently explore the Pareto front in case of multiobjective design space exploration problems. Simcenter 3D, comprising the FE-preprocessor and the Transmission Builder functionalities, was used because its hybrid FE–analytical approach offers the best-in-class numerical efficiency in terms of gear contact problem simulations. As part of this study, ad hoc procedures were developed to eliminate human intervention from the evaluation pipeline, leading to a fully automated evaluation process.

The proposed methodology was tested by considering the multiobjective design space exploration of helical gear transmission. Three levels of torque were also used to analyze the impact of the loading conditions on the obtained results. For each subcase, a Pareto front appeared at the bottom-left corner of each chart, featuring the hyperbolic shape typical of competing goals scenarios.

As summarized in Table 3, solutions PF-A and PF-B extracted from each of the Pareto fronts, corresponded to realizations of the same design parameters, suggesting that the optimal results obtained for one torque level were robust enough to maintain optimality within the considered torque variations. Further investigations are required to assess the validity of such findings on a more methodological level.

Future works will further expand the proposed approach by considering different types of lightweight gears, eventually combining micro- and macrogeometrical parameters with topological choice variables, such as the type and number of holes in the gear body.

Author Contributions: Conceptualization, F.C., R.A., L.M., A.R. and D.M.; methodology, F.C., L.M., A.R. and D.M.; software, F.C.; validation, F.C., R.A. and L.M.; formal analysis, F.C., R.A., L.M., A.R. and D.M.; investigation, F.C., R.A., L.M., A.R. and D.M.; resources, A.R. and D.M.; data curation, F.C., R.A., L.M., A.R. and D.M.; writing—original draft preparation, F.C., R.A. and D.M.; writing—review and editing, F.C., R.A., A.R. and D.M.; visualization, F.C. and R.A.; supervision, A.R. and D.M.; project administration, A.R. and D.M.; funding acquisition, A.R. and D.M. All authors have read and agreed to the published version of the manuscript.

Funding: This research was funded by VLAIO (Flemish government agency Flanders Innovation and Entrepreneurship), R&D project BE QUIET no. HBC.2019.

Institutional Review Board Statement: Not applicable.

Informed Consent Statement: Not applicable.

Data Availability Statement: Not applicable.

Acknowledgments: The authors kindly acknowledge VLAIO (Flemish government agency Flanders Innovation and Entrepreneurship) for their support of the R&D project BE QUIET (BElgian consortium for QUIEt Transmissions, no. HBC.2019.2323), coordinated by Siemens Digital Industries Software, Belgium.

Conflicts of Interest: The authors declare no conflict of interest.

References

1. Dobrzański, L.; Tański, T.; Čížek, L.; Brytan, Z. Structure and properties of magnesium cast alloys. *J. Mater. Process. Technol.* **2007**, *192–193*, 567–574. [CrossRef]
2. New Equipment Digest. NASA's Revolutionary Hybrid Gear Lightens Your Load. Available online: https://www.newequipment. com/research-and-development/article/22058083/nasas-revolutionary-hybrid-gear-lightens-your-load (accessed on 10 June 2022).
3. Ohrn, K.E. Aircraft Energy Use. In *Encyclopedia of Energy Engineering and Technology*, 2nd ed.; CRC Press: Boca Raton, FL, USA, 2014; pp. 32–38. [CrossRef]
4. Shweiki, S.; Palermo, A.; Mundo, D. A Study on the Dynamic Behaviour of Lightweight Gears. *Shock Vib.* **2017**, *2017*, 7982170. [CrossRef]
5. Benaïcha, Y.; Mélot, A.; Rigaud, E.; Beley, J.-D.; Thouverez, F.; Perret-Liaudet, J. A decomposition method for the fast computation of the transmission error of gears with holes. *J. Sound Vib.* **2022**, *532*, 116927. [CrossRef]
6. Catera, P.G.; Mundo, D.; Gagliardi, F.; Treviso, A. A comparative analysis of adhesive bonding and interference fitting as joining technologies for hybrid metal-composite gear manufacturing. *Int. J. Interact. Des. Manuf.* **2020**, *14*, 535–550. [CrossRef]
7. Yang, J.; Zhang, Y.; Lee, C.-H. Multi-parameter optimization-based design of lightweight vibration-reduction gear bodies. *J. Mech. Sci. Technol.* **2022**, *36*, 1879–1887. [CrossRef]

8. Ramadani, R.; Belsak, A.; Kegl, M.; Predan, J.; Pehan, S. Topology Optimization Based Design of Lightweight and Low Vibration Gear Bodies. *Int. J. Simul. Model.* **2018**, *17*, 92–104. [CrossRef]

9. Hou, L.; Lei, Y.; Fu, Y.; Hu, J. Effects of Lightweight Gear Blank on Noise, Vibration and Harshness for Electric Drive System in Electric Vehicles. *Proc. Inst. Mech. Eng. Part K J. Multi-Body Dyn.* **2020**, *234*, 447–464. [CrossRef]

10. Shweiki, S.; Rezayat, A.; Tamarozzi, T.; Mundo, D. Transmission Error and strain analysis of lightweight gears by using a hybrid FE-analytical gear contact model. *Mech. Syst. Signal Process.* **2019**, *123*, 573–590. [CrossRef]

11. Vivet, M.; Tamarozzi, T.; Desmet, W.; Mundo, D. On the modelling of gear alignment errors in the tooth contact analysis of spiral bevel gears. *Mech. Mach. Theory* **2021**, *155*, 104065. [CrossRef]

12. Cappellini, N.; Tamarozzi, T.; Blockmans, B.; Fiszer, J.; Cosco, F.; Desmet, W. Semi-analytic contact technique in a non-linear parametric model order reduction method for gear simulations. *Meccanica* **2017**, *53*, 49–75. [CrossRef]

13. Simcenter. HEEDS. Available online: https://www.plm.automation.siemens.com/global/en/products/simcenter/simcenter-heeds.html (accessed on 22 June 2022).

14. Siemens PLM Software. Boosting Productivity in Gearbox Engineering Using a Revolutionary Approach to Multibody Simulation. Available online: https://resources.sw.siemens.com/en-US/white-paper-gearbox-multibody-simulation (accessed on 22 June 2022).

15. Simcenter 3D Motion: Siemens PLM Software. Advanced FE-Preprocessor. Available online: https://www.plm.automation.siemens.com/global/en/products/simcenter/simcenter-3d.html (accessed on 22 June 2022).

16. Cai, Y.; Hayashi, T. The Linear Approximated Equation of Vibration of a Pair of Spur Gears (Theory and Experiment). *J. Mech. Des.* **1994**, *116*, 558–564. [CrossRef]

17. Cai, Y. Simulation on the Rotational Vibration of Helical Gears in Consideration of the Tooth Separation Phenomenon (A New Stiffness Function of Helical Involute Tooth Pair). *J. Mech. Des.* **1995**, *117*, 460–469. [CrossRef]

18. Parker, R.G.; Vijayakar, S.M.; Imajo, T. Non-linear dynamic response of a spur gear pair: Modelling and experimental comparisons. *J. Sound Vib.* **2000**, *237*, 435–455. [CrossRef]

19. Shweiki, S.; Korta, J.; Palermo, A.; Adduci, R.; Mundo, D. Combining Finite Element Analysis and Analytical Modelling for Efficient Simulations of Non-Linear Gear Dynamics. In Proceedings of the 7th European Congress on Computational Methods in Applied Sciences and Engineering, Crete Island, Greece, 5–10 June 2016.

20. ISO. ISO 6336-1:1996—Calculation of Load Capacity of Spur and Helical Gears—Part 1: Basic Principles, Introduction and General Influence Factors. Available online: https://www.iso.org/standard/12632.html (accessed on 20 June 2022).

21. Shen, Y.; Yang, S.; Liu, X. Nonlinear dynamics of a spur gear pair with time-varying stiffness and backlash based on incremental harmonic balance method. *Int. J. Mech. Sci.* **2006**, *48*, 1256–1263. [CrossRef]

22. Vivet, M.; Mundo, D.; Tamarozzi, T.; Desmet, W. An analytical model for accurate and numerically efficient tooth contact analysis under load, applied to face-milled spiral bevel gears. *Mech. Mach. Theory* **2018**, *130*, 137–156. [CrossRef]

23. Blockmans, B.; Adduci, R.; Fiszer, J.; Cappellini, N.; Desmet, W. Finite Element Based Simulation of Planetary Gearboxes Using Model-Order Reduction. In Proceedings of the International Gear Conference, Lyon, France, 29 August 2018; Chartridge Books: Oxford, UK, 2018; pp. 1024–1033.

24. Korta, J.A.; Mundo, D. Multi-objective micro-geometry optimization of gear tooth supported by response surface methodology. *Mech. Mach. Theory* **2017**, *109*, 278–295. [CrossRef]

25. Blockmans, B. *Model Reduction of Contact Problems in Flexible Multibody Dynamics with Emphasis on Dynamic Gear Contact Problems*; KU Leuven: Leuven, Belgium, 2018.

26. Vivet, M.; Verhoogen, J.; Jiránek, P.; Tamarozzi, T. A New Gear Contact Method for the Tooth Contact Analysis of Spiral Bevel Gear Drives in Multibody Simulations. In Proceedings of the NAFEMS International Seminar Advances in Structural Dynamic Simulation, Online, 29–30 March 2022.

27. Tamarozzi, T.; Blockmans, B.; Desmet, W. Dynamic Stress Analysis of the High-Speed Stage of a Wind Turbine Gearbox Using a Coupled Flexible Multibody Approach. In Proceedings of the Conference for Wind Power Drives 2015, Aachen, Germany, 10–12 March 2015; Books on Demand: Canfield, OH, USA, 2018; p. 87.

28. Palermo, A.; Mundo, D.; Hadjit, R.; Desmet, W. Multibody element for spur and helical gear meshing based on detailed three-dimensional contact calculations. *Mech. Mach. Theory* **2013**, *62*, 13–30. [CrossRef]

29. Litvin, F.L.; Fuentes, A. *Gear Geometry and Applied Theory*, 2nd ed.; Cambridge University Press: Cambridge, UK, 2004; ISBN 9780521815178.

30. Siemens. *White Paper: Boosting Productivity in Gearbox Engineering*; 2019; Available online: http://www.fp7demetra.eu/images/files/Siemens-PLM-Boosting-Productivity-in-Gearbox-Engineering-wp-65217-A20.pdf (accessed on 30 July 2022).

31. Shabana, A.A. *Dynamics of Multibody Systems*, 5th ed.; Cambridge University Press: Cambridge, UK, 2020; ISBN 1108485642.

32. Andersson, A.; Vedmar, L. A dynamic model to determine vibrations in involute helical gears. *J. Sound Vib.* **2003**, *260*, 195–212. [CrossRef]

33. Heirman, G.; Cappellini, N.; Tamarozzi, T.; Toso, A. Contact Modeling between Objects. US Patent 10,423,730, 24 September 2019.

34. Weber, C.; Banaschek, K.; Niemann, G. *Formänderung und Profilrücknahme Bei Gerad-und Schrägverzahnten Rädern*; F. Vieweg: Brunswick, Germany, 1955.

35. Craig, R.R.; Bampton, M.C.C. Coupling of substructures for dynamic analyses. *AIAA J.* **1968**, *6*, 1313–1319. [CrossRef]

36. Agrawal, O.P.; Shabana, A.A. Dynamic analysis of multibody systems using component modes. *Comput. Struct.* **1985**, *21*, 1303–1312. [CrossRef]
37. Blockmans, B.; Tamarozzi, T.; Naets, F.; Desmet, W. A nonlinear parametric model reduction method for efficient gear contact simulations. *Int. J. Numer. Methods Eng.* **2015**, *102*, 1162–1191. [CrossRef]
38. Ascher, U.M.; Petzold, L.R. *Computer Methods for Ordinary Differential Equations and Differential-Algebraic Equations*; Siam: Philadelphia, PA, USA, 1998; Volume 61.
39. Palermo, A.; Britte, L.; Janssens, K.; Mundo, D.; Desmet, W. The measurement of Gear Transmission Error as an NVH indicator: Theoretical discussion and industrial application via low-cost digital encoders to an all-electric vehicle gearbox. *Mech. Syst. Signal Process.* **2018**, *110*, 368–389. [CrossRef]
40. Choi, C.; Ahn, H.; Park, Y.J.; Lee, G.H.; Kim, S.C. Influence of gear tooth addendum and dedendum on the helical gear optimization considering mass, efficiency, and transmission error. *Mech. Mach. Theory* **2021**, *166*, 104476. [CrossRef]
41. Kim, S.-C.; Moon, S.-G.; Sohn, J.-H.; Park, Y.-J.; Choi, C.-H.; Lee, G.-H. Macro geometry optimization of a helical gear pair for mass, efficiency, and transmission error. *Mech. Mach. Theory* **2020**, *144*, 103634. [CrossRef]
42. Garambois, P.; Perret-Liaudet, J.; Rigaud, E. NVH robust optimization of gear macro and microgeometries using an efficient tooth contact model. *Mech. Mach. Theory* **2017**, *117*, 78–95. [CrossRef]
43. Chase, N.; Rademacher, M.; Goodman, E.; Averill, R.; Sidhu, R. A Benchmark Study of Multi-Objective Optimization Methods. *BMK-3021 Rev.* **2009**, *6*, 1–24.

Article

Squeeze Film Damper Modeling: A Comprehensive Approach

Edoardo Gheller *, Steven Chatterton, Andrea Vania and Paolo Pennacchi

Department of Mechanical Engineering, Politecnico di Milano, Via G. la Masa 1, 20156 Milan, Italy
* Correspondence: edoardo.gheller@polimi.it

Abstract: Squeeze film dampers (SFDs) are components used in many industrial applications, ranging from turbochargers to jet engines. SFDs are applied when the vibration levels or some instability threatens the safe operation of the machine. However, modeling these components is difficult and somewhat counterintuitive due to the multiple complex phenomena involved. After a thorough investigation of the state of the art, the most relevant phenomena for the characterization of the SFDs are highlighted. Among them, oil film cavitation, air ingestion, and inertia are investigated and modeled. The paper then introduces a numerical model based on the Reynolds equation, discretized with the finite difference method. Different boundary conditions for oil feeding and discharging are implemented and investigated. The model is validated by means of experimental results available in the literature, whereas different designs and configurations of the feeding and sealing system are considered. Eventually, an example of the application of a SFD to a compressor rotor for the reduction of vibration and correction of the instability is proposed. The paper provides an insight regarding the critical aspects of modeling SFDs, underscoring the limits of the numerical model, and suggesting where to further develop and improve the modeling.

Keywords: squeeze film damper; seal instability; rotor dynamics; lubrication

Citation: Gheller, E.; Chatterton, S.; Vania, A.; Pennacchi, P. Squeeze Film Damper Modeling: A Comprehensive Approach. *Machines* **2022**, *10*, 781. https://doi.org/10.3390/machines10090781

Academic Editors: Marco Ceccarelli, Giuseppe Carbone and Alessandro Gasparetto

Received: 25 July 2022
Accepted: 4 September 2022
Published: 7 September 2022

Publisher's Note: MDPI stays neutral with regard to jurisdictional claims in published maps and institutional affiliations.

1. Introduction

Vibrations represent an intrinsic problem in all fields of mechanical engineering including rotordynamics. Rotating machines are subject to remarkable loads and, with the development of machines that operate above some critical speeds, the control of vibrations is fundamental to guarantee long time operation. The typical problems in this field are excessive steady state synchronous vibration levels and subsynchronous rotor instabilities. The first one usually arises from excessive unbalance or due to operation close to a critical speed. The second one may depend on the presence of instability sources, connected to cross-coupling effects present in bearing systems and seals, among others. In some cases, the increase of the vibration, when crossing a critical speed during a runup or a rundown, can be harmful for the operation of the machine and the addition of some damping to the system is often required.

To this aim, squeeze film dampers remain one of the most effective components used because they offer the advantage of dissipating vibration energy when the shaft is supported by rolling element bearings. In addition, SFDs can improve the dynamic stability characteristics of rotor-bearing systems.

The most common design for these components is the one coupled with a rolling element bearing, as shown in Figure 1. The shaft is supported by a rolling element bearing and the coupling is often referenced as journal. The shaft vibration is transferred to the external ring of the bearing that "squeezes" the lubricant film, placed between the housing and the outer surface of the journal, generating high dynamic pressures. Therefore, dynamic forces counteract the lateral displacement of the shaft generating the damping effect. The anti-rotation pin is often applied to avoid any spinning motion of the journal, so that only translational displacements are possible, i.e., the journal can only translate or orbit without spinning about its axis of symmetry. The shaft spinning is decoupled from the journal motion thanks to the presence of the bearing.

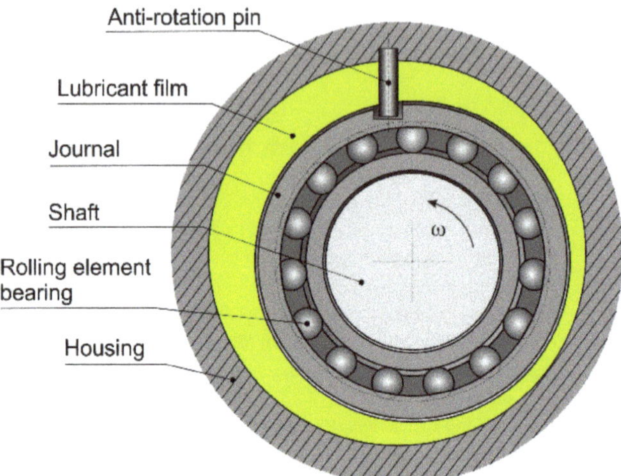

Figure 1. Simple squeeze film damper without centering mechanism, shaft rotation indicated by the arrow.

This configuration is characterized by strong non-linearities due to the "bottoming-out": the journal remains in contact with the casing surface at the run-up; when the level of the vibration is increased, the detachment of the two components happens resulting in a discontinuous change of the properties of the system. To reduce the non-linearity and the risk of collision between the static element and the whirling one in the case of large journal displacements, different supports are used, such as O-rings and squirrel cages. The selection of the proper stiffness of the support is fundamental for the correct operation of the SFD. If the support is too stiff, no relative motion between the shaft and the cage will be possible, i.e., no squeezing of the oil film; whereas if the stiffness is too low, the SFD can behave like a non-supported one [1,2].

Damping is the design parameter for all SFDs, and an optimal value for each application must be obtained. As a matter of fact, the utilization of a device whose damping capability is not aligned with the one requested by the system is useless if not dangerous. If the level of damping is too high, the SFD will dynamically behave as a rigid connection. Conversely, if the level of damping is too small, nothing will change in the dynamic response of the machine.

There are many studies in the literature that provide guidelines to determine the correct damping needed by a machine. In general, it depends on the dynamic characteristics of the machine itself, the typical operating conditions, and the kind of excitations [3,4].

Different models with different levels of complexity have been developed to predict the dynamic characteristics of SFDs. The first ones were based on the 1 D Reynolds equation for short plain journal bearings. This approximation is legitimate when the length to diameter ratio is lower than 0.25 and if no sealing mechanism is adopted, [5]. The effect of the SFD on the journal is modeled by means of linearized stiffness and damping coefficients likewise oil-film bearings. If no spinning motion is considered, no stiffening effect is obtained from the SFD. On the contrary, the long bearing approximation can be adopted when the length to diameter ratio tends to infinity or if seals limiting the oil flow are applied. In both cases, an analytical solution is possible. For this reason, many estimations of the coefficients are present in the literature [2,3].

The motion of the shaft is modelled, for convenience, as i) circular synchronous precessions, centered or with a static eccentricity, or ii) small amplitude motions about a static displaced center. The first model is usually applied when the response to unbalance is investigated, the second one is used for critical speed and stability analyses as shown by San Andrés in [6].

From these early works, it is possible to understand that the clearance and the length to diameter ratio are two important parameters influencing the operation of the bearing together with the amplitude of the vibration, [1–3].

The 1-D Reynolds equation model has the advantage of simplicity, but the predictions can be considered reliable only for very simple geometries and for a limited range of operating conditions.

The main phenomena affecting the dynamic performance of SFDs are the fluid inertia, the liquid cavitation, the air ingestion, and the geometrical features.

Inertia is usually neglected in the derivation of the Reynolds equation, but for large clearances and amplitudes of motion, associated to higher vibrational frequencies, the added mass produced by the oil dynamic pressurization found experimentally has a value comparable to the mass of the entire SFD as highlighted by San Andrés and Vance in [7]. Different models that consider the effect of inertia can be found in the literature. As also reported by San Andrés and Vance in [8], for moderate values of the squeeze Reynolds number ($Re = \frac{\rho \omega c_l^2}{\mu} \leq 10$ with ρ and μ being the density and dynamic viscosity of the oil respectively, ω the vibration frequency and c_l the SFD clearance) the fluid inertia can be assumed not to affect the shape of the fluid purely viscous velocity profiles and consider the fluid temporal inertia in the modeling of SFDs. In [9], the effects of convective inertia and temporal inertia are considered together. In [10], a detailed description of the equations necessary to include the inertial contribution is presented together with an application.

Cavitation is stated as one of the principal reasons why predictions on the force coefficients, made with the simple model used in [1,5], do not fit the experimental results. For this reason, relevant effort has been put in the investigation and modeling of cavitation. In [11], Zeidan and Vance experimentally recognized five different cavitation regimes: un-cavitated film, cavitation bubble following the journal, oil–air mixture, vapor cavitation, vapor and gaseous cavitation. The second regime is considered as a transient condition, steady only for reduced whirling frequencies, that evolves in the third one with the shaft acceleration. The most common regimes are the third and fourth that sometimes combine with each other. Diaz and San Andrés in [12] concentrated mostly on vapor cavitation and air entrainment. They tested a bearing in open-ends and in fully flooded configuration, changing whirling frequencies and pressure of supply oil, and measuring the dynamic pressure generated. The authors showed the difference between the pressure evolution in time for the two-cavitation mechanism. For the vapor cavitation, the pressure profile is nearly identical for every cycle, while for air entrainment the pressure measurements showed great variability from one cycle to the other. Similar conclusions regarding the gaseous cavitation can be found in [13].

Due to the differences measured between the two phenomena, vapor cavitation and air ingestion are treated and modeled differently. Different vapor cavitation models and algorithms have been developed. The first cavitation model that was introduced is the so called π-film model, also known as Gumbel condition. Here, the relative pressure is considered zero in the region where it assumes negative values. According to this hypothesis, the ruptured film extends over half the angular length of the bearing. One of the most used algorithms is the so-called Elrod's cavitation algorithm, [14]. An evolution of this approach consists in the adoption of the linear complementarity problem (LCP), [15].

In [8,16,17] the effect of air ingestion and bubbly mixture is experimentally investigated. Air is "sucked" inside the SFD, and, after some cycles, the bubbles of air are finely dispersed in the mixture and persist also in the high-pressure zone. The presence of a compressible foamy mixture can explain the variability of the pressure's peak values. Different models that take into account the air ingestion are present in the literature. Among them, Diaz [18] provided a detailed procedure, supported by a series of experimental results, to include the air ingestion effect in the 2D Reynolds equation, based on the hypothesis of a homogeneous bubbly mixture. To correctly determine the percentage of air inside of the mixture, a reference value is needed. In the experimental campaign, the air volume fraction is controlled at the feeding system. In industrial applications the SFD is fed with pure oil and air is ingested from the discharge locations. It is therefore necessary to predict the reference value of ingested air. In [19], the authors introduced a model to evaluate the air entrainment in open-ends short SFDs. Some years later, Mendez et al. [20] adapted Diaz's model to finite length bearings. Both the models presented in [18,20] are based on a simplified form of the Rayleigh–Plesset equation to model the presence of air bubbles in the oil in open-ends SFDs. Gehannin et al. in [21] considered instead the complete form of the equation and proposed a comparison with experimentally derived measures to evaluate the impact of these two different forms of the equation on the accuracy of the model.

Regarding the geometrical characteristics of the SFD, in [22], San Andrés et al. reported an extensive experimental campaign that thoroughly investigates the effect of different geometrical features on the dynamic properties of the SFDs. Six different configurations are tested, and the focus is set on the effect on the force coefficients of film clearance, length of the SFD, groove feeding and hole feeding, sealing ends and open ends, whirl orbit amplitude, shape of orbit, and number and disposition of feed holes.

In this paper, a comprehensive model based on the 2D Reynolds equation is introduced: The different phenomena described above are taken into considerations and discussed. The model is then validated with experimental and numerical data taken from the literature. The modeling of the different phenomena describing the dynamic behavior of SFDs is taken from several past works found in the literature. A simplified approach is considered to reduce the level of the difficulty and the parameters to be controlled. The goal of this work is to obtain a model that can be easily replicated and adapted.

In the literature there are more refined models based on the bulk-flow equations [23], and computational fluid dynamics [24–27]. Both approaches guarantee higher precision of the results, but the modeling and computational effort is higher than the one required by the model proposed in this work. The latter one gives the opportunity of investigating different phenomena in an approachable and straightforward way.

Eventually, an example of application of a SFD to a centrifugal compressor rotor for the reduction of vibration is proposed and a parametric investigation on the different parameters influencing the dynamic behavior of SFDs is performed. Moreover, the effect of the application of a SFD on the correction of an instability is also presented. In future works, the model proposed will be revised and improved to increase the accuracy.

2. Materials and Methods

The proposed model is based on the 2D Reynolds equation discretized with the finite difference approach. The inertia and air ingestion are modeled as extra terms of the Reynolds equation.

2.1. Oil Film Modeling

The approach to the analysis of the dynamic performance of SFDs is to simulate circular orbits of the shaft, whether centered (see Figure 2a) or not (Figure 2b), or small perturbations around the position of equilibrium. For simplicity, the proposed model is developed for centered circular orbits (CCOs), but it can easily be adopted for non-centered circular orbits or even noncircular orbits and oscillations around the equilibrium position if it is possible to identify a function that describes the behavior of the oil film thickness as a function of the time.

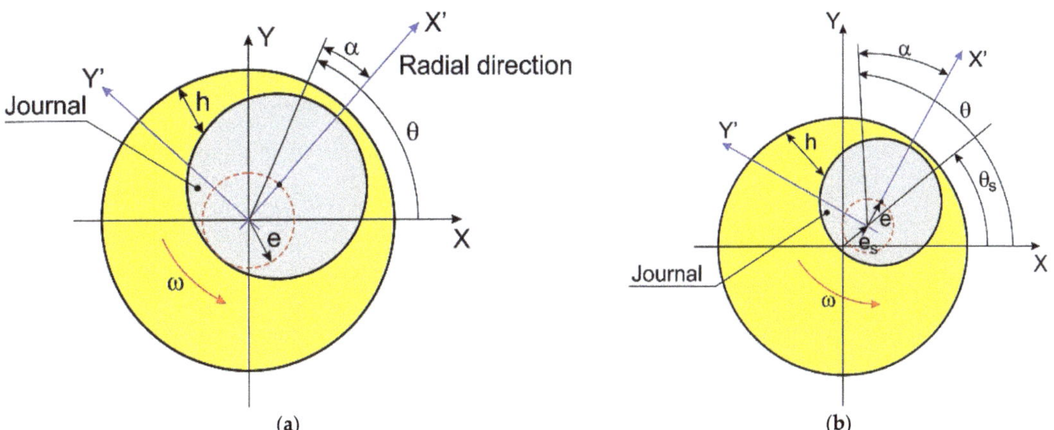

Figure 2. (**a**) Representation of circular centered orbit, (**b**) representation of circular eccentric orbit. Journal orbit represented by the red circle.

The rotation θ in the absolute frame of reference (X-Y) and the rotation α in the relative frame of reference (X'-Y') are related as follows:

$$\theta = \alpha + \omega t, \tag{1}$$

By considering the fixed reference system, it is possible to write the variation of the oil film thickness h in time and space domains:

$$h(\theta, t) = c_l - (e \cos \omega t + e_s \cos \theta_s) \cos \theta - (e \sin \omega t + e_s \sin \theta_s) \sin \theta, \tag{2}$$

where e is the orbit radius and, e_s and θ_s are the amplitude and phase of the static eccentricity respectively (see Figure 2).

Considering Equation (1), at each time instant:

$$\frac{\partial}{\partial t} = -\omega \frac{\partial}{\partial \vartheta} = -\omega \frac{\partial}{\partial \theta}. \tag{3}$$

This assumption is valid only in case of CCOs and if the pressure field can be assumed to remain constant along the orbiting motion, no feedholes, discharge holes and piston ring seals. Moreover, if the orbiting frequency remains constant in time, Equation (3) allows to simplify every time derivative as a spatial one. This transformation allows to reduce the calculation time. In fact, it is not necessary to develop a time transient simulation since the simulation at one time instant is representative of the behavior of the oil for the entire orbit of the shaft.

2.2. Reynolds Equation

The general equations to describe the dynamic behavior of a viscous Newtonian fluid are the 3-D Navier–Stokes equations:

$$\frac{\partial \rho}{\partial t} + \nabla \cdot \left(\rho \vec{V} \right) = 0, \tag{4}$$

$$\rho \left(\frac{\partial \vec{V}}{\partial t} + \vec{V} \cdot \nabla \left(\vec{V} \right) \right) = -\nabla P + \nabla \cdot \left(\mu \nabla \vec{V} \right) + \nabla \left(-\frac{2\mu}{3} \nabla \cdot \vec{V} \right) + \rho g, \tag{5}$$

where (4) is the continuity equation and (5) are the conservation of momentum equations within the flow boundary.

Taking into consideration the SFD application it is possible to adopt some simplifying hypotheses, such as: (i) fluid density ρ is considered constant, valid if cavitation is not present, (ii) fluid kinematic viscosity is constant, valid if temperature can be considered almost constant, (iii) inertia and body forces are neglected, (iv) fluid flow is considered laminar.

Finally, considering the SFD geometry the classical Reynolds equation can be obtained, [5]:

$$\frac{\partial}{\partial x}\left(h^3\frac{\partial P}{\partial x}\right) + \frac{\partial}{\partial y}\left(h^3\frac{\partial P}{\partial y}\right) = 12\mu\frac{\partial}{\partial t}(h). \tag{6}$$

In the case of constant whirling frequency, Equation (3) can be substituted inside Equation (6).

2.3. Fluid Inertia

In general, the fluid inertia forces are negligible if the value of the squeeze film Reynolds number is lower than 1. In case of high vibration frequencies, or SFDs with larger clearance, for example in case of inlet and outlet grooves, this value is greater than one and is usually lower than 50, [6]. As reported in [28,29], models that include inertia's effect give results closer to ones obtained experimentally for both force coefficients.

In this work, an approach similar to the one proposed in [30], a single Reynolds-like equation, is considered in which the effect of temporal inertia is added. Convective inertia terms are considered negligible as in [28]. Using cylindrical coordinates, the equation used in the model is:

$$\frac{\partial}{R\partial\theta}\left(\frac{h^3}{12\mu}\frac{\partial P}{R\partial\theta}\right) + \frac{\partial}{\partial y}\left(\frac{h^3}{12\mu}\frac{\partial P}{\partial y}\right) = \frac{\partial}{\partial t}(h) + \frac{\rho h^2}{12\mu}\frac{\partial^2 h}{\partial t^2}. \tag{7}$$

2.4. Air Ingestion

To fully consider the effect of air entrainment, the same approach adopted by the authors in [19] has been adopted. The Reynolds equation must be modified to consider the compressibility of the fluid and the effect of the presence of air bubbles on density and dynamic viscosity.

$$\frac{\partial}{R\partial\theta}\left(\frac{\rho h^3}{12\mu}\frac{\partial P}{R\partial\theta}\right) + \frac{\partial}{\partial y}\left(\frac{\rho h^3}{12\mu}\frac{\partial P}{\partial y}\right) = \frac{\partial}{\partial t}(\rho h) + \frac{\rho h^2}{12\mu}\frac{\partial^2 \rho h}{\partial t^2}, \tag{8}$$

$$\rho = (1-\beta)\rho_L, \tag{9}$$

$$\mu = (1-\beta)\mu_L, \tag{10}$$

$$\beta = \frac{1}{1 + \frac{P(x,t)-P_v}{P_{G\sigma}}\left(\frac{1}{\beta_0}-1\right)}. \tag{11}$$

where β is the air–mixture volume fraction, β_0 is the reference value for β, $P_{G\sigma}$ is the pressure of the air bubble for the critical radius, P_v is the vapor cavitation pressure, and μ_L and ρ_L are the dynamic viscosity and density for the pure oil.

In [19], a model to evaluate β_0 is presented for short SFDs. However, the short-length bearing approximation is not always applicable. For example, it is limited to $L/D < 0.25$. In [20], it is proposed to numerically evaluate the volumetric inflow of air at the sides of finite length SFD and evaluate the new reference value of volume air fraction. The pressure cycle is then repeated with the updated value of β_0. This procedure is continued until the convergence on β_0 is reached. A procedure like this one has been adopted in the present work and the starting value of reference volume fraction of ingested air will be considered zero.

2.5. Negative Pressure Zone

As previously mentioned, different models and algorithms that deal with vapor cavitation have been adopted in the literature. In this work, the approach presented by Fan and Behdinan in [31,32] is applied. The vapor cavitation is solved as a linear complementarity problem as suggested in [15]. The algorithm proposed in [33] is applied in the solution of the *LCP*.

2.6. Geometrical Discretization

The cylindrical geometry of the SFD il flattened in a 2D plane, as shown in Figure 3.

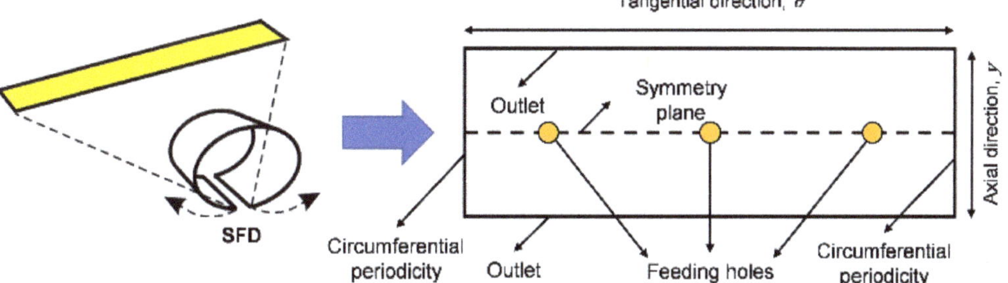

Figure 3. 2D geometrical transformation of the SFD.

A structured mesh is considered for the spatial discretization. The approach followed for the discretization of the holes is explained in the following sub-section.

2.7. Inlet Boundary Conditions

Different types of boundary conditions can be adopted at the inlet port. If the feeding holes are not considered in the modeling, it is possible to consider only half of the SFD by the symmetry boundary condition:

$$\left.\frac{\partial P}{\partial y}\right|_{Symmetry\ plane} = 0. \tag{12}$$

Conversely, if the feeding system is considered, the flowrate is imposed at each hole. If the laminar flow is assumed and no central groove is present, the flow rate is as follows:

$$Q_{inlet} = C_i\left(P_{supply} - P(x_h, z_h)\right)\left[\frac{m^3}{s}\right], \tag{13}$$

where $P(x_h, y_h)$ is the pressure of the oil at the hole location and C_i is a coefficient that includes the orifice area and flow coefficient. In a 3-D model, this flow rate would be directed radially but, since this model is planar, it will be considered in the axial and tangential directions. A more detailed description can be found in [34].

The circular geometry of the hole is simplified as a rectangle, as shown in Figure 4. At the edges of the boundary, the pressure is constant and equal to the feeding one imposed at the center. Considering that the flow from the hole is delivered in both axial and tangential directions, the whole geometry of the SFD must be considered.

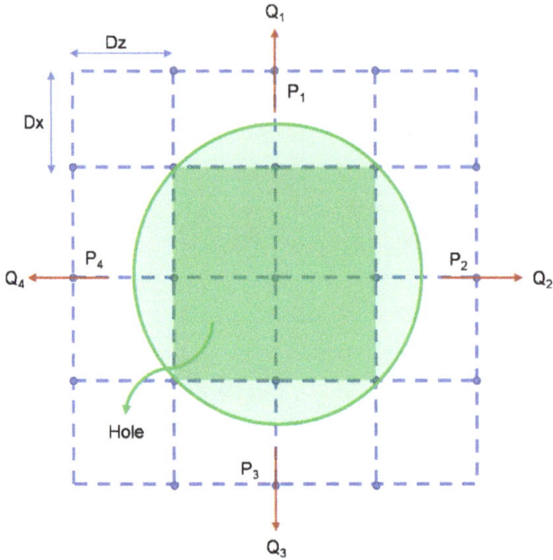

Figure 4. Mesh representation for the feeding hole boundary condition.

Considering the flow entering from one side of the discretized hole:

$$q_z = -\frac{h^3}{12\mu}\frac{\partial P}{\partial z}\left[\frac{m^2}{s}\right],$$ (14)

It is possible to solve

$$Q_1 = -\frac{\overline{h_1}^3}{12\mu}\left(\frac{P_1 - P_{supply}}{Dx}\right)2Dz,$$ (15)

where P_1 is the interpolation of the three points above the boundary of the hole, and $\overline{h_1}$ is the oil film height at $\frac{Dx}{2}$ from the side of the rectangle. Similar expressions can be written for the rest of the flow vectors. The same holds for all other sides P_2, P_3, and P_4.

In general, when the pressure of the oil inside the SFD in the vicinity of the feeding hole is higher than the supply pressure, a backflow happens: a flow rate of oil exits the land of the SFD and enters the supply circuit. As reported in [35], in practical application, check valves are applied to the feeding ducts to avoid backflows and to reduce the effect of pulsating pressure in the supply circuit. A detailed description of the application of check valve is present in [36]. For this reason, when the presence of feeding orifices is simulated in this model, Equation (13) will be used at the nodes where the orifices are located. If the pressure at the hole location is higher than the supply pressure, no boundary condition will be assigned.

In many applications, central grooves are applied as shown in Figure 5a.

In [34,37] it is proposed to model the feeding groove as a reservoir of oil at the feeding pressure. In [22], San Andrés et al. report instead that large values of dynamic pressure in the groove region occurred proving the previous assumption to be wrong. The same approach introduced in [28] is considered when modeling the presence of grooves. As shown in Figure 5b, the flow inside the groove is divided into two regions: a recirculating one and a through-flow close to the journal. Only the second one is active in the dynamic pressure generation, and therefore an effective groove depth is considered. Moreover, the feeding orifices are considered.

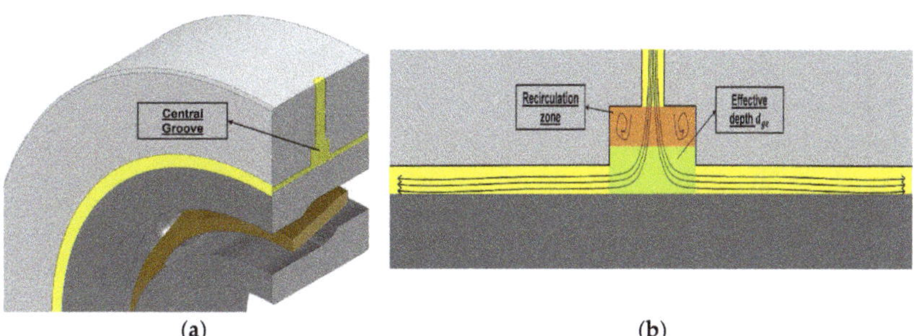

(a)   (b)

Figure 5. 3D representation of SFD equipped with central groove (**a**); Schematic representation of the flow paths passing through the central groove (**b**). Adapted from [28].

The value of the effective groove depth (d_{ge}) is usually optimized using as benchmark the force coefficients obtained experimentally [28].

2.8. Outlet Boundary Condition

As reported in [5], many different boundary conditions can be assigned for the outlet section. In general, the SFD can be exposed to ambient pressure air, and in this case the boundary condition to be assigned is:

$$P(L,t) = P_{air},\tag{16}$$

where P_{air} is the ambient pressure at the outlet.

In this case the SFD is subjected to high air entrainment, comporting a reduction of the damping capacity of the device. With an open-ends configuration, the exiting flow rate is higher, a condition that will require a higher inlet flow rate of oil. For this reason, SFDs are usually sealed at the ends. The sealing is usually not complete otherwise. Due to the oil heating, the damping capacity would decrease. In the scientific literature, it is possible to find many types of sealing to reduce the leakage of the SFD. One of the most common is the piston ring shown in Figure 6 [5,35,38].

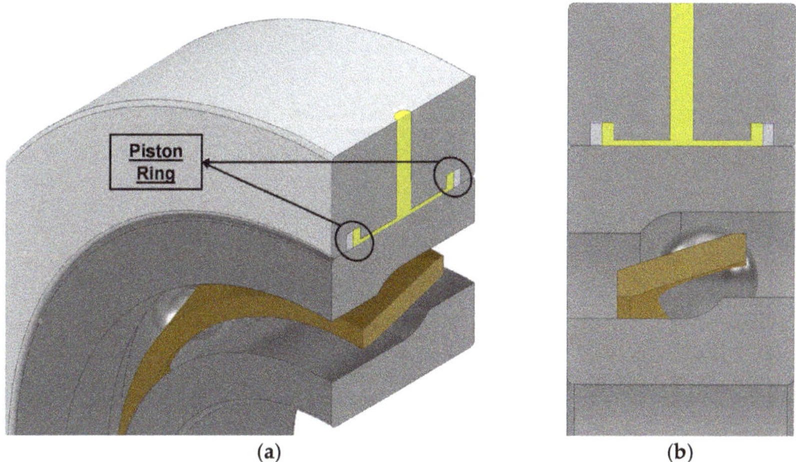

(a)   (b)

Figure 6. 3D representation of SFD equipped with piston-ring as sealing mechanism at the outlet (**a**); Frontal view of SFD equipped with piston-ring as sealing mechanism at the outlet (**b**).

The piston ring seal represents a limitation on the outlet flowrate, and it can be defined as follows [34]:

$$q_{out} = \frac{C_p(P(\theta, L) - P_{out})h_p^3}{12\mu w_p} \left[\frac{m^2}{s}\right], \tag{17}$$

where C_p is the piston ring loss coefficient, $0 < C_p < 1$, P_{out} is the pressure outside the seal, usually ambient pressure, h_p and w_p are the piston ring radial gap and axial dimension respectively. C_p has a main impact on the evaluation of the outlet flowrate. Moreover, it may not be straightforward to be obtained. In this application is considered as a tuning parameter to evaluate the effect of the seals.

Substituting Equation (17) into Equation (15):

$$\left.\frac{h^3}{12\mu}\frac{\partial P}{\partial y}\right|_L + \frac{C_p P(\theta, L)h_p^3}{12\mu w_p} = \frac{C_p P_{out}h_p^3}{12\mu w_p}. \tag{18}$$

The modeling of the inlet and outlet boundary conditions proposed in this work is simplified with respect to what can be found in other references [35,39]. The future improvement of the proposed model will re-evaluate the modeling of these boundary conditions, especially if different configurations of sealing and feeding mechanisms are considered.

2.9. Circumferential Periodicity

The circumferential periodicity must be satisfied at the edges of the flattened geometry (Figure 3). To maintain the continuity, the pressure and the circumferential gradient of the pressure along the axial direction must be equal on both sides. It was noted that the assignment of the pressure boundary condition is enough for the circumferential periodicity: the pressure gradient calculated on the two edges is equal.

2.10. Forces and Force Coefficients

Once the geometry and the mesh are defined and the boundary conditions are assigned, the Reynolds equation can be solved, and the pressure distribution obtained. The forces acting on the journal can be obtained by integrating the pressure profile along the circumferential and axial directions as follows:

$$\begin{bmatrix} F_x \\ F_y \end{bmatrix} = -\int_0^L \int_0^{2\pi} P(\theta, z, t) \begin{bmatrix} \cos\theta \\ \sin\theta \end{bmatrix} R \, d\theta dy. \tag{19}$$

where F_x and F_y are the horizontal and vertical force in the absolute reference frame respectively (Figure 2).

The dynamic behavior of the SFD is represented by the force coefficients. As reported in many sources, among them [6,22], the SFD itself does not generate any kind of stiffness because, without the journal spinning, no pressure is generated at a given static displacement if there is no precession. The SFD forces are represented in the linearized form as follows:

$$\begin{bmatrix} F_x \\ F_y \end{bmatrix} = -\begin{bmatrix} C_{xx} & C_{xy} \\ C_{yx} & C_{yy} \end{bmatrix}\begin{bmatrix} v_x \\ v_y \end{bmatrix} - \begin{bmatrix} M_{xx} & M_{xy} \\ M_{yx} & M_{yy} \end{bmatrix}\begin{bmatrix} a_x \\ a_y \end{bmatrix}, \tag{20}$$

where v_x and v_y are the instantaneous journal velocities and a_x, a_y are the instantaneous journal accelerations.

Damping and added mass coefficients along the x and y directions are typical of small shaft orbiting around the static equilibrium position. In case of circular centered orbits, the SFD generates a constant reaction film force in a relative frame rotating with frequency ω. In most rotodynamic applications, linearized force coefficients are considered. They represent changes in bearing reaction forces to infinitesimal amplitude motions about an equilibrium position. As the definition states, these coefficients are applicable only in the case of small motions around an equilibrium position. As reported in [40], in SFDs the orbit radius can go to half the clearance, defining an orbit far from being close to the equilibrium position and thus violating the main hypothesis behind linearized force coefficients. For this reason, an orbit-based model, such as the one proposed in [40], is adopted in this work. The counterclockwise orbit of the SFD is divided into points where the forces are evaluated. The equation of motion is then written in the frequency domain by applying the Fourier transform to both the orbit points and the forces:

$$\begin{bmatrix} F_x(\Omega) \\ F_y(\Omega) \end{bmatrix} = -\left(i\Omega \begin{bmatrix} C_{xx} & C_{xy} \\ C_{yx} & C_{yy} \end{bmatrix} - \Omega^2 \begin{bmatrix} M_{xx} & M_{xy} \\ M_{yx} & M_{yy} \end{bmatrix} \right) \begin{bmatrix} X(\Omega) \\ Y(\Omega) \end{bmatrix}. \tag{21}$$

Equation (21) can be rewritten as follows:

$$\begin{bmatrix} F_x(\Omega) \\ F_y(\Omega) \end{bmatrix} = -H(\Omega) \begin{bmatrix} X(\Omega) \\ Y(\Omega) \end{bmatrix}, \tag{22}$$

where H_{ij} coefficients are the four unknowns in Equation (22), but only two equations are available. For this reason, the same procedure is applied to the clockwise orbit, obtained by applying a negative value of ω. So, the final system to be solved is

$$\begin{bmatrix} F_x^{cc}(\Omega) \\ F_y^{cc}(\Omega) \\ F_x^{c}(\Omega) \\ F_y^{c}(\Omega) \end{bmatrix} = -H(\Omega) \begin{bmatrix} X^{cc}(\Omega) \\ Y^{cc}(\Omega) \\ X^{c}(\Omega) \\ Y^{c}(\Omega) \end{bmatrix}, \tag{23}$$

where the apex c stands for clockwise and the apex cc stands for counterclockwise.

Once the matrix of complex stiffness H is obtained, the single coefficients can be calculated as:

$$C_{ij} = \frac{Imag(h_{ij})}{\omega}, \tag{24}$$

$$M_{ij} = -\frac{Real(h_{ij})}{\omega^2}. \tag{25}$$

3. Model Validation

The model was validated with both numerical and experimental data available in the literature. The numerical and experimental results presented in [22] were considered due to the different geometrical configuration tested. In this work, four configurations (SFD A, B, E and F) were selected as reference for the validation. They differ in terms of clearance, SFD length, as well as the presence of a central groove and an end seal. The tested diameter is constant and equal to 127 mm. In [22], the oil has density $\rho_L = 805$ kg/m^3 and dynamic viscosity $\mu_L = 0.0265$ Pa·s. The geometrical characteristics of the SFDs considered for the validation are listed in Table 1.

Table 1. Geometrical characteristics of SFDs from [22]. d_G and L_G represent the physical depth and length of the central groove, not present in SFD E and F. d_E and L_E represent the depth and length of the grooves at the discharge, not present in SFD E and F. Piston ring seals are applied only for SFD B.

		SFD A	SFD B	SFD E	SFD F
Clearance	c_l [mm]	0.141–0.251	0.138	0.122	0.267
Length	L [mm]	2×25.4	2×12.7	25.4	25.4
Central groove depth	d_G [mm]	9.5	9.5	no	no
Central groove length	L_G [mm]	12.5	12.5	no	no
End groove depth	d_E [mm]	3.5	3.5	no	no
End groove length	L_E [mm]	2.5	2.5	no	no
Seal	-	yes	yes	no	no

SFDs E and F are tested with a fixed static eccentricity and by changing the amplitude of the circular orbit vibration. The considered e/cl ratios are: 0.05, 0.14, 0.29, and 0.43.

The tested frequencies are $10 \div 250$ Hz for SFD E and $10 \div 100$ Hz for SFD F. The obtained coefficients are constant for the whole frequency range, therefore only the values at 100 Hz and 50 Hz are shown respectively. The evolution of both the mass and damping coefficients for SFD F is shown in Figure 7, where it is possible to see that the results obtained with the model presented in this paper agree well with both the experimental and numerical results in [22].

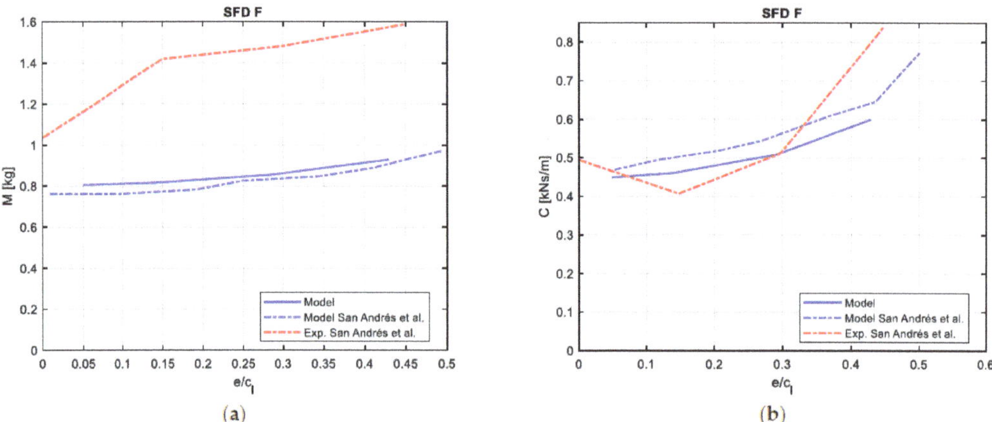

Figure 7. SFD F: numerical and experimental results from [22]. (**a**) Evolution of mass coefficient as a function of the orbit radius; (**b**) Evolution of damping coefficient as a function of the orbit radius.

Both the numerical results for the mass coefficient shown in Figure 7a underestimate the experimental results. In [22], the authors attribute the discrepancy to the high value of the feeding pressure that determines a higher value of the radial component of the force, directly responsible for the mass coefficient.

Similarly, the evolution of the mass and damping coefficient for SFD E is shown in Figure 8. In this case only the experimental results are available. It is possible to notice an acceptable agreement for the damping coefficients, the maximum difference between the experimental and numerical results is lower than the 25%. On the other hand, an important discrepancy between for the mass coefficients is shown. A possible explanation could be the high level of the feeding pressure that strongly affects the dynamic pressure distribution.

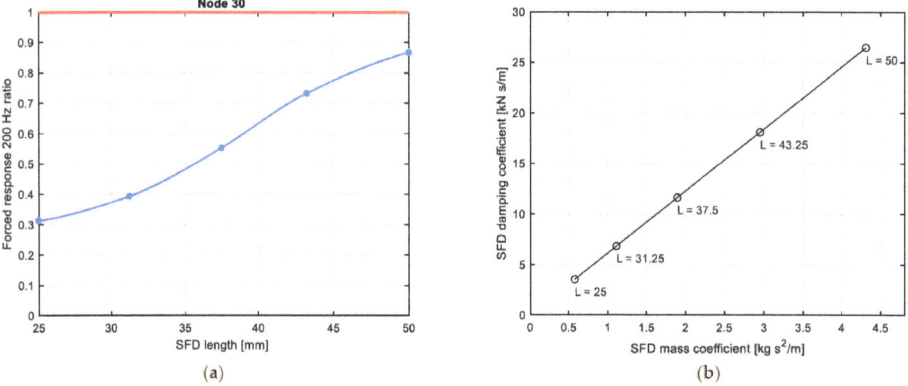

Figure 8. SFD E: experimental results from [22]. (**a**) Evolution of mass coefficient with orbit radius; (**b**) Evolution of damping coefficient with orbit radius.

SFDs B and A are tested in [22] at different static eccentricities with a constant orbit radius $e = 0.055c_l$ and for the frequency range $110 - 210$ Hz. Moreover, for these configurations, there is no variation of the force coefficients with the frequency and the only frequency considered is 150 Hz. For both configurations, the effective groove depth is tuned to match the results presented in [22]. The evolution of the dynamic coefficients with the static eccentricity for the open-ends configuration of SFD B is shown in Figure 9. Similarly to SFD F, the numerical results agree well with the experimental ones. A similar trend was obtained for SFD A. The results are not reported for the sake of brevity. In Figure 9, the values of the force coefficients are adimensionalized considering the same reference values reported in [22].

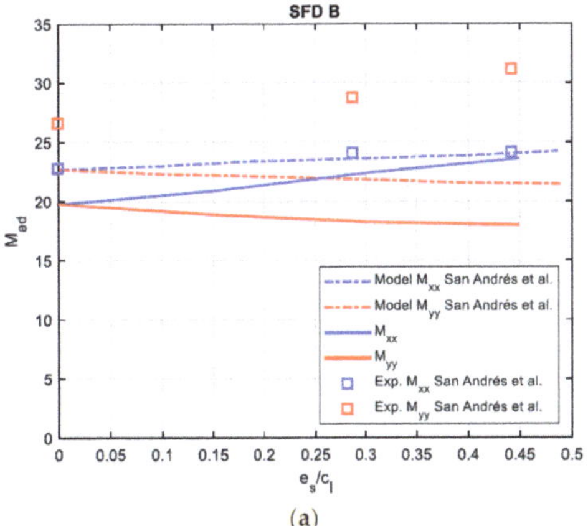

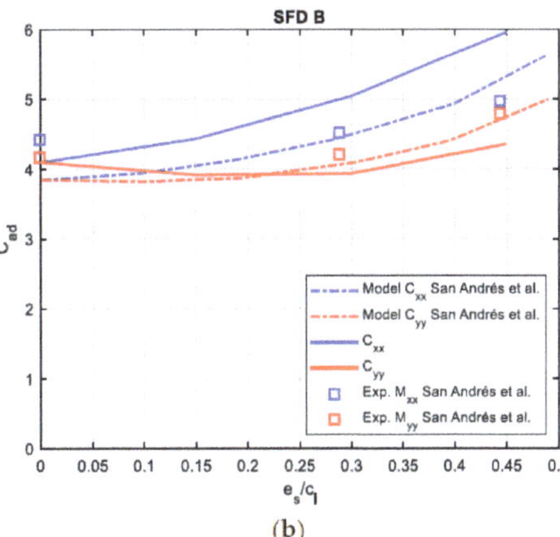

Figure 9. SFD B open ends: experimental and numerical results from [22]. (**a**) Evolution of mass coefficient with static eccentricity; (**b**) Evolution of damping coefficient with static eccentricity.

For every SFD configuration, the mesh independency check was performed considering a structured grid. Both rectangular and squared grids were considered, and the final number of elements adopted was selected with a trade-off between numerical accuracy and computational time. The evolution of the radial and tangential force relative error with the number of mesh points is reported.

For the sake of brevity, only the evaluation conducted for SFD F is reported. The number of axial points N_z is selected and the tangential point N_x are evaluated as $N_x = kN_z \frac{2\pi R}{L}$ with $0 < k \leq 1$. When $k = 1$, the elements are squared.

The evolution of the radial and tangential forces as a function of the number of mesh points and for some values of k (0.05, 0.125 0.5, 1) is shown in Figure 10. Increasing the number of mesh points both errors reach an asymptote. When k is reduced, i.e., when for the same number of axial points, the number of tangential points is reduced, the shape of the error evolution is flat. Generally, a relative error below 1% can be considered acceptable. To keep the alculation time low, for SFD F, the mesh configuration selected has $k = 0.125$ and approximately 1×10^4 mesh points.

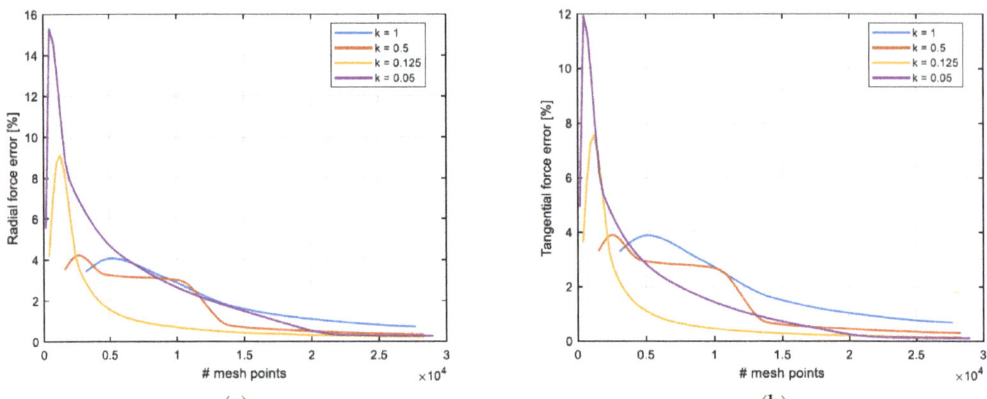

Figure 10. (**a**) Radial force error evolution with number of mesh points for SFD F; (**b**) tangential force evolution with number of mesh points for SFD F.

4. Application

The proposed model has been integrated in the finite beam element analysis of a high-speed centrifugal compressor coupled with a gear element. The shaft of the machine is long 0.7 m and the nominal diameter is 50 mm. The impeller is 70 mm long and has a maximum diameter of 140 mm, while the minimum one is 33 mm. The finite element discretization of the structure, with a total of 34 nodes, is shown in Figure 11.

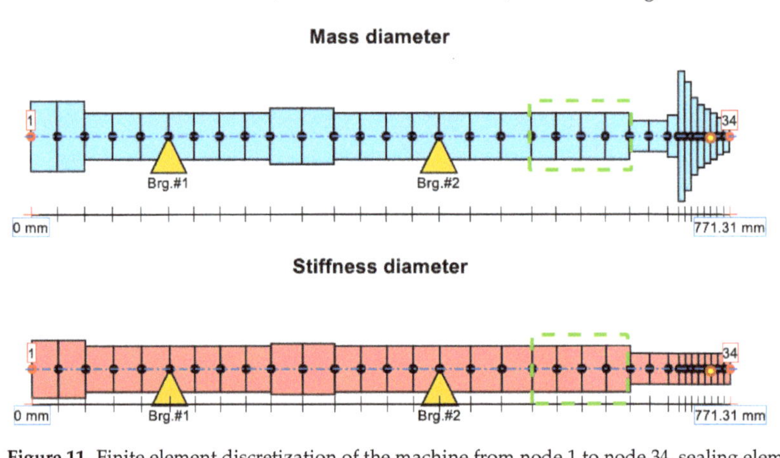

Figure 11. Finite element discretization of the machine from node 1 to node 34, sealing element in green rectangle.

As shown in Figure 11, the stiffness and mass diameter are different for the different elements. The yellow triangles represent the two roller element bearings. The green rectangle represents the region where a sealing element is placed. The scheme of the machine represents an actual application while the application investigated in the next pages is a hypothesis. In the analysis an unbalance force of 3×10^{-6} [Kg·m] is placed in the yellow node of the impeller (node 31). The effect of the seal is not taken into consideration while the attention is focused on the reduction of the vibration of the machine, focusing on the impeller. The operational speed range of the compressor goes from $0 - 300$ Hz and 200 Hz is considered as the operating frequency. The forced responses to the unbalance at three nodes of the impeller are shown in Figure 12. It is possible to see that, due to the characteristics of the bearings, the system is barely damped and when crossing the natural frequency, at 186 Hz, the vibration's amplitude is, in the last node, higher than 2×10^{-4} m. Due to the small gaps between the impeller and the cage and to reduce the aerodynamic losses, it is important to reduce as much as possible the level of the vibration.

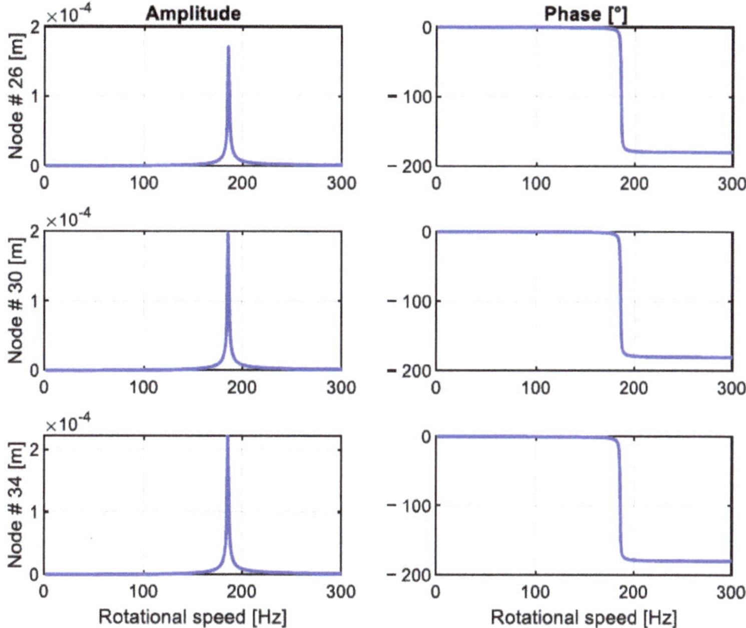

Figure 12. Amplitude and phase of vibration at nodes 26, 30, and 34.

To reduce the vibration peak, a SFD is applied in parallel with the first bearing. The new structure is shown in Figure 13. The SFD is supposed to be supported by an external squirrel cage defined by its own mass (m_{cage}) and stiffness (k_{cage}), respectively. Moreover, the squirrel cage acts as a centering mechanism. The SFD introduces an external source of damping (c_{SFD}) and added mass (m_{SFD}).

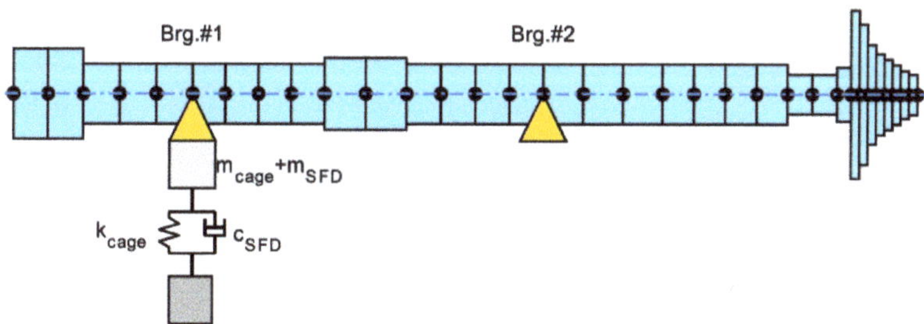

Figure 13. Finite element discretization with SFD.

For simplicity, a plain SFD without grooves, feeding system, and seals is considered. The geometrical characteristics of the SFD and the properties of the ISO VG 46 oil considered are listed in Table 2.

Table 2. Geometrical characteristics and oil properties of the SFD.

		SFD
Clearance	c_l [mm]	0.3
Length	L [mm]	30
Diameter	D [mm]	100
Oil dynamic viscosity	μ_L [Pa·s]	0.0775
Oil density	ρ_L [Kg/m^3]	870
Cage mass	m_{cage} [kg]	1.16
Cage stiffness	k_{cage} [N/m]	2×10^7

At first, the forced response of the configuration with the squirrel cage but without considering the presence of the oil is performed. The comparison between the two forced responses for the same impeller nodes considered in Figure 12 is shown in Figure 14.

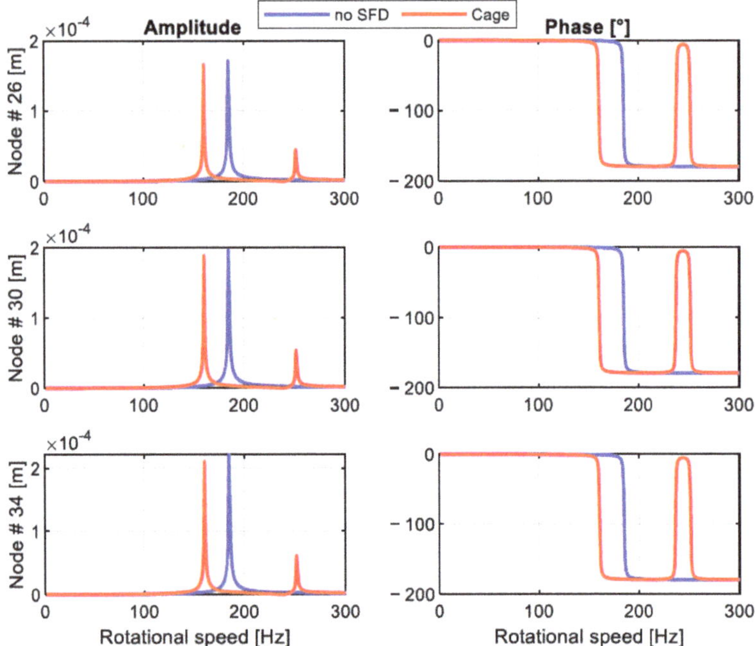

Figure 14. Forced response comparison for the original configuration and the configuration with the squirrel cage.

As it is possible to see from Figure 14, the introduction of the squirrel cage significantly changes the forced response. The introduction of the squirrel cage has the effect of a tuned mass damper. The resonance peak at 182 Hz is moved to 161 Hz. Moreover, a second resonance peak is present at 252 Hz.

Then, the previously mentioned SFD is considered for the forced response. The comparison between the forced response between the original configuration, the configuration with the squirrel cage, and the SFD configuration is shown in Figure 15. The introduction of the SFD is strongly effective in the reduction of the level of the vibration peak.

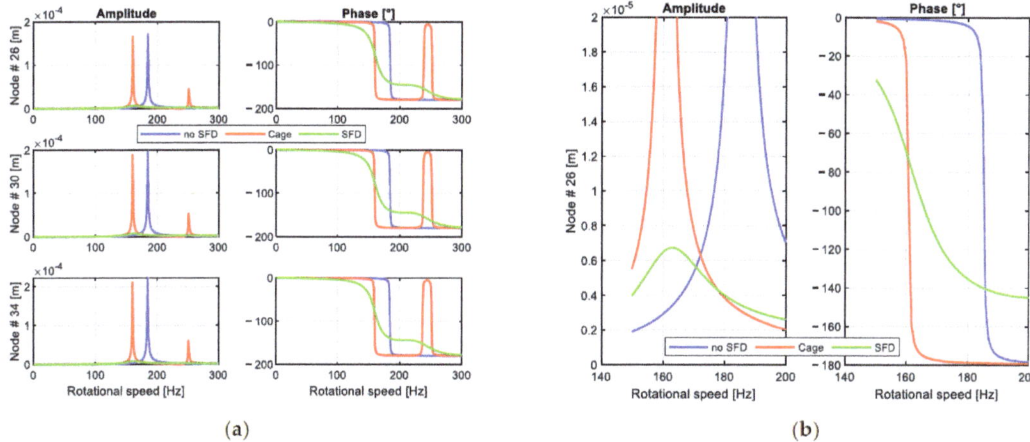

(a)

(b)

Figure 15. Forced response comparison for the original configuration, the configuration with the squirrel cage, and the configuration with the SFD (**a**). Forced response comparison for the original configuration, the configuration with the squirrel cage, and the configuration with the SFD for node 34 (**b**).

Then, the effect of some geometrical parameters on the forced response of the system is evaluated. As previously mentioned, the operating frequency considered is 200 Hz. Therefore, considering the evolution of the forced responses shown in Figure 15, the frequency range from 150 Hz to 200 Hz is considered for the following analysis.

4.1. SFD Clearance

The first parameter to be investigated is the clearance of the SFD. The ratio between the forced response at 200 Hz of the configuration with the SFD and the original one is shown in Figure 16. The forced response decreases with the increase of the SFD clearance even though the damping coefficients increases when the SFD clearance is decreased. This behavior is related to the increase of the resonance frequency when the SFD clearance is reduced. Therefore, a higher level of vibration is obtained at 200 Hz (see Figure 17).

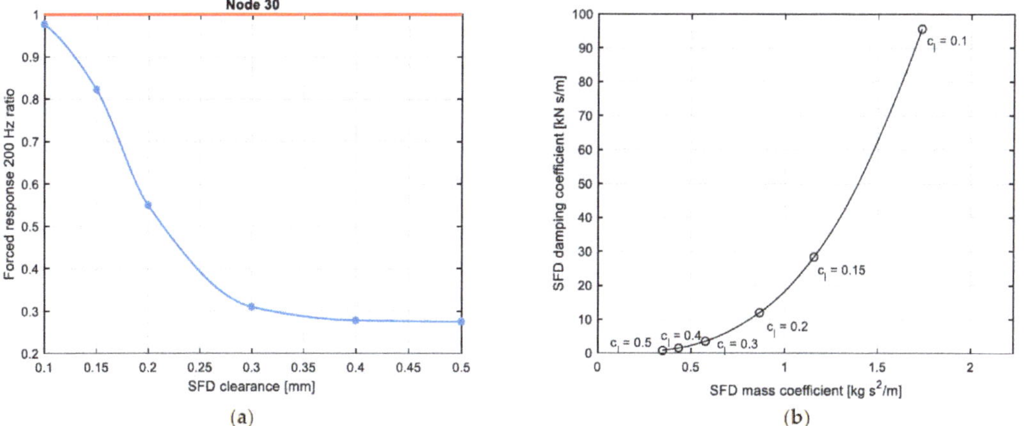

(a)

(b)

Figure 16. Forced response ratio at 200 Hz for node 30 for different values of SFD clearance (**a**), evolution of SFD force coefficients with clearance (**b**).

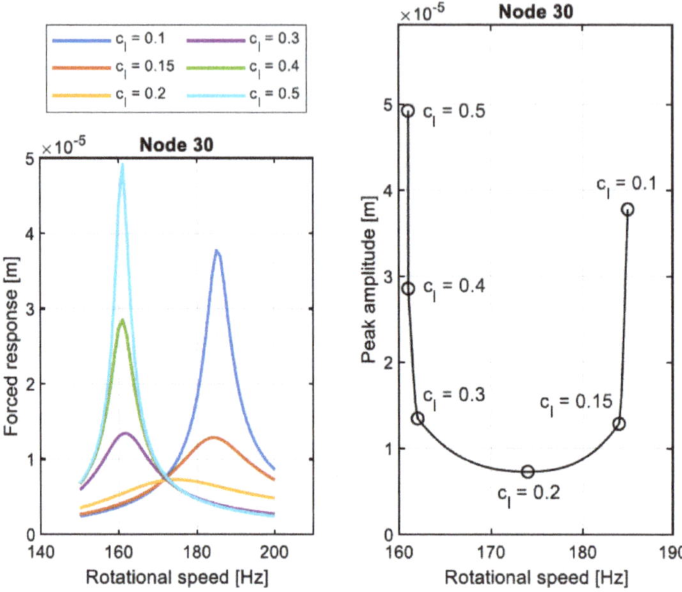

Figure 17. Comparison of forced response for different values of SFD clearance, circle markers indicate the clearance value.

4.2. SFD Length

The effect of the length of the SFD on the forced response is investigated. For this analysis, the selected SFD clearance is 0.3 mm because it minimizes the vibration level at 200 Hz and guarantees an acceptable level for the resonance peak vibration. The ratio of the forced response at 200 Hz for the original configuration and the configuration with the SFD is shown in Figure 18a. The minimum forced response is obtained when the shortest SFD is considered. The force coefficients of the SFD increase with the SFD length, see Figure 18b. Therefore, also in this case, the most suitable SFD to reduce the vibration level at 200 Hz is the one characterized by the lowest force coefficients.

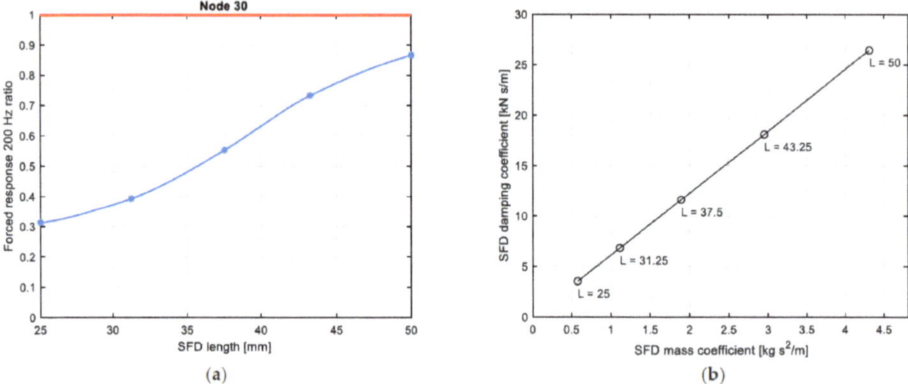

Figure 18. Forced response ratio at 200 Hz for node 30 for different values of SFD length (**a**), evolution of SFD force coefficients with length (**b**). Red line for system without SFD, blue line indicates system with SFD of (**a**). Markers highlights tested configurations.

The comparison between the forced responses obtained considering the different values of length of SFD is shown in Figure 19. Moreover, in this case, the results are reported in the frequency range of interest. It is possible to see that the minimum forced response at 200 Hz is obtained with the shortest damper. On the contrary, the minimum of the vibration peak is obtained when $L = 37.5$ mm, as shown in the right part of Figure 19. Therefore, the proper SFD configuration must be selected according to the optimization required.

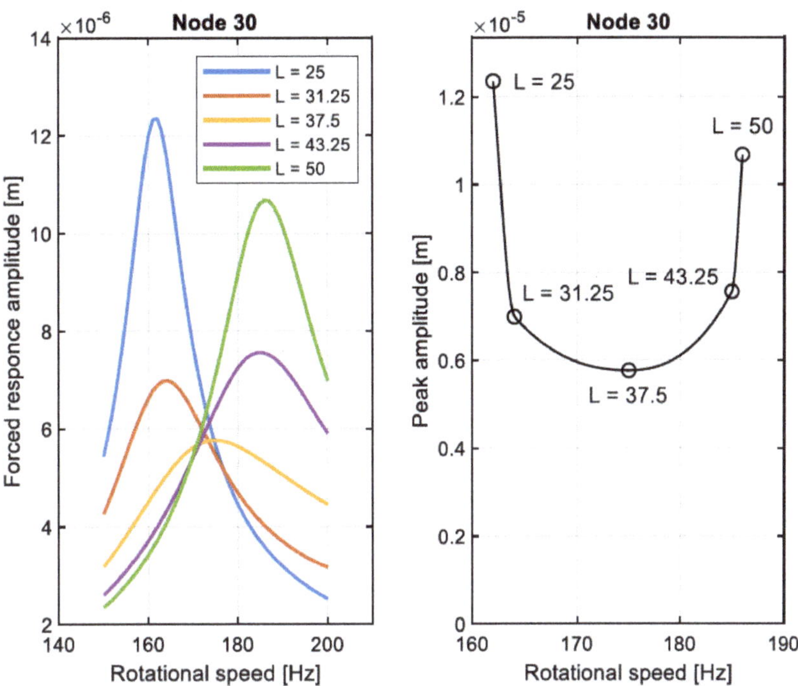

Figure 19. Comparison of forced response for different values of SFD length at node 30, markers highlight configurations tested.

4.3. Groove Effective Depth

Another tuning parameter that can be selected for the geometry of the SFD is the effective depth of the groove. For this reason, different values of d_{ge} have been investigated. For convenience, the relative value of the effective groove depth is considered (d_{ge}/cl). The ratio of the forced response at 200 Hz for the original configuration and the configuration with the grooved SFDs is shown in Figure 20a. The clearance considered is 0.3 mm and the lands of the SFD have a length of 12.5 mm. The groove length considered is 5mm and it is placed in the center of the SFD. When the effective groove depth is one, the damper geometry results in a grooveless damper of length 30 mm. From the analysis shown in Figure 20a, the higher the groove depth, the lower the forced response at 200 Hz. The evolution of the ratio between the SFD force coefficients for the different values of the relative effective groove depth and the values obtained when $d_{ge}/cl = 1$ is shown in Figure 20b. Increasing the groove depth, the damping coefficient is reduced while the mass coefficient is highly increased. The evolution of the forced responses for the considered frequency range is shown in Figure 21. Moreover, in this case, the configuration that minimizes the forced response at 200 Hz is not the one that minimizes the amplitude of the peak.

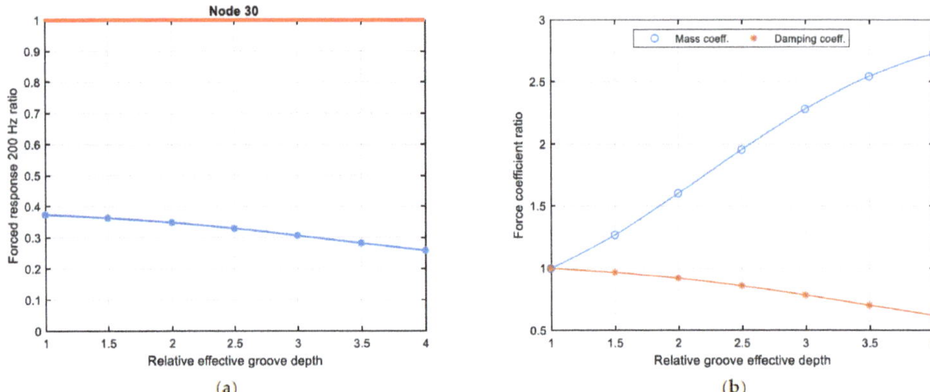

Figure 20. Forced response ratio at 200 Hz for node 30 for different values of relative effective groove depth (**a**), evolution of SFD force coefficients ratio with effective groove depth (**b**). Red line for system without SFD, blue line indicates system with SFD of (**a**). Markers highlights tested configurations.

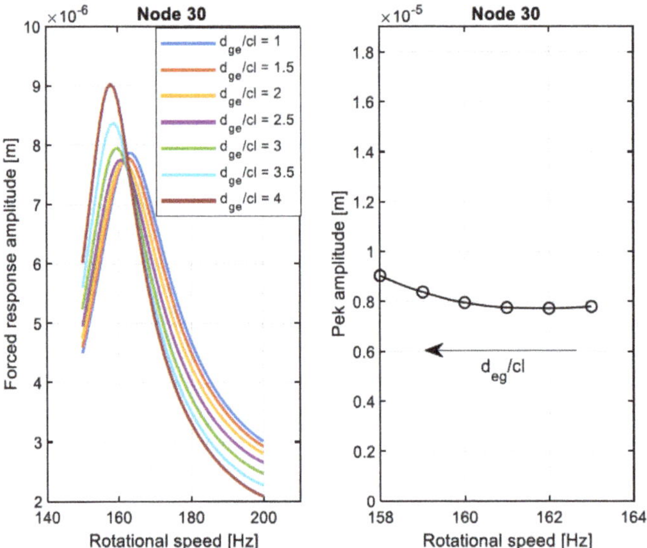

Figure 21. Comparison of forced response for different values of relative groove effective depth at node 30. Markers highlight tested configurations, arrow highlights growing direction of d_{eg}/cl.

4.4. Feeding Pressure

In this section, the feeding system is considered. The SFD considered has a length of 30 mm and clearance equal to 0.3 mm, and the diameter of the holes is considered equal to 2 mm. Moreover, the coefficient C_i is considered as $1 \times 10^{-9} \frac{m^3}{s\,Pa}$. The effect of the feeding pressure on the forced response has been investigated. The ratio of the forced response at 200 Hz for the original configuration and the configuration with SFDs is shown in Figure 22. When the feeding pressure is considered zero, the feeding system is not included in the modeling. The presence of the feeding system seems to have a small impact on the forced response at 200 Hz, and the inlet pressure is not influencing the forced response.

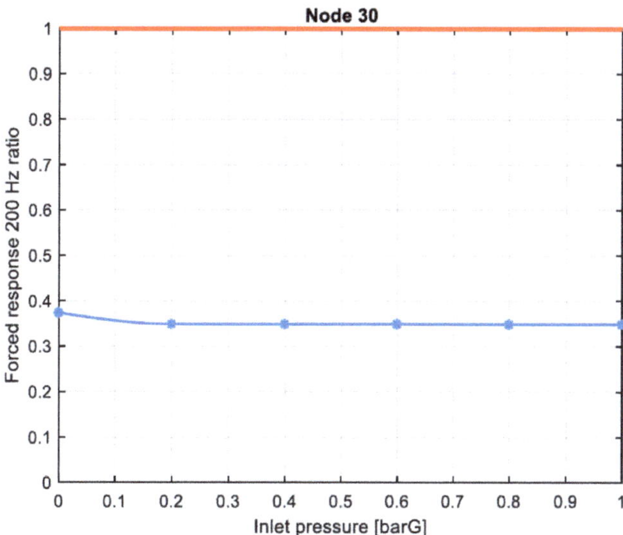

Figure 22. Forced response ratio at 200 Hz for node 30 for different values of feeding pressure depth. Red line for system without SFD, blue line indicates system with SFD.

If the feeding system is not considered in the modeling and no cavitation or air ingestion is present, both the direct SFD force coefficients are equal and constant with the tested frequencies. On the contrary, when the feeding system is modeled, the xx and yy force coefficients are slightly different. Moreover, the mass coefficients show a dependency with the frequency which is more evident when the feeding pressure is increased, as shown in Figure 23.

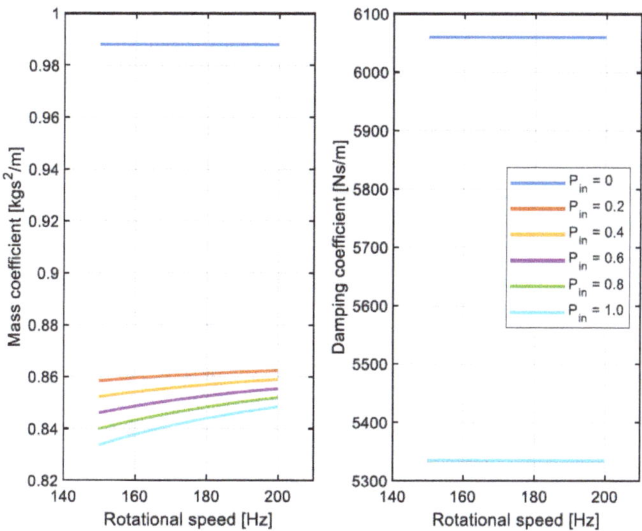

Figure 23. Evolution of xx mass and damping coefficient with the rotational speed considering different inlet pressure values.

The differences in the force coefficients shown in Figure 23 determine different forced response between the modeling with and without the feeding system (Figure 24). On the contrary, the evolution of the mass coefficient with the rotational speed does not have an impact on the forced response.

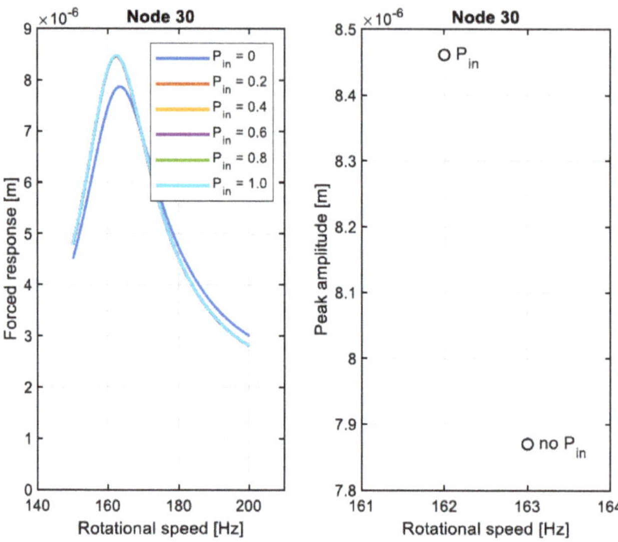

Figure 24. Comparison of forced response for different values of inlet pressure at node 30.

4.5. End Seals

For convenience, the effect of the sealing mechanism is shown considering the overall seal coefficient $C_{seal} = C_p h_p^3 / w_p \mu$. The highest value of C_{seal} corresponds to the open ends condition while the lowest value of C_{seal} corresponds to the ideal condition of complete sealing. The ratio of the forced response at 200 Hz for the original configuration and the configuration with the sealed SFDs is shown in Figure 25a. Increasing the sealing effect determines an increase of the forced response at 200 Hz at node 30. The evolution of the force coefficients with C_{seal} at 200 Hz is shown in Figure 25b. Increasing the sealing effect determines an increase of both the force coefficients of the SFD.

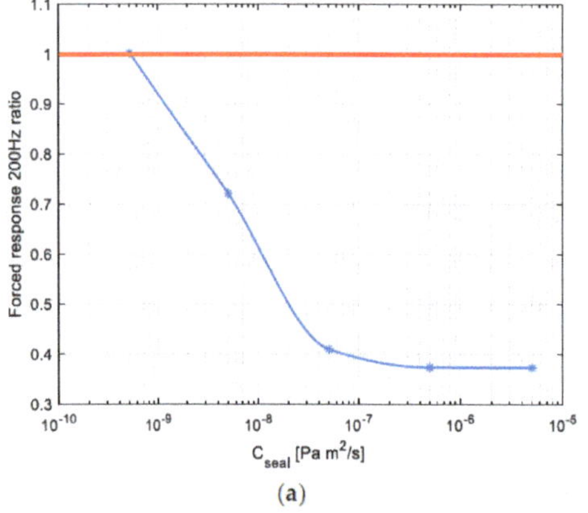

(a)

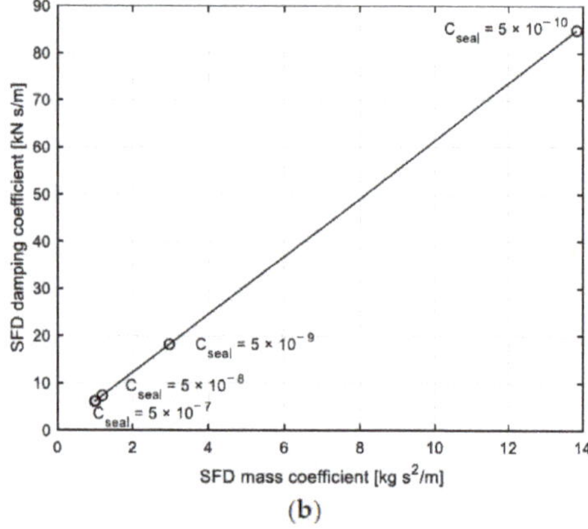

(b)

Figure 25. Forced response ratio at 200 Hz for node 30 for different values of C_{seal} (a), evolution of SFD force coefficients for different values of C_{seal} (b). Red line for system without SFD, blue line indicates system with SFD of (a). Markers highlights tested configurations.

The evolution of the forced responses for the considered frequency range is shown in Figure 26. Moreover, in this case, when the force coefficients are increased, the frequency of the peak of the forced response increases.

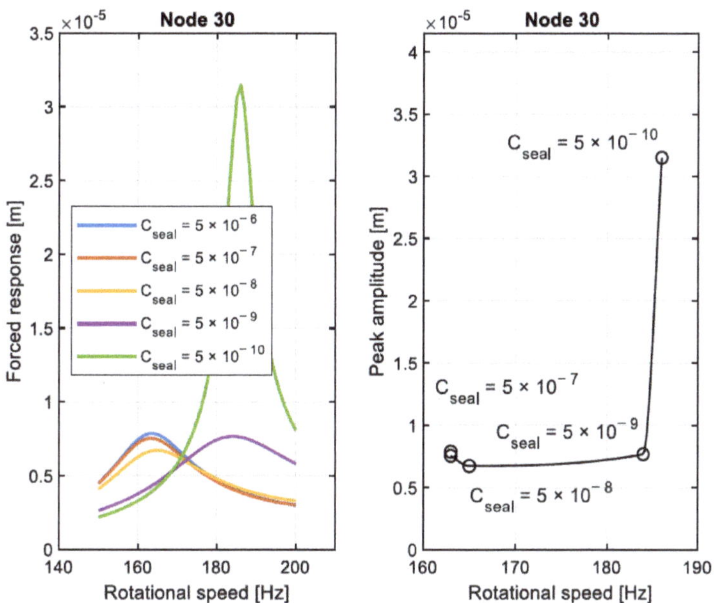

Figure 26. Comparison of forced response for different values of C_{seal} at node 30.

4.6. Correction of an Instability

In this section the effect of the seal placed before the impeller is considered as source of instability. The stiffness matrix at the nodes of the seal is introduced as follows:

$$K_{seal} = \begin{bmatrix} 0 & k_{seal} \\ -k_{seal} & 0 \end{bmatrix},$$ (26)

The parameter k_{seal} determines whether the compressor is affected by instability. The effect of k_{seal} on the stability of the system is investigated. The first instability is present at $k_{seal} = 15,000 \, \text{N/m}$. However, the system is unstable for the whole frequency range only when $k_{seal} \geq 17,500 \, \text{N/m}$. For this reason, our analysis is focused at $k_{seal} = 17,500 \, \text{N/m}$.

The same architecture shown in Figure 13 is considered. The damping introduced in the system by the SFD tends to have a stabilizing effect. The dimensionless damping factor is studied as an indication of the stabilizing effect. This indicator is defined as:

$$\eta_i = -\frac{Real(\lambda_i)}{Imag(\lambda_i)},$$ (27)

The SFD considered is similar to that described in Table 2 but now the clearance is set to 0.5 mm. The dimensionless damping factor for the original system, system with the cage, and the system with the SFD is shown in Figure 27. Both the original system and the system with the cage are affected by instability. On the contrary, when the SFD is added to the system the correction of the instability is achieved.

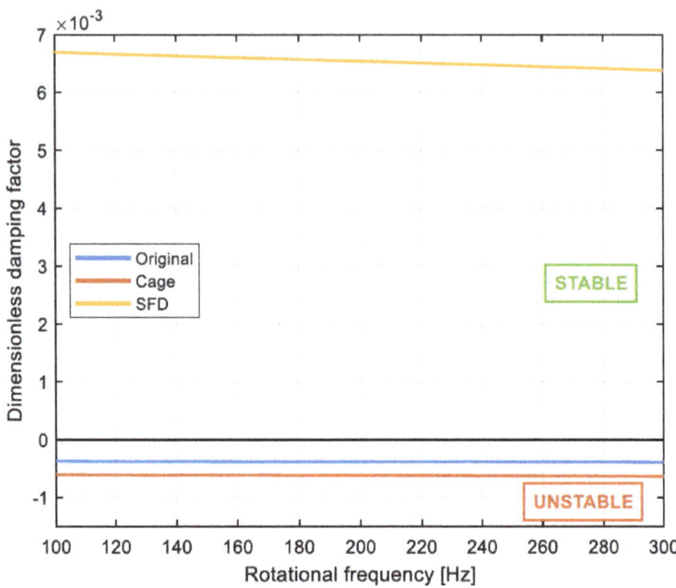

Figure 27. Evolution of dimensionless damping factor for unstable modes with rotational frequency of original system, system with cage, and system with SFD.

5. Conclusions

In this paper, a thorough investigation on the state of the art of SFDs is reported. The most critical features are highlighted and an evaluation of how to treat them is given. A comprehensive model based on the 2D Reynolds equation is presented. The classic Reynolds equation is modified to include an extra term to take into consideration the effect of the temporal inertia and the air ingestion modeling is also considered. The equation is numerically solved with the finite difference approach.

The proposed model is then validated with numerical and experimental results available in the literature. At first, the effect of the air ingestion is considered. Secondly, the overall effect of different geometrical configurations is considered. The results obtained show an acceptable agreement for the evaluated configurations.

A finite element code to simulate the dynamic behavior of turbomachines was developed. The effect of the SFD is included considering the force coefficients calculated from the finite difference solution of the Reynolds equation. A slightly unbalanced centrifugal compressor was considered and a parametric investigation on several parameters of the SFD was performed to test the effectiveness of the SFD in the vibration reduction. In general, the application of the SFD is effective in reducing the level of the vibration. From the investigation, it is highlighted that the selection of the most appropriate design of the SFD is highly influenced by the application. Moreover, the application of the SFD also proved effective in the correction of the instability.

The model derived in this paper has proven effective and efficient in the prediction of the dynamic properties of SFDs and can therefore be considered a useful tool in the initial design of these critical components. More accurate and precise models based on CFD simulations are present in the literature, but these are characterized by a higher level of complexity and longer simulation times.

Author Contributions: Conceptualization, E.G.; methodology, E.G. and S.C.; software, E.G. and S.C.; validation, E.G.; formal analysis, E.G.; investigation, E.G.; resources, P.P.; data curation, E.G.; writing—original draft preparation, E.G.; writing—review and editing, A.V.; visualization, E.G.; supervision, A.V.; project administration, P.P.; funding acquisition, P.P. All authors have read and agreed to the published version of the manuscript.

Funding: This research received no external funding.

Data Availability Statement: Data used are confidential and cannot be shared.

Conflicts of Interest: The authors declare no conflict of interest.

References

1. Zeidan, F.Y.; San Andrés, L.; Vance, J.M. Design and Application of Squeeze Film Dampers in Rotating Machinery. In *Proceedings of the 25th Turbomachinery Symposium, Houston, TX, USA, 13–15 September 2022*; Texas A&M University, Turbomachinery Laboratories: College Station, TX, USA, 1996; pp. 169–188.
2. Vance, J.M.; Zeidan, F.Y.; Murphy, B. *Machinery Vibrations and Rotordynamics*; John Wiley & Sons Inc.: Hoboken, NJ, USA, 2010.
3. Gunter, E.J., Jr.; Barrett, L.E.; Allaire, P.E. *Design and Application of Squeeze Film Dampers for Turbomachinery Stabilization*; Texas A&M University, Gas Turbine Laboratories: College Station, TX, USA, 1975; pp. 127–141. [CrossRef]
4. Chen, W.J.; Rajan, M.; Rajan, S.D.; Nelson, H.D. Optional Design of Squeeze Ilm Dampers for Flexible Rotor Systems. *J. Mech. Transm. Autom. Des.* **1988**, *110*, 166–174. [CrossRef]
5. della Pietra, L.; Adiletta, G. The Squeeze Film Damper over Four Decades of Investigations. Part I: Characteristics and Operating Features. *Shock Vib. Dig.* **2002**, *34*, 3–26.
6. San Andrés, L. Squeeze Film Damper: Operation, Models and Theoretical Issues. In *Modern Lubrication Theory*; Texas A&M University: College Station, TX, USA, 2012.
7. San Andrés, L.; Vance, J.M. Effects of Fluid Inertia and Turbulence on the Force Coefficients for Squeeze Film Dampers. *J. Eng. Gas Turbine Power* **1986**, *108*, 332–339. [CrossRef]
8. San Andrés, L.A.; Vance, J.M. Effects of Fluid Inertia on Finite-Length Squeeze-Film Dampers. *ASLE Trans.* **1987**, *30*, 384–393. [CrossRef]
9. Hamzehlouia, S.; Behdinan, K. Squeeze Film Dampers Executing Small Amplitude Circular-Centered Orbits in High-Speed Turbomachinery. *Int. J. Aerosp. Eng.* **2016**, *2016*, 5127096. [CrossRef]
10. Chen, X.; Ren, G.; Gan, X. Dynamic Behavior of a Flexible Rotor System with Squeeze Film Damper Considering Oil-Film Inertia under Base Motions. *Nonlinear Dyn.* **2021**, *106*, 3117–3145. [CrossRef]
11. Zeidan, F.Y.; Vance, J.M. Cavitation Regimes in Squeeze Film Dampers and Their Effect on the Pressure Distribution. *Tribol. Trans.* **1990**, *33*, 447–453. [CrossRef]
12. Diaz, S.E.; San Andrés, L.A. Air Entrainment vs. Lubricant Vaporization in Squeeze Film Dampers: An Experimental Assessment of Their Fundamental Differences. *J. Eng. Gas Turbine Power* **2001**, *123*, 871–877. [CrossRef]
13. Reinke, P.; Ahlrichs, J.; Beckmann, T.; Schmidt, M. High-Speed Digital Photography of Gaseous Cavitation in a Narrow Gap Flow. *Fluids* **2022**, *7*, 159. [CrossRef]
14. Elrod, H.G. Cavitation Algorithm. *J. Lubr. Technol.* **1981**, *103*, 350–354. [CrossRef]
15. Almqvist, A.; Fabricius, J.; Larsson, R.; Wall, P. A New Approach for Studying Cavitation in Lubrication. *J. Tribol.* **2014**, *136*, 011706. [CrossRef]
16. Diaz, S.E.; Andres, L.A. Reduction of the Dynamic Load Capacity in a Squeeze Film Damper Operating with a Bubbly Lubricants. *J. Eng. Gas Turbine Power* **1999**, *121*, 703–709. [CrossRef]
17. San Andrés, L.; Diaz, S.E. Flow Visualization and Forces from a Squeeze Film Damper Operating with Natural Air Entrainment. *J. Tribol.* **2003**, *125*, 325–333. [CrossRef]
18. Diaz, S.E. *An Engineering Model for Prediction of Forces in SFD's and Experimental Validation for Operation with Entrainment*; Texas A&M University: College Station, TX, USA, 1999.
19. Diaz, S.; San Andrés, L. A Model for Squeeze Film Dampers Operating with Air Entrainment and Validation with Experiments. *J. Tribol.* **2001**, *123*, 125–133. [CrossRef]
20. Méndez, T.H.; Torres, J.E.; Ciaccia, M.A.; Díaz, S.E. On the Numerical Prediction of Finite Length Squeeze Film Dampers Performance with Free Air Entrainment. *J. Eng. Gas Turbine Power* **2010**, *132*, 012501. [CrossRef]
21. Gehannin, J.; Arghir, M.; Bonneau, O. Evaluation of Rayleigh-Plesset Equation Based Cavitation Models for Squeeze Film Dampers. *J. Tribol.* **2009**, *131*, 024501. [CrossRef]
22. San Andrés, L.; Jeung, S.; Den, S.; Savela, G. Squeeze Film Dampers: An Experimental Appraisal of Their Dynamic Performance. In *Proceedings of Asia Turbomachinery Pump Symposium, Kuala Lumpur, Malaysia, 23–26 May 2022*; Turbomachinery Laboratory, Texas A&M Engineering Experiment Station: Singapore, 2016.
23. Gehannin, J.; Arghir, M.; Bonneau, O. Complete Squeeze-Film Damper Analysis Based on the "Bulk Flow" Equations. *Tribol. Trans.* **2010**, *53*, 84–96. [CrossRef]

24. Lee, G.J.; Kim, J.; Steen, T. Application of Computational Fluid Dynamics Simulation to Squeeze Film Damper Analysis. *J. Eng. Gas Turbine Power* **2017**, *139*, 102501. [CrossRef]

25. Xing, C.; Braun, M.J.; Li, H. A Three-Dimensional Navier-Stokes-Based Numerical Model for Squeeze-Film Dampers. Part 1-Effects of Gaseous Cavitation on Pressure Distribution and Damping Coefficients without Consideration of Inertia. *Tribol. Trans.* **2009**, *52*, 680–694. [CrossRef]

26. Xing, C.; Braun, M.J.; Li, H. A Three-Dimensional Navier-Stokes-Based Numerical Model for Squeeze Film Dampers. Part 2-Effects of Gaseous Cavitation on the Behavior of the Squeeze Film Damper. *Tribol. Trans.* **2009**, *52*, 695–705. [CrossRef]

27. Zhou, H.-L.; Chen, X.; Zhang, Y.-Q.; Ai, Y.-T.; Sun, D. An Analysis on the Influence of Air Ingestion on Vibration Damping Properties of Squeeze Film Dampers. *Tribol. Int.* **2020**, *145*, 106168. [CrossRef]

28. Delgado, A.; San Andrés, L. A Model for Improved Prediction of Force Coefficients in Grooved Squeeze Film Dampers and Oil Seal Rings. *J. Tribol.* **2010**, *132*, 032202. [CrossRef]

29. Hamzehlouia, S.; Behdinan, K. Thermohydrodynamic Modeling of Squeeze Film Dampers in High-Speed Turbomachinery. *SAE Int. J. Fuels Lubr.* **2018**, *11*, 129–146. [CrossRef]

30. Fan, T.; Hamzehlouia, S.; Behdinan, K. The Effect of Lubricant Inertia on Fluid Cavitation for High-Speed Squeeze Film Dampers. *J. Vibroeng.* **2017**, *19*, 6122–6134. [CrossRef]

31. Fan, T.; Behdinan, K. The Evaluation of Linear Complementarity Problem Method in Modeling the Fluid Cavitation for Squeeze Film Damper with Off-Centered Whirling Motion. *Lubricants* **2017**, *5*, 46. [CrossRef]

32. Fan, T.; Behdinan, K. Investigation into the Effect of Piston Ring Seals on an Integrated Squeeze Film Damper Model. *J. Mech. Sci. Technol.* **2019**, *33*, 559–569. [CrossRef]

33. Almqvist, A. A Pivoting Algorithm Solving Linear Complementarity Problems 2020. Available online: https://it.mathworks.com/matlabcentral/fileexchange/41485-a-pivoting-algorithm-solving-linear-complementarity-problems (accessed on 6 September 2022).

34. Marmol, R.A.; Vance, J.M. Squeeze Film Damper Characteristics for Gas Turbine Engines. *J. Mech. Des. Trans. ASME* **1978**, *100*, 139–146. [CrossRef]

35. San Andrés, L.S.; Koo, B.; Jeung, S.-H. Experimental Force Coefficients for Two Sealed Ends Squeeze Film Dampers (Piston Rings and O-Rings): An Assessment of Their Similarities and Differences. *J. Eng. Gas Turbine Power* **2019**, *141*, 021024. [CrossRef]

36. Andrés, L.S.; Rodríguez, B. On the Experimental Dynamic Force Performance of a Squeeze Film Damper Supplied Through a Check Valve and Sealed With O-Rings. *J. Eng. Gas Turbine Power* **2021**, *143*, 111011. [CrossRef]

37. Lund, J.W.; Smalley, A.J.; Tecza, A.J.; Walton, J.F. Squeeze-Film Damper Technology: Part 1-Prediction of Finite Length Damper Performance. In *Turbo Expo: Power for Land, Sea, and Air, Phoenix, AR, USA, 27–31 March 1983*; American Society of Mechanical Engineers: Houston, TX, USA, 1983. [CrossRef]

38. San Andrés, L.; Koo, B. Effect of Lubricant Supply Pressure on SFD Performance: Ends Sealed with O-Rings and Piston Rings. In *Proceedings of the 10th International Conference on Rotor Dynamics–IFToMM, online*; Springer Science and Buisness Media B.V.: Dordrecht, The Netherlands, 2019; Volume 60.

39. Jeung, S.-H.; San Andrés, L.; Den, S.; Koo, B. Effect of Oil Supply Pressure on the Force Coefficients of a Squeeze Film Damper Sealed with Piston Rings. *J. Tribol.* **2019**, *141*, 061701. [CrossRef]

40. San Andrés, L.; Jeung, S.-H. Orbit-Model Force Coefficients for Fluid Film Bearings: A Step beyond Linearization. *J. Eng. Gas Turbine Power* **2016**, *138*, 022502. [CrossRef]

Article

Conceptualization and Implementation of a Reconfigurable Unmanned Ground Vehicle for Emulated Agricultural Tasks

Raza A. Saeed [1,*], **Giacomo Tomasi** [1], **Giovanni Carabin** [1], **Renato Vidoni** [1,2,*] **and Karl D. von Ellenrieder** [1]

1   Field Robotics South Tyrol (FiRST) Lab, Libera Università di Bolzano, 39100 Bolzano, Italy
2   Competence Center for Plant Health, Libera Università di Bolzano, 39100 Bolzano, Italy
*   Correspondence: raza.saeed@unibz.it (R.A.S.); renato.vidoni@unibz.it (R.V.)

Abstract: Small-to-medium sized systems able to perform multiple operations are a promising option for use in agricultural robotics. With this in mind, we present the conceptualization and implementation of a versatile and modular unmanned ground vehicle prototype, which is designed on top of a commercial wheeled mobile platform, in order to test and assess new devices, and motion planning and control algorithms for different Precision Agriculture applications. Considering monitoring, harvesting and spraying as target applications, the developed system utilizes different hardware modules, which are added on top of a mobile platform. Software modularity is realized using the Robot Operating System (*ROS*). Self- and ambient-awareness, including obstacle detection, are implemented at different levels. A novel extended Boundary Node Method is used for path planning and a modified Lookahead-based Line of Sight guidance algorithm is used for path following. A first experimental assessment of the system's capabilities in an emulated orchard scenario is presented here. The results demonstrate good path-planning and path-following capabilities, including cases in which unknown obstacles are present.

Keywords: UGVs; reconfigurable robots; mechatronic design; field robotics; path and trajectory planning

Citation: Saeed, R.A.; Tomasi, G.; Carabin, G.; Vidoni, R.; von Ellenrieder, K.D. Conceptualization and Implementation of a Reconfigurable Unmanned Ground Vehicle for Emulated Agricultural Tasks. *Machines* **2022**, *10*, 817. https://doi.org/10.3390/machines10090817

Academic Editors: Marco Ceccarelli, Giuseppe Carbone and Alessandro Gasparetto

Received: 30 July 2022
Accepted: 13 September 2022
Published: 16 September 2022

Publisher's Note: MDPI stays neutral with regard to jurisdictional claims in published maps and institutional affiliations.

1. Introduction

In modern agriculture, concepts like precision agriculture, proximal monitoring, and sustainable agriculture are currently important, if not fundamental, for answering the need for increased food production, fighting climate change, and alleviating labor shortages. Indeed, digitalization is impacting agriculture through technologies and advanced data processing techniques for, e.g., land assessment, soil–crop suitability, weather information, crop growth, biomass and productivity, and precision farming [1,2]. It is anticipated that precision agriculture (PA), also known as precision farming or smart farming [3], will increase production with fewer resources by permitting farmers to continuously monitor and manage crops [4]. Therefore, there is a demand for quantitatively establish the effectiveness of *PA* in common agricultural applications by testing baseline automated platforms with integrated sensors, controls, information technologies, and algorithms. In this regard, agricultural robotics represents an important part of agri-digitalization [5]. Existing solutions mostly include specialized, task-based, agricultural robots, i.e., robots that can perform a particular task for a specific set of field conditions, but are not suitable for other tasks. Recent literature reports different robotic and mobile robotic applications in agriculture for, e.g., seeding [6], spraying [7], mowing [8], weeding [9,10], pruning [11,12], monitoring and inspection [13–15], and harvesting [16]. The existing literature has been reviewed and summarized by [17,18]. Thus, when considering the complexity of agricultural environments resulting from disparate operating conditions, e.g., farm size, orchard topology, and crops, growers must use different machines for different crops and production methods. This is not cost-effective, especially for small farms. To address this issue, some researchers and companies have developed multipurpose robotic platforms that can be used for different production methods [19–22]. Indeed, reconfigurability and modularization can represent a possible solution when incorporated into the design of field robots [23–29]. A reconfigurable robot is thus composed of many modules with different functions which can be quickly reconfigured to operate under new circumstances, perform different tasks, or recover from damage [28,29]. In keeping with the philosophy that the whole can be greater than the sum of its parts, individual modules generally have limited sensing, perception, control, computing, and motion capabilities. However, when assembled, the modules should act as a single robotic system. Typically, some sort of uniform docking interface is used between each module to permit the transfer of mechanical power, electrical power, and communication. The complete robot is usually composed of a primary unit, or main platform, and additional specialized modules, such as grippers, wheels, cameras, payload, and energy storage and generation units. Valuable examples of contemporary prototypes and commercial solutions following such an approach are Thorvald [25], MARS [30], GARotics [31], SAGA [32], GRAPE [33], and CATCH [34]. Based on the referenced literature, developing a fully autonomous system like these is still an open research challenge, in particular when the costs and complexity of reconfiguration are considered. By trying to make these platforms useful in numerous tasks such as seeding, spraying, weeding, harvesting, and monitoring, various challenges must be addressed, e.g., harvesting speed, disease detection, path-following/tracking accuracy, field navigation, obstacle avoidance, protection from accidents while operating, human–robot collaboration and multiple-robot collaboration to complete even more complex tasks. Thus, the effective combination of all the involved technologies and the implementation of a complex and extendable infrastructure to support every task of modern cultivation are the results to be targeted.

Therefore, the availability of a modular and reconfigurable platform from the hardware and software point of view in a research lab unlocks the opportunity to also emulate different situations and test different solutions from a basic and applied research standpoint in order to advance the overall state of the art. Indeed, sensors, actuators, and other equipment are needed to experimentally reproduce the main features of different applications (e.g., generating external and interacting forces, creating weight shifts that affect trajectory tracking, localizing the ground vehicle, and permitting safe remote human interaction). The successful integration and implementation of this hardware for applications in precision agriculture requires the development of new algorithms and technical solutions.

In this approach in mind, a reconfigurable system for lab and field activities is conceived in this work starting from hardware already available at the FiRST-Field Robotics South-Tirol Lab of the Free University of Bozen–Bolzano (Italy, see [35]). Regarding the three main configurations and cases related to precision agriculture tasks to be addressed or emulated in the future, i.e., monitoring, harvesting, and spraying, the following considerations and design guidelines have been adopted. Specific functions are enabled using different Hardware (*HW*) and Software (*SW*) modules in order that their combination can enhance the functionality of the entire robot, as well as ensure robustness in case of failure, e.g. redundant sensors. Given an orchard scenario, the robot has to be controlled to safely reach goal points, as well as navigate within the rows and the orchard passages; a map-based reliable path planning algorithm that minimizes the distance travelled to reach the goal points must be implemented. Given the fact that the Unmanned Ground Vehicle (UGV) has to maintain a safe distance from obstacles, e.g., trees, the edges of the orchard, fixed and dynamic obstacles, safety-margins and boundary zones must be implemented to allow safe motion of the robot. The UGV system state should be described by its position and orientation (pose) with respect to the environment. Due to the steering mechanism, e.g. skid steering, and environmental characteristics, the wheels are often subjected to slipping and uneven ground that frequently generate disturbances. Since linear and angular speed commands, that can be used separately or combined, are used to actuate the wheels a path-tracking algorithm supported by a proper sensory feedback must be developed and implemented. A combination of feedback navigation sensors is needed for motion control, as well for both self- and ambient-awareness. To act on the environment, manipulation capabilities are to be allowed through at least a 6 degree of freedom manipulator. Finally, good payload capabilities and sufficient clearance between devices must be ensured to accommodate additional modules, or to reconfigure modules for different working scenarios. Overall, the *HW* must permit the emulation of sensors, actuators, and other equipment needed to experimentally reproduce the main features of the three test scenarios/applications. At the same time the *SW* must support the real-time implementation of: (i) a complete navigation system that can be customized to address the different emulated agricultural scenarios and (ii) a reliable path-following controller.

The main contributions of this work include:

1. the functional mechatronic design of a reconfigurable mobile robot to be build on top of an available mobile platform;
2. the implementation and prototyping of the modular mechanical and electronic systems based on the design concepts;
3. software modules exploiting *ROS* that can perform automatic tasks combining path planning, obstacle detection, and path-following;
4. the implementation of customized path-planning as well as path-following algorithms based on recent literature results;
5. preliminary tests of the mobile robot performance in an emulated agricultural scenario.

The rest of the paper is organized as follows: Section 2 describes the proposed functional design concept of hardware and software modules. Section 3 presents the implementation and experimental testing results. Finally, the main conclusions drawn from this study are provided in Section 4.

2. Functional Design Concept and System Configuration

The main platform on which the modular system has been conceived and outfitted for PA is the Husky Unmanned Ground Vehicle (UGV) from Clearpath Robotics, already available at our premises. It presents a skid steering mechanism, which relies on wheel slippage. This simple and robust driving mechanism allows high mobility and assures large traction for manoeuvring on rough surfaces relying on few actuators [36]. On top of it, the proposed functional design concept of hardware and software modules is presented in Figure 1 together with the main flow of information between them. The blocks with the solid line are considered to be standard/required modules. The blocks with a dashed line represent modules that might be added and connected to the main robotic platform, depending on the tasks to be performed. This is conceptualized through the opportunity to add self- and ambient-awareness sensors, a robotic manipulator installed in different locations, a second control box and power unit to extend the robot capabilities, and an adjustable frame to support devices in different ways. As is common in field robotics, the electronic and software architecture of the system are configured so that the *high-level control* tasks (e.g., path planning, trajectory planning, and task planning and other computationally intensive processes, such as image processing for LiDAR and stereo vision) are separated from *low-level control* tasks, such as trajectory tracking control, path following control, and state estimation. By doing so the overall control of the mobile platform is more efficient and somewhat modular.

The following subsections introduce and describe the hardware and software modules.

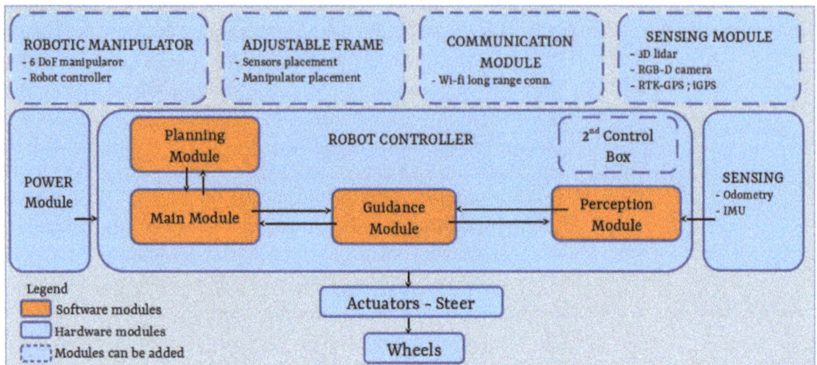

Figure 1. Functional architecture of the reconfigurable mobile platform. The main *SW* modules are depicted in orange. Arrows represent the main flow of information and the communication between modules.

2.1. HW Configuration

A mobile robotic base represents the ground-layer of the system, and the overall modular platform has been conceived and realized on top. The base is a Husky unmanned ground vehicle (UGV) from Clearpath Robotics. The robot is a four-wheeled skid-steering platform, and it is fully supported by Robot Operating System (*ROS*) [36]. The robot's main external dimensions are 990 mm in length, 670 mm in width, and 390 mm in height. Its weight is 50 kg, it has a payload of 75 kg, and it can reach a maximum speed of 1 m per second. Two kinds of sensors were integrated with the robot to provide information about the robot's state in terms of the body's angular orientation and odometry: onboard wheel encoders with a resolution of 78,000 ticks/m and a CH Robotics UM6 nine-DoF inertial measurement unit (*IMU*). The latter provides Euler angles with a resolution of 0.01 degree at 500 Hz rate. Several different vehicle configurations were explored to ensure:

1. the vision-based systems have good fields-of-view for detecting and then avoiding obstacles, in particular in front of the vehicle, i.e., recognize obstacles to be avoided at a minimum of 2 m in front of the vehicle;
2. the robotic manipulator has sufficient clearance over the entire range of its motion, i.e., keep the same footprint, and there would be minimal electronic signal noise from the *UHF* and WiFi transmitters;
3. the positioning systems are appropriately placed to receive satellite and *UHF* signals, i.e., not occluded or disturbed by other devices.

For the vision-based systems, we adopted a Velodyne VLP-16 LiDAR (range = 100 m, 360 × 30 deg field-of-view, and +/− 15 degree variation) and am Intel Realsense D455 RGB-D camera (range = 0.4 to 6 m, 1280 × 720 resolution, and 86 × 57 degree depth field-of-view).

We decided to mount these devices on a single platform so that they can be rotated together to shift their vertical fields of view. This makes the data more accessible and reliable for both systems by minimizing the amount of re-calibration necessary when repositioning them. Concerning the manipulator, a Universal Robots UR5e manipulator [37] has been considered. As shown in the system layout, it is placed in the front part of the vehicle to ensure adequate clearance. In addition, a pyramid-like structure has been designed for the front of the Husky. The form of this structure permits the base of the robotic manipulator to be mounted in different orientations, which allows us to test our system's ability to re-tune itself when it is reconfigured automatically. A detailed design configuration for the concept is shown in Figure 2. The designed robot frame (Figure 2) provides the mounting points for all *HW* modules. We have decided to use the onboard computer provided with the Husky UGV for the low-level control processes and use a second computer in a newly designed control box (high-level control box) to support electronics. The battery onboard the Husky UGV can provide electrical power to additional sensors that may be required, e.g., LiDAR, stereo camera, and RTK-GPS (outdoor), or ArduSimple simpleRTK2B [38] or iPS (indoor) positioning system, e.g., Pozyx [39]. Fieldwork can severely limit the ability to collect data, as no main power is typically available for recharging. The battery in the second control box is included to increase the time the robot can be operated without needing to stop to recharge. The system can either be run for long periods when attached to the main power line for debugging in the lab or run automatically for several hours in the field. In addition to the batteries, additional connectors, cables, a Ethernet bridge, a USB hub, and RS232 serial interfaces (designed through a set of Arduino Teensy micro-controllers) are used to integrate the high-level computer and different sensors, as well as to interface with the low-level computer in the Husky and the UR5 robotic manipulator. Lastly, an emergency stop push button has been added to ensure that the system can be rapidly disabled during field testing in case of unexpected behavior of the vehicle control system. Block diagrams of the overall electronics configuration and the high-level control box can be seen in Figure 3. In general, the heavy components (i.e., the UR5e manipulator, the UR5e manipulator control box, and the high-level control box with internal batteries) are distributed uniformly across the vehicle. Moreover, the center of gravity is placed as low as possible to maintain the stability of the ground vehicle on sloped terrain.

Once finalized, the mechanical and electronic systems used for our experiments were prototyped, as shown later in the experimental setup (see Section 3).

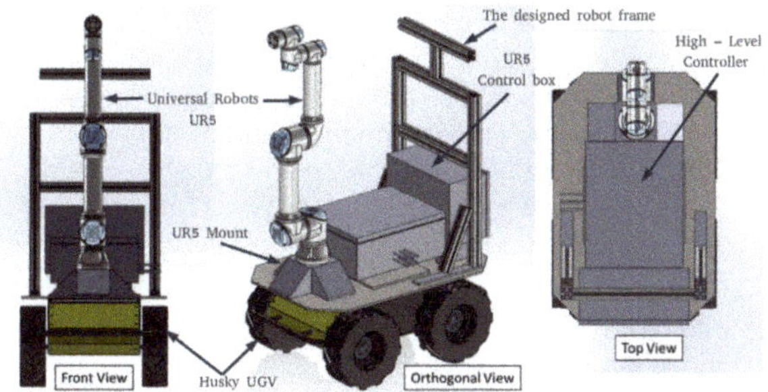

Figure 2. The design concept of the reconfigurable unmanned ground vehicle (UGV).

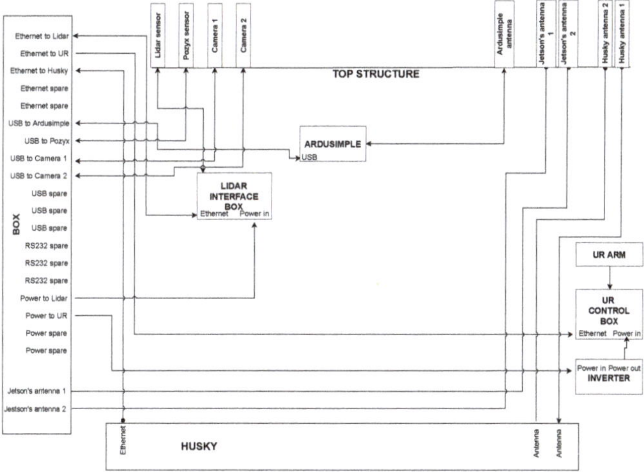

Figure 3. *Cont.*

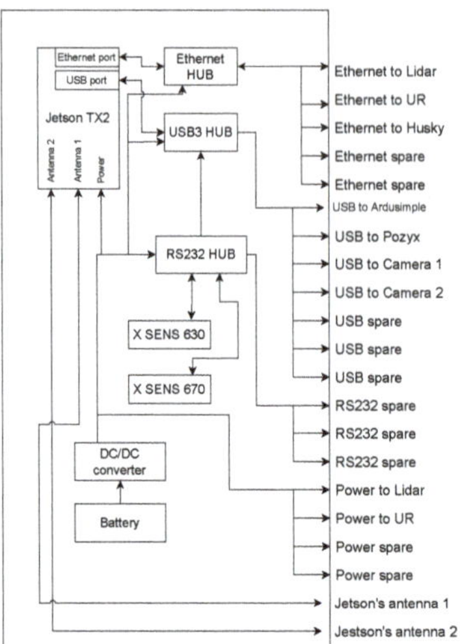

Figure 3. Configuration of the electronics and of the high-level control box.

2.2. SW Configuration

In this work, a complete automatic navigation system with a multimodal sensor setup is conceived and implemented. The system consists of different software modules, including guidance, planning, perception, and main modules. Essential requirements for fully automatic operation of the robot in outdoor environments are mapping, a collision-free path-planning algorithm, path-following control, and detection and localization of obstacles. Each software module provides specific functions. i.e., the *Guidance module* was designed to control the mobile robot's motion, the *Planning module* was developed to generate the shortest collision-free path, and the *Perception module* was created to detect and localize objects. We use *ROS* as the central framework to implement a proposed automatic navigation system, and we developed *ROS* packages to implement each software module. Coordination between software modules is illustrated in Figures 1 and 4. In the following subsections, each coordinated system module, as well as the main block and developed algorithms, are explained in more detail.

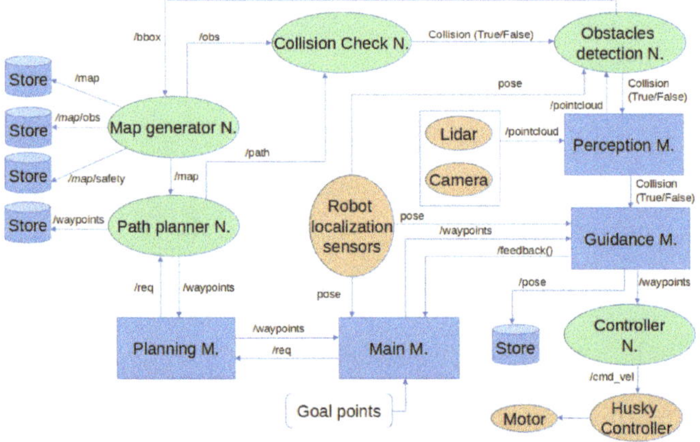

Figure 4. The *ROS* nodes scheme shows the system architecture in the first stage.

2.2.1. Main Module

The software modules are coordinated with each other through the *Main module*, the central module of the system. It receives user inputs, outputs commands, and calls upon other modules to perform a specific task. The main user input is the list of goals that the robot should visit. The goal points are expressed with respect to a pre-defined map of the environment that comes from previous knowledge of the environment. It can be made available as a grid-map of the orchard or, in future, thanks to the post-processing of a geo-referenced raster obtained through a previous survey with, e.g., a mapping drone. The grid-map of the environment is divided into a number of small square grid cells of the same size. Each grid cell can either correspond to a navigable area or to a space occupied by obstacles, e.g., trees.

The objective is to plan an optimal or sub-optimal route and then visit the goal points while avoiding obstacles, if needed. The *Main module* communicates with the *Planning module* to generate a path and with the *Perception module* to avoid obstacles. Given an initial pose g_0 and a sequence of goal points ($goal_{points} = g_1, g_2, \ldots, g_n$), the *Planning module* is then in charge of computing the shortest collision-free path connecting goal-to-goal points. Then, the *Main module* provides the reference waypoints to the *Guidance module*, which creates velocity commands to actuate the robot.

More in detail, as illustrated in Figure 4, the *Main module* receives a collision-free point-to-point path of waypoints between each pair of goal points from the *Planning module*, i.e., the *Main module* subscribes to the topic /*waypoints*. Moreover, the *Main module* has information both on the map and on the proprioceptive sensor-related data, i.e., it subscribes to the topics /*map* and /*pose*, which are computed according to the connected sensors, for localizing the robot on the map. It then sends the generated waypoints to the *Guidance module* to create a velocity command for the controller to move the robot towards the target point. The *Perception module* detects any unexpected obstacle, i.e., not present in the available map, that may appear along the path. For each newly detected object along the path, the *Perception module* evaluates whether it is occluding (blocking) the robot's path or not. If the object blocks the path, it is considered an obstacle, and the *SW* cancels the planned path and waypoints. When this happens, the *Main module* receives an alert (through the /*feedback()*) and sends a request (/*req*) to the *Planning module* to generate a new route considering the initial map updated with the obstacles, the robot's current location as the initial pose, and the remaining goal points as locations to be visited. All the *Store* blocks represent the data-storage to handle different variables in different situations when needed, such as during the experimental tests.

2.2.2. Perception Module

The main purpose of the *Perception module* is to enable the robot to safely navigate through the environment while avoiding collisions with other objects. The *Perception module* can use the information either from a stereo RGB-D camera or LiDAR sensors to detect and localize objects around the robot that may eventually become obstacles. Both sensors provide point cloud data that must be filtered and processed to give an understanding of the surrounding environment. Since the two devices sense the environment differently, the provided point clouds have different properties. The LiDAR has a 360° view, whereas the camera is constrained to its horizontal field of view; the 3D LiDAR returns points in grayscale, while the camera can add red–green–blue (RGB) color information to the points. Therefore, to permit modularity and fast reconfiguration, two different packages for the two sensors have been developed. They share the same structure and class architecture.

The *Perception module* architecture is shown in Figure 5. There are two main nodes, called the *Detector_node* and *BoundingBox_node*, that are used to identify, localize, and estimate the size of possible obstacles. The *Detector_node* takes a point cloud as input and generates an array of point-cloud clusters as outputs. A filtered point cloud is used for debugging and visualization the scenario in Rviz, the 3D ROS visualizer. The *Bounding-Box_node* takes as input the array of point-cloud clusters generated by the *Detector_node* and finds an axis-aligned bounding box for each cluster. The overall module can be decomposed into three main steps: pre-processing, clustering, and bounding-box definition.

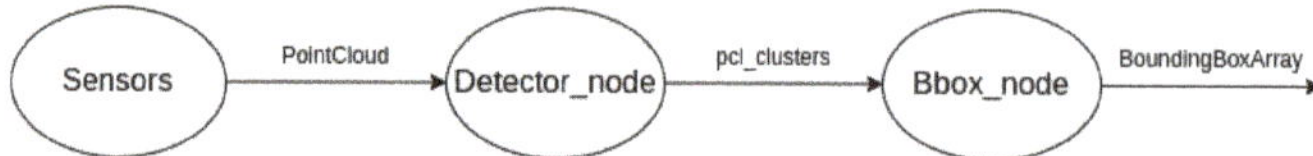

Figure 5. Perception module architecture.

The pre-processing step consists of a series of algorithms to prepare the data for the clustering step. Figure 6 presents an example of the perception module steps. Particularly, the point cloud is downsampled to reduce the computational load by using a voxel grid and pass-through filters (see Figure 6a) that mainly merge and approximate sets of points and cut the data to consider a reduced xyz-range of points. After downsampling the data, the *Detector_node* performs ground-removal using a plane-model segmentation algorithm [40] with a distance threshold parameter of 0.01 m. In the next step, since the ground is not perfect and the sensor is noisy, sometimes it happens that some points survive the segmentation step; an outlier-removal function is run to remove isolated points that remain after ground removal. Then, a DBSCAN algorithm [41] (see Figure 6b) is used to group all points belonging to the same object. A single point cloud is generated and added to an array for each object detected. The last step of the *Perception module* framework takes the array of clusters and finds an axis-aligned bounding box for each array element. When the system starts, the map frame is fixed in a known position in the environment and is used as the fixed-reference frame. We use this frame to generate axis-aligned bounding boxes for all the detected obstacles. When point cloud data is received from the sensor, it is filtered and the resulting point cloud is transformed with respect to the map frame. At that point, a bounding-box node generator to find the bounding boxes of the obstacles with coordinates expressed in the map frame (fixed frame) is run. The bounding box obtained is described with three coordinates representing the position of its center and three values representing the dimensions along coordinate axes. The final output of the perception module is published and accessible in the */BoundingBoxArray* topic (/bbox in Figure 4).

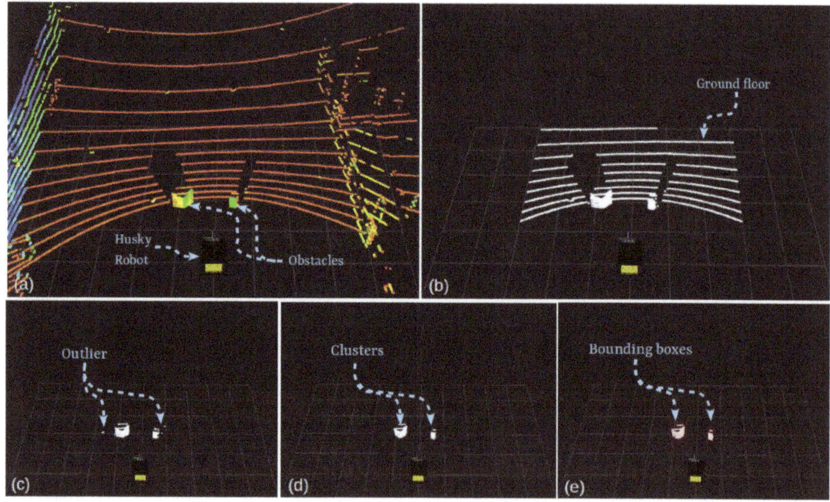

Figure 6. Perception module steps: (**a**) raw PointCloud, (**b**) pass-through, (**c**) plane removal, (**d**) outlier removal and clustering, and (**e**) bounding-box generation.

2.2.3. Dynamic Map Implementation and Management

The definition and exploitation of the environmental map is fundamental for the implemented framework both for defining the goal points as well as for searching and finding the suitable path to be traveled. Suppose dynamic environments like the ones targeted in this research activity are considered. In that case, the map should be updated, and the path can be replanned if needed. The requirements for the map module can be defined as two sub-tasks.

The first one generates a dynamic map updated in real-time. The second sub-task checks if the computed path is still safely achievable (with no collisions) after the map is updated with the detected obstacle(s). The package developed to accomplish these tasks is called *nav_map* and consists of two nodes named *MapGenerator_node* and *CollisionCheck_node*. *MapGenerator_node* creates a first map of the known environment from a file as soon as the node starts setting its fixed frame in a know location, then it subscribes to the */BoundingBoxArray* topic and updates the map every time a new message is received. Whenever a new message is received, it updates the map by removing or adding objects. The map is an OccupancyGrid map in which each cell can have three different values, respectively, it the cell represents a free area, obstacle area, or safety area. Besides the map, *MapGenerator_node* publishes an additional topic containing the coordinates of the cells containing an obstacle area. This information is used by *CollisionCheck_node* to check if the objects detected around the robot have to be considered as possible collision obstacles. *CollisionCheck_node* subscribes to the path topic */path* and compares the obstacle's cell coordinates and the path's cell coordinates. If there is at least one overlap, *CollisionCheck_node* will trigger a collision message as true. This strategy was implemented to avoid that the robot considers as obstacles objects that are not in its path and so, avoiding planning a new path when the previously computed path is still valid and collision-free. An example is shown in Figure 7.

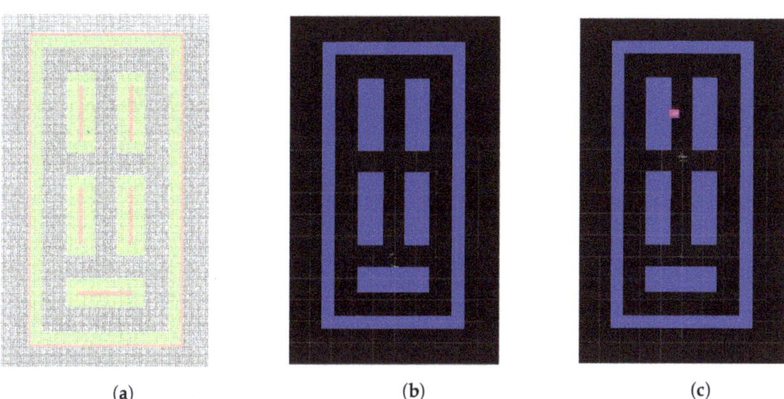

(a)   (b)   (c)

Figure 7. Example of a known map provided in (**a**) csv, (**b**) OccupancyGrid free map, and (**c**) updated map with detected obstacles.

2.2.4. Planning Module

In the targeted applications, several given goal points that the robot needs to visit are fixed a priori, e.g., proximal monitoring of specific locations/plants. The *Planning module* based on *ROS* is created to connect these goal points with the shortest collision-free path in the robot's working environment. This module gets a request from the *Main module* to generate a path between two goal points (*/req/*). The request is composed of the first goal point (current position of the robot) and the second goal point. The developed path-planning method, which we discuss in the following paragraph, is implemented to generate a collision-free path between every pair of sequenced goal points. For this purpose, a path planner *ROS* node is created to compute a continuous obstacle-free path, where the path is assumed to go through a set of successive waypoints (x_k, y_k) for $k = 1, \ldots, N$; see Figure 4. This node gets the starting and end points (from /*req*) as well as info from the map and the localization sensors and strategies adopted. Based on these data, the planner can generate waypoints representing a collision-free path between goal points. The *Planning module* is then in charge of sending the waypoints to the *Guidance module*. The path planner *ROS* node is in charge of checking for the shortest path and waypoints from the robot's current position to the goal position within the created map.

The path planner node uses an extended version of the authors' developed path planning methods, called boundary-node method (*SW*) and path enhancement method (PEM) [42–44]. The path-planning method calculates the shortest path considering obstacle avoidance to reach the destination point safely with the minimum distance traveled. The shortest path is generated in a two-step procedure. First, the *SW* Method generates the initial feasible path (IFP) between goal points. The IFP is generated from a sequence of waypoints w that the robot has to travel as it moves toward the destination point without colliding with obstacles.

An example of path planning and obstacle avoidance for a mobile robot in a static environment using *SW* is illustrated in Figure 8a. Based on the extended *SW*, the robot is simulated by a nine-node quadrilateral element. If the nodes are denoted by a vector $p(q), (q = 1 \ldots 9)$, the robot's location is represented by the centroid node $p(5)$, and nodes $p(1 \rightarrow 4)$ with $p(6 \rightarrow 9)$ represent the eight boundary nodes that help the robot move forward and avoid obstacles. The contour line represents the potential function utilized to direct the robot toward the goal point. It has the lowest potential value at the final destination point, and increases as the robot moves away. As shown in Figure 8a, the line color represents the potential value, i.e., red corresponds to the lowest potential value, and dark gray corresponds to the highest potential value. The sequence of the red circles represents the best solution to IFP. The simulated robot can only move in eight possible directions. In each iteration, the current location of the robot and boundary nodes move in one particular direction. Additionally, this method uses an optimization technique based on the lowest potential value to let the robot find the path and yield fast convergence. The node with the lowest potential value is chosen as the best position among all boundary nodes, and the robot updates its position to the best position.

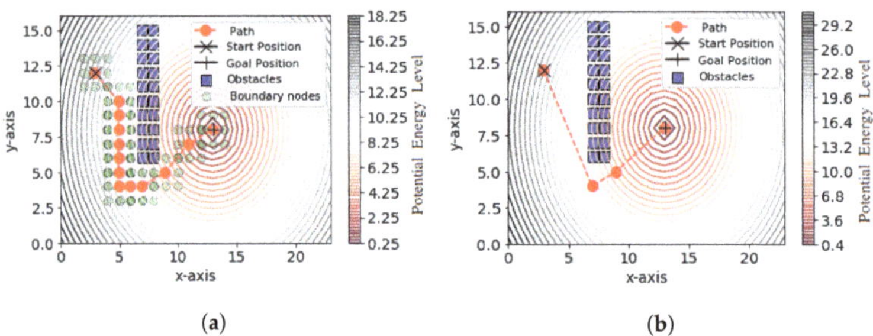

(a) (b)

Figure 8. Example of path planning for a mobile robot [44]: (**a**) The obtained solution to IFP by using *SW*, where the sequence of the red circles represents the IFP. (**b**) The shortest path found by using PEM, where the solid red line represents the shortest path.

The obtained IFP between goal points is not an optimal path in terms of total path length. Therefore, in the second step, the PEM is used to construct an optimal or near-optimal path from IFP by reducing the number of waypoints and the overall path length. Figure 8b illustrates the computed shortest path from waypoints. Waypoints defining the path are marked with red circles, while the red dashed line represents the shortest path. A more-detailed description of how the environment is created can be found in [43]. If the path provided by the path planner fails, or any unexpected obstacle is detected along the path, the path planner is contacted to compute a new path to the destination. This is the case when unexpected obstacles are detected in the working environment along the path.

In this case, the path-planning method computes a new path for the robot to avoid collision with static and dynamic obstacles. Once the path is found, the *Planning module* forwards the waypoints to the *Guidance module*.

2.2.5. Guidance Module

This module aims to control the robot to follow the path, that consists of a series of waypoints joined by line segments. As reported in [45,46], the path-following method is one of the typical control methods for autonomous vehicles. This method allows a robot to follow a predefined path independent of time, and thus without any restriction on the time-propagation along the path. An accurate path-following method is an essential aspect of the automatic navigation of robots in farm environments.

This study employs and adapts a line-of-sight (*LOS*) guidance algorithm for path-following to drive the robot to follow a predefined path, which is based on a lookahead-based LOS guidance algorithm. The main advantages of lookahead LOS guidance are the simplicity and ease of implementation [46]. Furthermore, the lookahead method is used to compute control inputs in real-time, which is advantageous when the given path is not smooth or when the path is specified using waypoints [47]. The method assumes that the robot moves at a constant desired forward speed, and uses the relative pose (position and heading angle) between the robot and the nearest path segment being followed to generate the desired heading angle. Since the lookahead-based steering method only generates the desired speed and heading angle rather than the control inputs, it is known as a guidance law [48]. The lookahead approach utilizes motion information about the position, orientation, velocity, and acceleration.

In the following paragraph, we describe the robot motion model with only three degrees of freedom (*DOF*). The robot's motion is assumed to be constrained to the horizontal plane, and a plane view of the robot is shown in Figure 9. The three *DOF* kinematic equations of the robot are reduced to [48]

$$\dot{\eta} = R(\psi)v, \tag{1}$$

where ψ is the heading angle of the vehicle. We used the global reference frame $G\{x, y\}$ and the body-coordinate frame $B\{x_b, y_b\}$ to describe the robot's motion, location, and orientation. As shown in Figure 9, the x axis of the global coordinate system $G\{x, y\}$ points toward the North, the y axis points toward the East, and the z axis indicates downward. The body-fixed frame $B\{x_b, y_b\}$ moves with the robot, the x axis points toward the head of the robot, the y axis points toward the right, the z axis indicates downward normal to the x–y surface, and U denotes the corresponding speed. The angle β represents the slide-slip angle, and the angle ψ represents the orientation angle of the mobile robot measured from the positive N axis of the body-fixed system. The point n denotes the center of the robot, which is a Cartesian coordinate about a global coordinate frame denoted by (x_b, y_b). $R(\psi)$ is the transformation matrix from $B\{x_b, y_b\}$ to $G\{x, y\}$, which is given by

$$R(\psi) := \begin{bmatrix} \cos\psi & -\sin\psi & 0 \\ \sin\psi & \cos\psi & 0 \\ 0 & 0 & 1 \end{bmatrix} \in SO(3), \tag{2}$$

and

$$\eta := \begin{pmatrix} x_n \\ y_n \\ \psi \end{pmatrix} \in \mathbb{R}^2 \times \mathcal{S}, \quad \text{and} \quad v := \begin{pmatrix} u \\ v \\ r \end{pmatrix} \in \mathbb{R}^3 \tag{3}$$

are the position and orientation (pose) vector and velocity vector (in body-fixed coordinates), respectively (see Figure 9). Here, the symbol $\mathbb{R}^n$ is the Euclidean space of dimension n, $\mathcal{S}$ is an Euler angle defined on the interval $[-\pi, \pi]$, and $SO(3)$ is the Special Orthogonal Group of order 3 [48]. Thus, η has three components representing two linear displacements and one angular rotation. The variables appearing in (3) include position x_n, y_n, orientation angle ψ, surge speed u, sway speed v, and yaw rate r. In this study, we developed a speed controller using the kinematic model only. The force-controller we used was the native one of the Husky robot developed by the company. The development of a lower-level controller based on a dynamic model is left for future work and improvements.

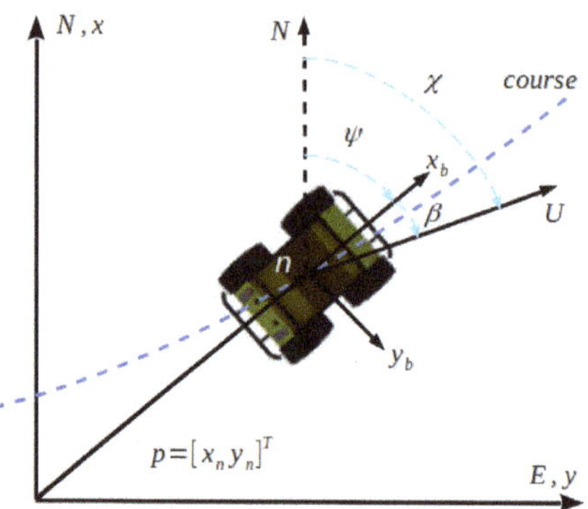

Figure 9. Three-DOF maneuvering coordinate system definitions.

The overall path to be followed consists of a set of n straight-line segments connected by $n+1$ waypoints. The automatic controller is constructed to steer the vehicle along a time-independent path, for example the path between waypoints p_k (x_k, y_k) and p_{k+1} (x_{k+1}, y_{k+1}). When the position of the robot is within a circle of acceptance with radius R around waypoint p_{k+1}, so that

$$(x_n - x_{k+1})^2 + (y_n - y_{k+1})^2 \leq R_{k+1}^2,\tag{4}$$

a switching mechanism is used to select the next waypoint p_{k+2}. The waypoints are fed sequentially to have a smooth path without *stop-and-go* behavior.

Line-of-Sight (LOS) guidance law:

Figures 9 and 10 show the geometry of the LOS guidance problem and main variables. As shown in the figures, the position of the robot in global coordinates can be written as $p = [x_n \, y_n]^T \in \mathbb{R}^2$, and the corresponding speed is defined as

$$U := \sqrt{\dot{x}_n^2 + \dot{y}_n^2} := \sqrt{u^2 + v^2} \in \mathbb{R}^+.\tag{5}$$

For an arbitrary waypoint on the path, the direction of the velocity vector with respect to the north axis is calculated by

$$\chi = \tan^{-1}\left(\frac{\dot{y}_n}{\dot{x}_n}\right) \in \mathcal{S} := [-\pi, \pi].\tag{6}$$

As the robot starts following the path to move towards the goal point, it firstly rotates towards the first goal direction, and then the robot accurately passes through all waypoints between each pair of goal points. However, based on the path planner, the angle between waypoints is always less than 90 degrees [43]. Consider a straight-line path defined by two consecutive waypoints at positions $p_k = [x_k, y_k]^T \in \mathbb{R}^2$ and $p_{k+1} = [x_{k+1}, y_{k+1}]^T \in \mathbb{R}^2$, respectively. The path makes an angle of

$$\alpha_k = \tan^{-1}\left(\frac{y_{k+1} - y_k}{x_{k+1} - x_k}\right) \in \mathcal{S}\tag{7}$$

with respect to the north axis of the NED frame.

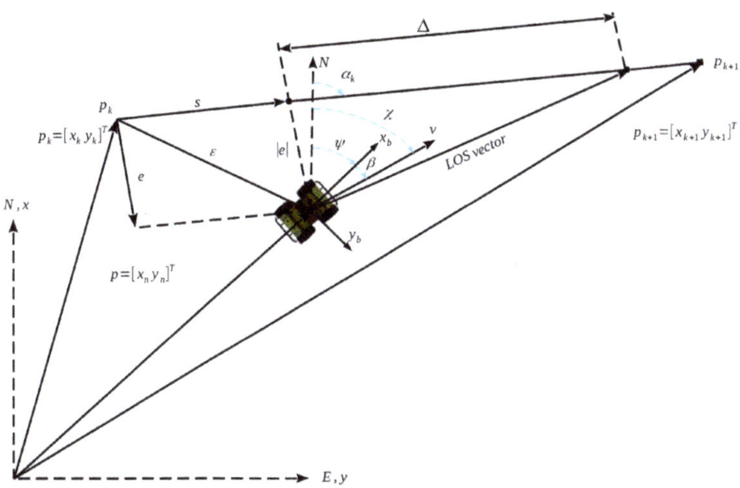

Figure 10. Line-of-sight path-following definitions.

The coordinates of the robot in the path-fixed reference frame are

$$\varepsilon = \begin{bmatrix} s \\ e \end{bmatrix} = R_\alpha^T(\alpha_k)(p - p_k) \tag{8}$$

where R_α^T is the transformation matrix from the inertial to the path-fixed frame given by

$$R_\alpha^T(\alpha_k) := \begin{bmatrix} \cos \alpha_k & \sin \alpha_k \\ -\sin \alpha_k & \cos \alpha_k \end{bmatrix}, \tag{9}$$

s is the along-track distance, and e is the cross-track error (see Figure 10). The cross-track error is defined as the orthogonal distance to the path tangential reference frame. From (8) and (9), the cross track error can be computed by

$$e = -(x_n - x_k)\sin \alpha_k + (y_n - y_k)\cos \alpha_k. \tag{10}$$

Its time derivative, which will be used later, is consequently given by

$$\dot{e} = -\dot{x}_n \sin \alpha_k + \dot{y}_n \cos \alpha_k. \tag{11}$$

From (1) and (2),

$$\begin{aligned} \dot{x}_n &= u\cos\psi - v\sin\psi \\ \dot{y}_n &= u\sin\psi + v\cos\psi \end{aligned} \tag{12}$$

so that

$$\dot{e} = -(u\cos\psi - v\sin\psi)\sin \alpha_k + (u\sin\psi + v\cos\psi)\cos \alpha_k. \tag{13}$$

The control objectives are formulated to drive the cross-track error to zero by steering the robot and controlling its forward speed. With the lookahead-based steering method, a fixed parameter, known as the lookahead distance Δ, which corresponds to a distance along the path ahead of point s, is used to define the LOS vector. The robot is steered so that its velocity vector is parallel to the LOS vector (Figure 10). The resulting velocity vector will have a component perpendicular to the path, driving the robot towards the path until the LOS vector is parallel to the path so that $e \to 0$. Then, the *Guidance module* calculates the linear and angular velocities, which are sent to the Husky controller and afterward to wheel motors to let the robot move automatically between goal points. The *Guidance module* sends the linear and angular velocity inputs to move the Husky robot automatically inside the working environment.

3. Prototyping and Preliminary Experimental Tests

The conceived and designed layout has been implemented, and first navigation tests have been performed. The ground robot platform is mechanically designed to handle suitable sensors. Thus, the mechanical and electronic systems used for our experiments were prototyped and integrated on top of the main platform. The final configuration of the mobile robot for the first experimental tests is shown in Figure 11a,b. The second control box arrangement is reported in Figure 11c. The main ambient awareness sensors installed on the robot are a 3D LiDAR Velodyne VPL16 and an Intel Realsense D455 camera. The robot is equipped with a mini-ITX computer (see [36]), while the second control box embedded computer is a Jetson TX2 platform. The preliminary test targeted an emulated orchard and a safe navigation along a collision-free path given multiple goal points.

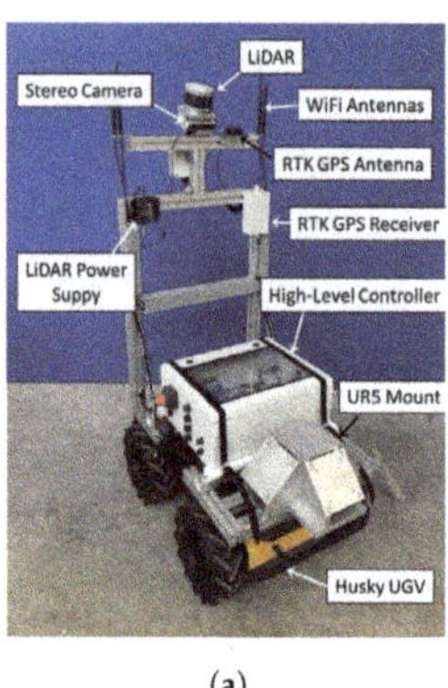

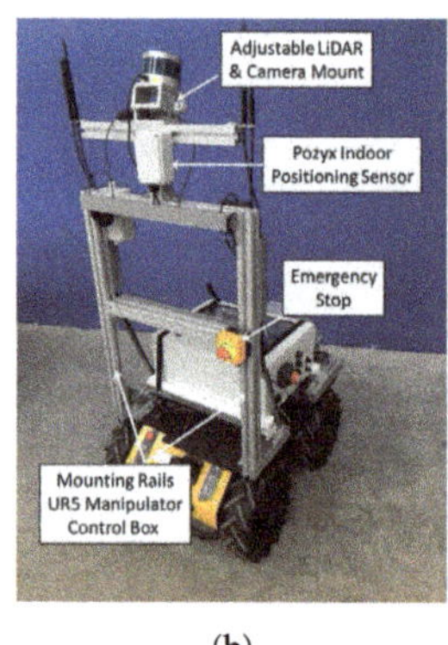

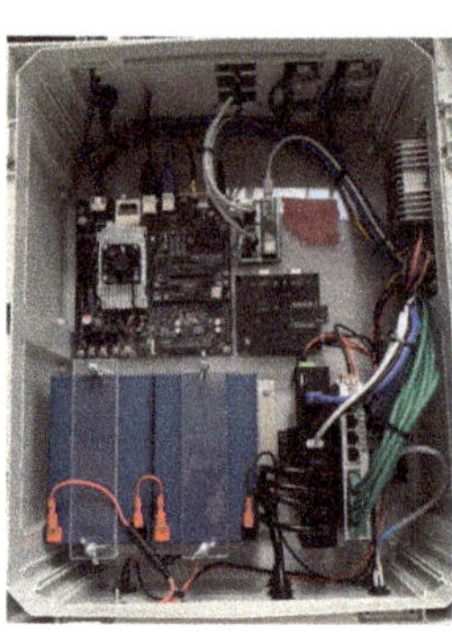

(a) (b) (c)

Figure 11. Mechanical and electronic configuration of the experimental platform (UR5 manipulator not shown): (**a**) front view, (**b**) back view, and (**c**) second control box.

3.1. Emulated Scenario and Test Inputs

For the initial evaluation of the implemented framework, a lab-emulated orchard was created in which parallel rows of tree were considered fixed obstacles on the map. The experimental emulated scenario is shown in Figure 12a, and the Gazebo digital twin is shown in Figure 12b.

The length and width of the environment are fixed to 13.2 m and 6.6 m, respectively. For simulating an orchard, the spacing between rows is set to 2.2 m (see [49]). The length of each row (static obstacles) is fixed at 2.2 m long. The robot's working space is decomposed into rectangular grid cells, and each grid cell represents 10 cm in the robot's working environment. The goal tolerance was set to 0.1 m, and the heading tolerance was set to 0.1 rad. The UGV chosen localization system was based on a filtered odometry approach; the Intel RealSense D455 $RGB\text{-}D$ camera or the Velodyne VLP-16 3D LiDAR were used (in different tests) for obstacle detection, and the extended BNM was adopted for path planning. The vision system was set to operate in front of the robot, setting a view size of $5 \times 6 \times 1.5$ m (xyz coordinates), and the obstacles were boxes of different dimension. The bounding boxes were generated with an offset of 0.2 m to ensure an additional safety. For path-tracking, the lookahead distance was set to 1 m for the LOS-guidance method, and the heading error was set to 0.05 rad.

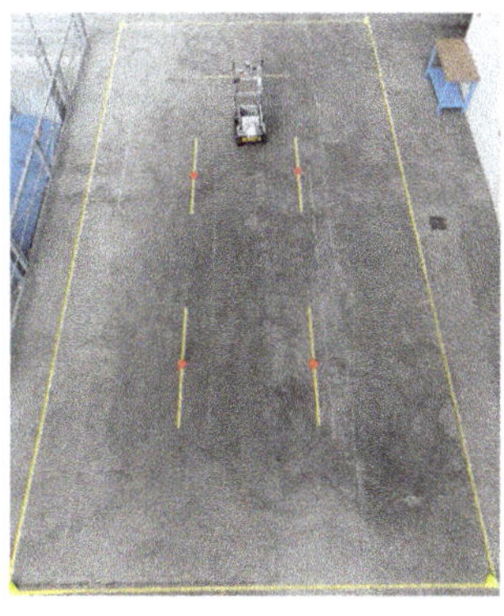
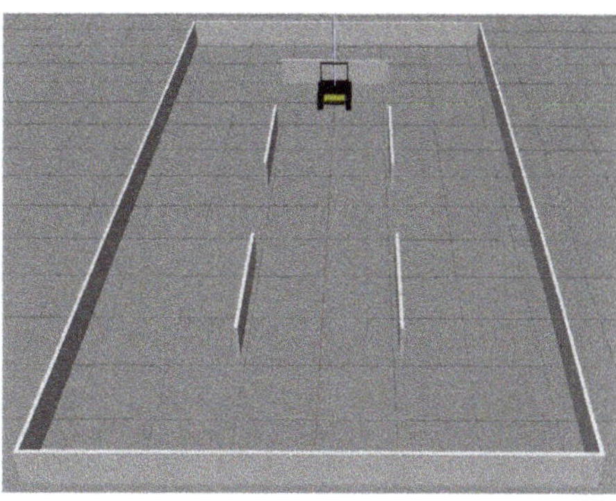

(a)   (b)

Figure 12. Robot working environment: (**a**) the real environment; (**b**) the simulated environment using Gazebo.

Given the input data to define the goal points and the working environment, the map generator node creates a map. Afterward, the map data are processed with the goal points through the path planner to compute a collision-free path between each pair of goal points.

The planned path between each two goal points formed by a discrete sequence of poses, i.e., waypoints, is fed into the *Guidance module*. At the same time, the output of the *Guidance module* is mapped to the wheel motors by a simulated low-level controller running onboard.

3.2. Experimental Tests and Discussion

As per the envisioned tasks, the mobile robot was sent to the proper orchard locations to perform predefined tasks. Then, we experimentally tested the system to evaluate the automatic navigation performance in a generated given-map environment. In the first test configuration, we evaluated the automatic navigation performance of the robot under static environmental conditions. Then, in the second test configuration, we evaluated the robot's navigation performance in a dynamic environment, i.e., with an unexpected obstacle along the path. During the experimental tests, we verified that the proposed framework generates an appropriate path, and the robot avoids static and dynamic obstacles without collisions.

Figures 13 and 14 show the planned path for a mobile robot using BNM in two testing scenarios as well as the performance of the navigation strategy in both testing scenarios. In the first test configuration (see Figure 13, the robot starts moving from initial position g_0 (0, 0) to visit three fixed goal points (g_1, g_2, and g_3) added to the pre-built map. The goal points were located at $(-2.2, -2.2)$, $(8.8, 2.2)$, and $(4.5, -2.2)$. The red cells express the fixed obstacles, blue dashed lines represent the planned path of the robot, yellow circles in the figure define the goal points, and green cells represent the area related to the safety margin introduced to allow the robot to safely navigate and steer. In this study, we adopted a safety margin around obstacles to avoid the possibility of overlapping the paths traced by the robot overlapping with obstacles. As it can be appreciated in Figure 13a, the robot might move very close to the obstacle. Therefore, a certain safety margin for anti-collision has to be assured. Considering the robot dimensions and footprint, it has been defined with a constant-size of six grid cells. All fixed obstacles are given in parallel lines, and the goal points are given between obstacles. The robot localization sensors are utilized to determine where a mobile robot is located within the environment. In this test, a filtered-odometry, i.e., encoders and IMU, approach has been adopted. The current coordinates of the robot are compared with the predefined ranges of the fixed goal point, and if the robot is within the range, the coordinates of the next destination point are considered.

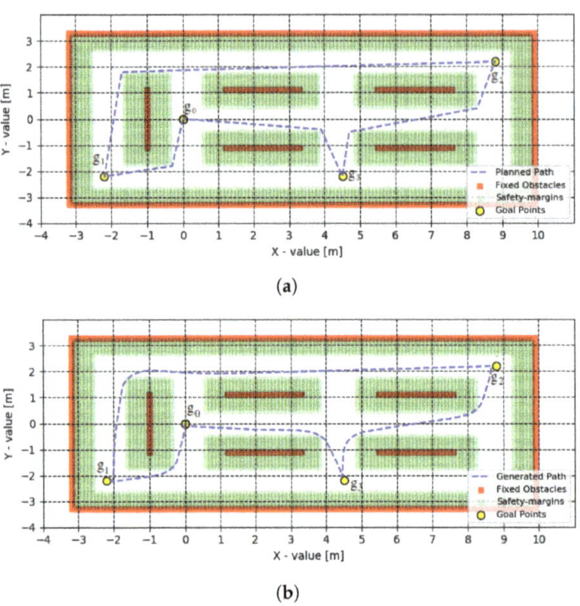

(a)

(b)

Figure 13. Simulated environment with fixed obstacles: (**a**) planned path using *BNM*; (**b**) path executed by the robot.

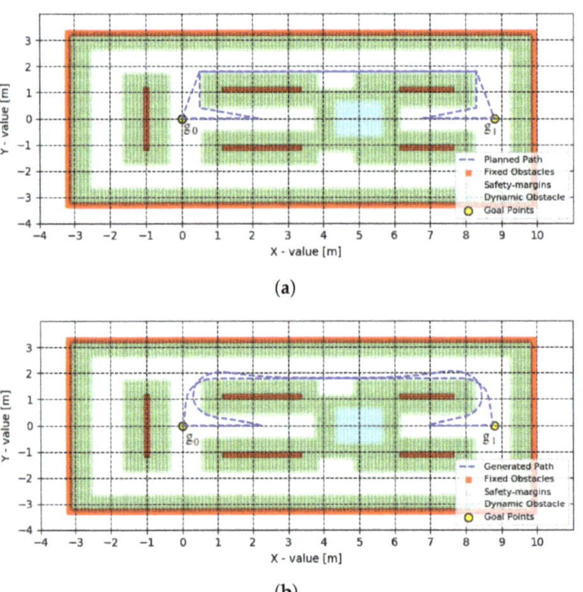

(a)

(b)

Figure 14. Simulated environment with fixed and dynamic obstacles: (**a**) planned and replanned path using *BNM*; (**b**) path executed by the robot.

In the second scenario, the experimental test was carried out with the addition of a dynamic obstacle, as shown in Figure 14. The robot starts moving from the initial position g_0 (0, 0) to visit a fixed goal point (g_1) located at (8.8, 0.0), and then the robot returns to the initial point. Since the robot sensors detect a new obstacle that occludes the pre-planned path, the *Planning module* generates a new path from the current position to the destination point (g_1).

The plots show that the robot's executed path is coherent with the planned one. The mobile robot shows appropriate navigation accuracy and performs smooth motion along the environment with an average error $\leq 10\,\text{cm}$. The traveled path mostly depends on the robot's kinematics, the accuracy of the sensors, and the environment. High navigation speeds lead to slightly higher errors. The proposed *ROS* package always directs the robot to follow the planned path, and the robot moves close to the calculated path. The path starts from the robot's initial position, passes through the intermediate goal points, and then returns to the initial point. A final path is assembled by connecting the paths between goal points in an iterative way until the path is completed, and the length of the path is the sum of the lengths of the goal-to-goal paths. From the first experimental scenario and reported experimental results, the generated path length is 24.3684 m, close to the 24 m of the ideal *BNM*-planned path, thus showing good performance. Screenshots of the experimental tests at different locations are presented in Figure 15.

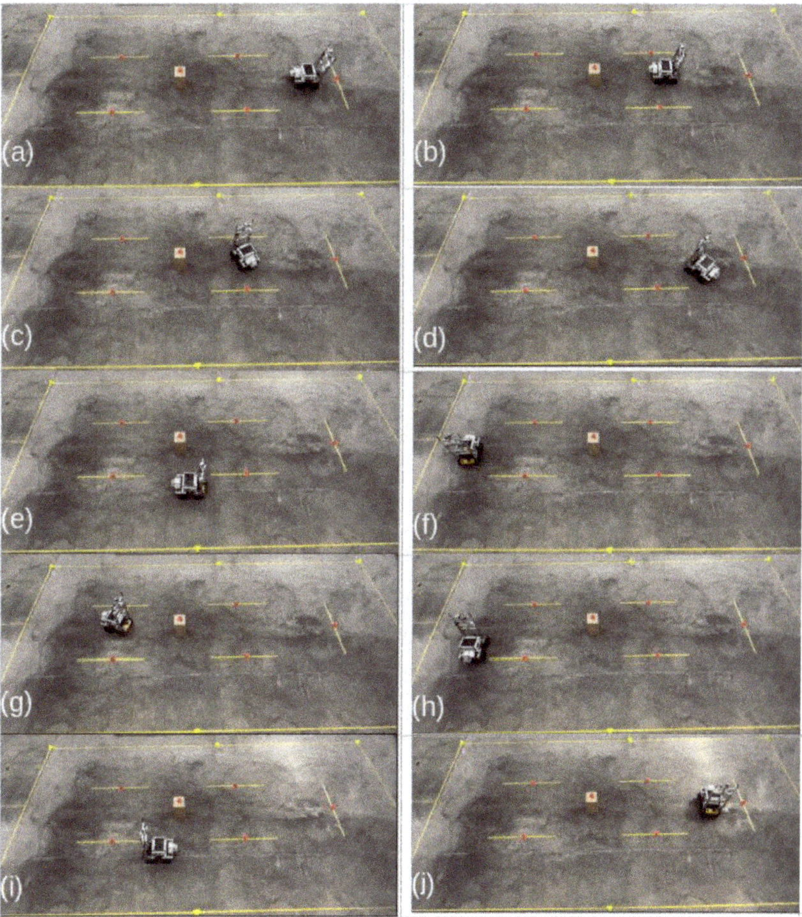

Figure 15. From (**a**) start to (**j**) end: sequence of screenshots of the simulated orchard environment with obstacles (*crop rows*) showing the navigation test of the robot prototype in the physical environment, with the fixed obstacles represented by yellow lines, and the dynamic obstacle, not present in the predefined map, represented by the box

Despite the overall good experimental results, the system may fail in certain circumstances. The perception module, for example, may fail detecting objects that have a size much smaller or much bigger than that of obstacles assumed during parameter tuning. This is due to the fact that when defining a cluster, the algorithm uses a *min* and *max* threshold value to constraint the number of points a cluster can contain. Additionally, because the planning module works based on the grid cells, increasing the cell size will reduce the system's accuracy. In the same way, the guidance model may fail to control the robot if we set a very small value for the acceptance radius. In this scenario, due to the noisy data from the localization sensors, the robot may never reach the goal area, and thus it would not switch to the next goal point.

4. Conclusions

Here, the conceptual design and experimental implementation of a modular mobile robotic system for agricultural field tasks is presented. Based on the functional design concept *HW* and *SW* modules are conceived, developed and integrated on a Husky unmanned ground vehicle. In particular, the software modules are implemented using *ROS*, following a high and low-level multi-layered approach, and allow the use of different sensors for localization and obstacle detection. Localization is implemented using a filtered-odometry method that can be integrated with indoor or outdoor positioning systems. Both an RGB-D camera and a 3D LiDAR sensor are configured for use on the platform, and either can be employed for obstacle avoidance. An extended Boundary Node Method, which has been specifically adapted for this work, is used for path planning. A modified Lookahead-based Line of Sight guidance algorithm is utilized for path following. Preliminary field tests in an emulated orchard scenario demonstrate that the system can perform path-following with a suitable accuracy, and obstacle detection, with path re-planning when needed. Planned future work will include intensive experimental testing, detailed data analyses, as well as performance assessments of the system's capabilities of executing the three *PA* scenarios (monitoring, harvesting, and spraying) in both emulated and real environments.

Author Contributions: Conceptualization and methodology, R.A.S., G.T., K.D.v.E. and R.V.; software, validation, and formal analysis, R.A.S. and G.T.; resources, K.D.v.E. and R.V.; writing—original draft preparation, R.A.S., G.T. and R.V.; writing—review and editing, all; supervision, G.C., K.D.v.E. and R.V.; project administration and funding acquisition, K.D.v.E. and R.V. All authors have read and agreed to the published version of the manuscript.

Funding: This research was supported in part by the "Reconfigurable Collaborative Agri-Robots (RE-COARO)" Südtirol/Alto Adige 4th Call (project #4122) and by the European Regional Development Fund (ERDF), FiRST Lab Project #FESR1084.

Institutional Review Board Statement: Not applicable.

Informed Consent Statement: Not applicable.

Data Availability Statement: Not applicable.

Acknowledgments: The authors are grateful to Matteo Malavasi, who designed and fabricated some of the electronic systems of the robot, and Josef Zelger, who designed and fabricated the mechanical superstructure.

Conflicts of Interest: The authors declare no conflict of interest.

References

1. Mondejar, M.E.; Avtar, R.; Diaz, H.L.B.; Dubey, R.K.; Esteban, J.; Gómez-Morales, A.; Hallam, B.; Mbungu, N.T.; Okolo, C.C.; Prasad, K.A.; et al. Digitalization to achieve sustainable development goals: Steps towards a Smart Green Planet. *Sci. Total Environ.* **2021**, *794*, 148539. [CrossRef] [PubMed]

2. Nasirahmadi, A.; Hensel, O. Toward the Next Generation of Digitalization in Agriculture Based on Digital Twin Paradigm. *Sensors* **2022**, *22*, 498. [CrossRef]

3. Monteiro, A.; Santos, S.; Gonçalves, P. Precision agriculture for crop and livestock farming—Brief review. *Animals* **2021**, *11*, 2345. [CrossRef]

4. Shafi, U.; Mumtaz, R.; García-Nieto, J.; Hassan, S.A.; Zaidi, S.A.R.; Iqbal, N. Precision agriculture techniques and practices: From considerations to applications. *Sensors* **2019**, *19*, 3796. [CrossRef] [PubMed]

5. Department, I.S. *World Robotics 2021—Service Robots*; VDMA Services GmbH: Frankfurt am Main, Germany, 2021.

6. Blender, T.; Buchner, T.; Fernandez, B.; Pichlmaier, B.; Schlegel, C. Managing a mobile agricultural robot swarm for a seeding task. In Proceedings of the IECON 2016-42nd Annual Conference of the IEEE Industrial Electronics Society, Florence, Italy, 23–26 October 2016; pp. 6879–6886.

7. Oberti, R.; Marchi, M.; Tirelli, P.; Calcante, A.; Iriti, M.; Tona, E.; Hočevar, M.; Baur, J.; Pfaff, J.; Schütz, C.; et al. Selective spraying of grapevines for disease control using a modular agricultural robot. *Biosyst. Eng.* **2016**, *146*, 203–215. [CrossRef]

8. Maini, P.; Gonultas, B.M.; Isler, V. Online coverage planning for an autonomous weed mowing robot with curvature constraints. *IEEE Robot. Autom. Lett.* **2022**, *7*, 5445–5452. [CrossRef]

9. McAllister, W.; Osipychev, D.; Davis, A.; Chowdhary, G. Agbots: Weeding a field with a team of autonomous robots. *Comput. Electron. Agric.* **2019**, *163*, 104827. [CrossRef]

10. Quan, L.; Jiang, W.; Li, H.; Li, H.; Wang, Q.; Chen, L. Intelligent intra-row robotic weeding system combining deep learning technology with a targeted weeding mode. *Biosyst. Eng.* **2022**, *216*, 13–31. [CrossRef]

11. Tinoco, V.; Silva, M.F.; Santos, F.N.; Rocha, L.F.; Magalhães, S.; Santos, L.C. A review of pruning and harvesting manipulators. In Proceedings of the 2021 IEEE International Conference on Autonomous Robot Systems and Competitions (ICARSC), Santa Maria da Feira, Portugal, 28–29 April 2021; pp. 155–160.

12. Botterill, T.; Paulin, S.; Green, R.; Williams, S.; Lin, J.; Saxton, V.; Mills, S.; Chen, X.; Corbett-Davies, S. A Robot System for Pruning Grape Vines. *J. Field Robot.* **2017**, *34*, 1100–1122. [CrossRef]

13. Kim, W.S.; Lee, D.H.; Kim, Y.J.; Kim, T.; Lee, W.S.; Choi, C.H. Stereo-vision-based crop height estimation for agricultural robots. *Comput. Electron. Agric.* **2021**, *181*, 105937. [CrossRef]

14. Vidoni, R.; Gallo, R.; Ristorto, G.; Carabin, G.; Mazzetto, F.; Scalera, L.; Gasparetto, A. Byelab: An agricultural mobile robot prototype for proximal sensing and precision farming. In Proceedings of the ASME International Mechanical Engineering Congress and Exposition, Proceedings (IMECE), Tampa, FL, USA, 3–9 November 2017; Volume 4A. [CrossRef]

15. Quaglia, G.; Visconte, C.; Scimmi, L.; Melchiorre, M.; Cavallone, P.; Pastorelli, S. Design of a UGV powered by solar energy for precision agriculture. *Robotics* **2020**, *9*, 13. [CrossRef]

16. Bac, C.W.; Van Henten, E.J.; Hemming, J.; Edan, Y. Harvesting robots for high-value crops: State-of-the-art review and challenges ahead. *J. Field Robot.* **2014**, *31*, 888–911. [CrossRef]

17. Moysiadis, V.; Sarigiannidis, P.; Vitsas, V.; Khelifi, A. Smart farming in Europe. *Comput. Sci. Rev.* **2021**, *39*, 100345. [CrossRef]

18. Oliveira, L.; Moreira, A.; Silva, M. Advances in agriculture robotics: A state-of-the-art review and challenges ahead. *Robotics* **2021**, *10*, 52. [CrossRef]

19. Kumar, A.; Deepak, R.S.; Kusuma, D.S.; Sreekanth, D. Review on multipurpose agriculture robot. *Int. J. Res. Appl. Sci. Eng. Technol.* **2020**, *8*, 1314–1318. [CrossRef]

20. Sowjanya, K.D.; Sindhu, R.; Parijatham, M.; Srikanth, K.; Bhargav, P. Multipurpose autonomous agricultural robot. In Proceedings of the 2017 International Conference of Electronics, Communication and Aerospace Technology (ICECA), Coimbatore, India, 20–22 April 2017; Volume 2, pp. 696–699.

21. Nandeesh, T.; M Kalpana, H. Smart Multipurpose Agricultural Robot. In Proceedings of the CONECCT 2021: 7th IEEE International Conference on Electronics, Computing and Communication Technologies, Bangalore, India, 9–11 July 2021. [CrossRef]

22. Tauze Zohora Saima, F.; Tamanna Tabassum, M.; Islam Talukder, T.; Hassan, F.; Sarkar, P.K.; Howlader, S. Advanced Solar Powered Multipurpose Agricultural Robot. In Proceedings of the 2022 3rd International Conference for Emerging Technology, INCET, Belgaum, India, 27–29 May 2022. [CrossRef]

23. Levin, M.; Degani, A. A conceptual framework and optimization for a task-based modular harvesting manipulator. *Comput. Electron. Agric.* **2019**, *166*, 104987. [CrossRef]

24. Le, A.V.; Arunmozhi, M.; Veerajagadheswar, P.; Ku, P.C.; Minh, T.H.Q.; Sivanantham, V.; Mohan, R.E. Complete path planning for a tetris-inspired self-reconfigurable robot by the genetic algorithm of the traveling salesman problem. *Electronics* **2018**, *7*, 344. [CrossRef]

25. Grimstad, L.; From, P.J. Thorvald II-a modular and re-configurable agricultural robot. *IFAC-PapersOnLine* **2017**, *50*, 4588–4593. [CrossRef]

26. Levin, M.; Degani, A. Design of a Task-Based Modular Re-Configurable Agricultural Robot. *IFAC-PapersOnLine* **2016**, *49*, 184–189. [CrossRef]

27. Denis, D.; Thuilot, B.; Lenain, R. Online adaptive observer for rollover avoidance of reconfigurable agricultural vehicles. *Comput. Electron. Agric.* **2016**, *126*, 32–43. [CrossRef]

28. Youchun, Z.; Gongyong, Z. Design of Multimodal Neural Network Control System for Mechanically Driven Reconfigurable Robot. *Comput. Intell. Neurosci.* **2022**, *2022*, 2447263. [CrossRef] [PubMed]

29. Xu, R.; Li, C. A modular agricultural robotic system (MARS) for precision farming: Concept and implementation. *J. Field Robot.* **2022**, *39*, 387–409. [CrossRef]

30. Lytridis, C.; Kaburlasos, V.G.; Pachidis, T.; Manios, M.; Vrochidou, E.; Kalampokas, T.; Chatzistamatis, S. An Overview of Cooperative Robotics in Agriculture. *Agronomy* **2021**, *11*, 1818. [CrossRef]

31. GARotics. Green Asparagus Harvesting Robotic System. Available online: http://echord.eu/garotics (accessed on 27 July 2022).

32. Albani, D.; IJsselmuiden, J.; Haken, R.; Trianni, V. Monitoring and mapping with robot swarms for agricultural applications. In Proceedings of the 2017 14th IEEE International Conference on Advanced Video and Signal Based Surveillance (AVSS), Lecce, Italy, 29 August–1 September 2017; pp. 1–6.

33. Reisch, B.I.; Owens, C.L.; Cousins, P.S. Grape. In *Fruit Breeding*; Springer: Berlin/Heidelberg, Germany, 2012; pp. 225–262.

34. Fernandez, R.; Montes, H.; Surdilovic, J.; Surdilovic, D.; Gonzalez-De-Santos, P.; Armada, M. Automatic detection of field-grown cucumbers for robotic harvesting. *IEEE Access* **2018**, *6*, 35512–35527. [CrossRef]

35. FiRST-Lab. Field Robotics South-Tyrol Lab. Available online: https://firstlab.projects.unibz.it/ (accessed on 24 July 2022).

36. Robotics, C. Husky Technical Specifications. Available online: https://www.clearpathrobotics.com/husky-unmanned-ground-vehicle-robot/ (accessed on 24 July 2022).

37. Universal Robots. Available online: https://www.universal-robots.com/ (accessed on 31 August 2022).

38. ArduSimple. Available online: https://www.ardusimple.com/ (accessed on 31 August 2022).

39. Pozyx. Available online: https://www.pozyx.io/ (accessed on 31 August 2022).

40. Fischler, M.A.; Bolles, R.C. Random sample consensus: A paradigm for model fitting with applications to image analysis and automated cartography. *Commun. ACM* **1981**, *24*, 381–395. [CrossRef]

41. Ester, M.; Kriegel, H.P.; Sander, J.; Xu, X. A density-based algorithm for discovering clusters in large spatial databases with noise. In Proceedings of the KDD, Portland, OR, USA, 2–4 August 1996; Volume 96, pp. 226–231.

42. Saeed, R.; Reforgiato Recupero, D.; Remagnino, P. The boundary node method for multi-robot multi-goal path planning problems. *Expert Syst.* **2021**, *38*, e12691. [CrossRef]

43. Saeed, R.; Recupero, D.; Remagnino, P. A Boundary Node Method for path planning of mobile robots. *Robot. Auton. Syst.* **2020**, *123*, 103320. [CrossRef]

44. Saeed, R.; Recupero, D. Path planning of a mobile robot in grid space using boundary node method. In Proceedings of the ICINCO 2019—16th International Conference on Informatics in Control, Automation and Robotics, Prague, Czech Republic, 29–31 July 2019; Volume 2, pp. 159–166. [CrossRef]

45. Lekkas, A.M.; Fossen, T.I. A time-varying lookahead distance guidance law for path following. *IFAC Proc. Vol.* **2012**, *45*, 398–403. [CrossRef]

46. Wang, X.; Wu, G. Modified LOS path following strategy of a portable modular AUV based on lateral movement. *J. Mar. Sci. Eng.* **2020**, *8*, 683. [CrossRef]

47. Ahn, J.; Shin, S.; Kim, M.; Park, J. Accurate Path Tracking by Adjusting Look-Ahead Point in Pure Pursuit Method. *Int. J. Automot. Technol.* **2021**, *22*, 119–129. [CrossRef]

48. von Ellenrieder, K.; Licht, S.; Belotti, R.; Henninger, H. Shared human–robot path following control of an unmanned ground vehicle. *Mechatronics* **2022**, *83*, 102750. [CrossRef]

49. Du, F.; Deng, W.; Yang, M.; Wang, H.; Mao, R.; Shao, J.; Fan, J.; Chen, Y.; Fu, Y.; Li, C.; et al. Protecting grapevines from rainfall in rainy conditions reduces disease severity and enhances profitability. *Crop Prot.* **2015**, *67*, 261–268. [CrossRef]

Article

An Intelligent Predictive Algorithm for the Anti-Rollover Prevention of Heavy Vehicles for Off-Road Applications

Antonio Tota [1,*], Luca Dimauro [1], Filippo Velardocchia [2], Genny Paciullo [3] and Mauro Velardocchia [1]

[1] Department of Mechanical and Aerospace Engineering, Politecnico di Torino, 10129 Torino, TO, Italy
[2] Department of Management and Production Engineering, Politecnico di Torino, 10129 Torino, TO, Italy
[3] Sezione Mobilità e Contromobilità—Caporeparto Mobilità, Centro Polifunzionale di Sperimentazione, 00010 Montelibretti, RM, Italy
* Correspondence: antonio.tota@polito.it

Abstract: Rollover detection and prevention are among the most critical aspects affecting the stability and safety assessment of heavy vehicles, especially for off-road driving applications. This topic has been studied in the past and analyzed in depth in terms of vehicle modelling and control algorithms design able to prevent the rollover risk. However, it still represents a serious problem for automotive carmakers due to the huge counts among the main causes for traffic accidents. The risk also becomes more challenging to predict for off-road heavy vehicles, for which the incipient rollover might be triggered by external factors, i.e., road irregularities, bank angles as well as by aggressive input from the driver. The recent advances in road profile measurement and estimation systems make road-preview-based algorithms a viable solution for the rollover detection. This paper describes a model-based formulation to analytically evaluate the load transfer dynamics and its variation due to the presence of road perturbations, i.e., road bank angle and irregularities. An algorithm to detect and predict the rollover risk for heavy vehicles is also presented, even in presence of irregular road profiles, with the calculation of the ISO-LTR Predictive Time through the Phase-Plane analysis. Furthermore, the artificial intelligence techniques, based on the recurrent neural network approach, is also presented as a preliminary solution for a realistic implementation of the methodology. The paper finally assess the efficacy of the proposed rollover predictive algorithm by providing numerical results from the simulation of the most severe maneuvers in realistic off-road driving scenarios, also demonstrating its promising predictive capabilities.

Keywords: rollover detection; heavy vehicles; off-road applications; predictive algorithms; artificial intelligence; load transfer ratio

Citation: Tota, A.; Dimauro, L.; Velardocchia, F.; Paciullo, G.; Velardocchia, M. An Intelligent Predictive Algorithm for the Anti-Rollover Prevention of Heavy Vehicles for Off-Road Applications. *Machines* **2022**, *10*, 835. https://doi.org/10.3390/machines10100835

Academic Editors: Marco Ceccarelli, Giuseppe Carbone and Alessandro Gasparetto

Received: 31 July 2022
Accepted: 8 September 2022
Published: 21 September 2022

Publisher's Note: MDPI stays neutral with regard to jurisdictional claims in published maps and institutional affiliations.

1. Introduction

Nowadays, vehicle rollover detection and prevention are two critical aspects that must be taken into account for the safety of car passengers and pedestrians to avoid dramatic fatal crashes [1] and accidents in the urban scenario. An analysis of the sequence critical events leading to loss of control, and hence to rollover of vehicles equipped with Electronic Stability Control (ESC) Systems is provided in [2], while typical scenarios and characteristics of rollover accidents are investigated in [3], referring to the test criteria of National Highway Traffic Safety Administration (NHTSA) [1,4]. Vehicles with a high position of the Centre of Gravity (CoG) are more prone to rollover, as it happens for buses [5,6], heavy commercial vehicles [7,8], and articulated heavy vehicles [9–14]. In a similar way, two-wheel vehicles are affected by stability issues [15,16]; hence, their components must be dynamically optimised [17], through a component to assembly dynamic analysis [18,19]. The rollover prevention is a typical challenge for the automotive sector, especially for fuel cell and hydrogen trucks, where the safety requirements demanded for their storage systems are stricter than for light-duty vehicle segments [20].

The root cause of rollover can be addressed to external factors, i.e., road irregularities and its bank angle, which also influence the powertrain management [21], and to aggressive input from the driver on the steering wheel. Considering these factors, rollovers are sometimes categorized, as conducted in [22–24], into two main types: un-tripped rollovers and tripped rollovers. The former [25] is related to driver fast maneuvering on smooth roads, while the latter [26] occurs due to sudden impacts that may apply lateral or vertical forces to the vehicle tyres [27,28], hence possible root causes are impacts with guardrails or hits with road objects as curbs and bumps [29]. In the last decades many researchers have studied the vehicle rollover risk, whose knowledge is mandatory for developing practical rollover prevention systems, which may be implemented using different control logic, through a two-stages process. Indeed, the first step is the detection of rollover risk, whereas the second is its mitigation and control, e.g., through an active handling suspension controller [30] or by means of a Model Predictive Control (MPC) strategy through active front steering [31], or using a braking system for vehicle dynamics control [32]. Several rollover indices (RIs) have been proposed in literature, starting from static rollover index, as the Static Stability Factor (SSF) [33–35], which only considers the geometrical parameters of vehicles. The vehicle dynamics [36,37] plays a key role in the rollover risk phenomena, and hence also for the indicators adopted to prevent it, which consider the most important vehicle states for rollover, such as roll angle and roll rate [38–40], lateral acceleration [41–44], side slip angle [45], yaw angle [46] and yaw rate [47,48]. The load transfer ratio (LTR) [49–51], which is based on the computation of vertical tire forces, is one of the widely used RIs for dynamic simulations, and shows a direct measure of how close the vehicle is to rollover [49]. A predictive LTR [52] has been proposed to provide a better prediction of rollover propensity w.r.t traditional LTR, while in [53] the contour line of load transfer ratio (CL-LTR) is proposed for an accurate prediction of vehicle rollover threat. Moreover, time to rollover (TTR) [39,43] and the rollover index (RI) [38,54] are used widely for the rollover risk detection. At a glance, many RIs have been proposed in literature, with slight differences on the factors affecting the rollover, thus some RIs are more adapt for specific rollover conditions and not work well in other situations. For example, an extensive research investigation is focused on the untripped rollover risks and only a minimal attention is reserved to tripped rollover conditions which represent the most critical aspect for off-road and autonomous driving applications where the disturbance from the external environment play an important role [55]. Additionally, the artificial intelligence (AI) techniques are also proposed in literature [56,57] to obtain better anti-rollover prevention features. In particular, [58] has identified the Recurrent Neural Networks (RNN) as potential rollover risk-detector algorithm. However, by considering time and costs constraints [59], the big experimental data required to generate an efficient algorithm is not always possible.

Based on the widespread and in-depth bibliographic research found in the literature, to the knowledge of the authors, there are still many open questions on the rollover prevention assessment that the present activity attempts to solve. This paper aims to cover some relevant aspects in the the detection and prevention of the incipient rollover, with the following contributions:

- The development of a non-linear three degrees-of-freedom mathematical model as simple as effective to catch the lateral load transfer dynamics even in presence of a banked road;
- The formulation of a model-based algorithm for the analytical identification of the critical rollover limits through the development of characteristic maps in the phase plane portrait, able to exhibit the influence of road local irregularities and global geometric factors, i.e., the bank angle;
- The formulation of a statistical algorithm, based on the recurrent neural network approach, for the estimation of the load transfer ratio in a realistic scenario, by considering typical measurable quantities available for an experimental implementation;

- The numerical assessment of a real-time algorithm able to predict in advance the time to reach a specific load transfer ratio considered as the incipient rollover limit.

It is also important to note that one of the most important advantages related to the methodology described in the paper is the versatility of application to different powertrain and vehicle architectures, by providing an algorithm that requires the typical measurements available onboard the vehicle.

The manuscript is organised as follows: Section 2 describes the set of equations used for modelling the non-linear vehicle roll dynamics; Section 3 presents the two algorithms adopted for the load transfer ratio estimation; the the phase plane analysis of load transfer characteristics is carried out in Section 4; the ISO-LTR predictive time is introduced in Section 5 and its simulation assessment is shown in Section 6; finally, some conclusions are drawn in Section 7.

2. Vehicle Model Description

The evaluation of the rollover condition, together with the definition of a predictive index to prevent the incipient risk, is firstly carried out by investigating the influence of the most relevant quantities on the load transfer dynamics. A straightforward methodology is proposed with a three degrees-of-freedom (dofs) vehicle model to obtain an analytical and a straightforward approach for analysing the vehicle roll dynamics on flat and banked roads. The 3-dofs of the vehicle model, whose scheme is reported in Figure 1, are the vertical and the lateral vehicle motions (with respect to the road plane) and the relative roll motion between the sprung and the unsprung masses. The virtual roll axis is R, G_s is the CoG of the sprung mass and G_{u_i} is the CoG of the $i_{th} = FL, FR, RL, RR$ unsprung mass. The hypothesis behind the model are:

1. All the bodies below the suspension system (tires, calipers, wheels carriers, suspension rods, etc. …) are considered as a unique rigid body, connected to the sprung mass m_s through the virtual roll axis R, and represented by four lumped masses m_{u_i}, where $i = FL, FR, RL$ and RR, each one placed in the the the Front Left (FL), Front Right (FR), Rear Left (RL) and Rear Right (RR) wheel rotational centers, respectively;
2. The roll moment of inertia of the unsprung mass is considered negligible;
3. The road bank angle ϕ_R is supposed to be equal for the front and the rear axles: absolute roll angle of the unsprung mass is ϕ_R;
4. The front and rear suspensions are represented as an equivalent torsional spring and damper system.

From Figure 1a, the following set of equilibrium equations for the whole vehicle is drawn :

$$
\begin{cases}
F_{z_R} + F_{z_L} - m_s a_{z_s} - \sum_i \left(m_{u_i} a_{z_{u_i}} \right) - mg \cos \phi_R = 0 \\
F_{y_R} + F_{y_L} - m_s a_{y_s} - \sum_i \left(m_{u_i} a_{y_{u_i}} \right) - mg \sin \phi_R = 0 \\
(F_{z_R} - F_{z_L})\frac{T}{2} - m_s a_{y_s}(h_R + h_s \cos \phi) - m_s g(h_R \sin \phi_R + h_s \sin (\phi_R + \phi)) + \\
-m_s a_{z_s} h_s \sin \phi - \sum_i \left(m_{u_i} a_{y_{u_i}} h_u \right) - m_u g h_u \sin \phi_R - G - I_s(\ddot{\phi}_R + \ddot{\phi}) = 0
\end{cases}
\tag{1}
$$

where $F_{z_R} = F_{z_{FR}} + F_{z_{RR}}$ and $F_{z_L} = F_{z_{FL}} + F_{z_{RL}}$ are the total vertical forces on the right and left tyres, respectively. $F_{y_R} = F_{y_{FR}} + F_{y_{RR}}$ and $F_{y_L} = F_{y_{FL}} + F_{y_{RL}}$ are the right and left side total lateral forces, respectively. m_s is the sprung mass, m_{u_i} is the $i_{th} = FL, FR, RL, RR$ lumped unsprung mass ($m_u = \sum_i m_{u_i}$) and $m = m_s + m_u$ is the vehicle total mass. a_{z_s} and $a_{z_{u_i}}$ are the vertical components of the sprung and i_{th} unsprung mass, respectively. a_{y_s} and $a_{y_{u_i}}$ are the lateral components of the sprung and i_{th} unsprung mass, respectivelly. Moreover, the two angle ϕ_R and ϕ are the bank angle and the roll angle of the sprung mass w.r.t. the unsprung mass. The absolute sprung mass roll angle is $\phi_A = \phi_R + \phi$. h_u and h_R are the heights of the unsprung mass and roll centre from the road, respectively, and h_s is the distance between the sprung mass CoG from the roll centre. g is the gravity acceleration,

T is the vehicle track width and I_s is the roll sprung mass moment of inertia. R_x and R_y represent the horizontal and vertical components of the internal reaction in the virtual roll axis, respectively.

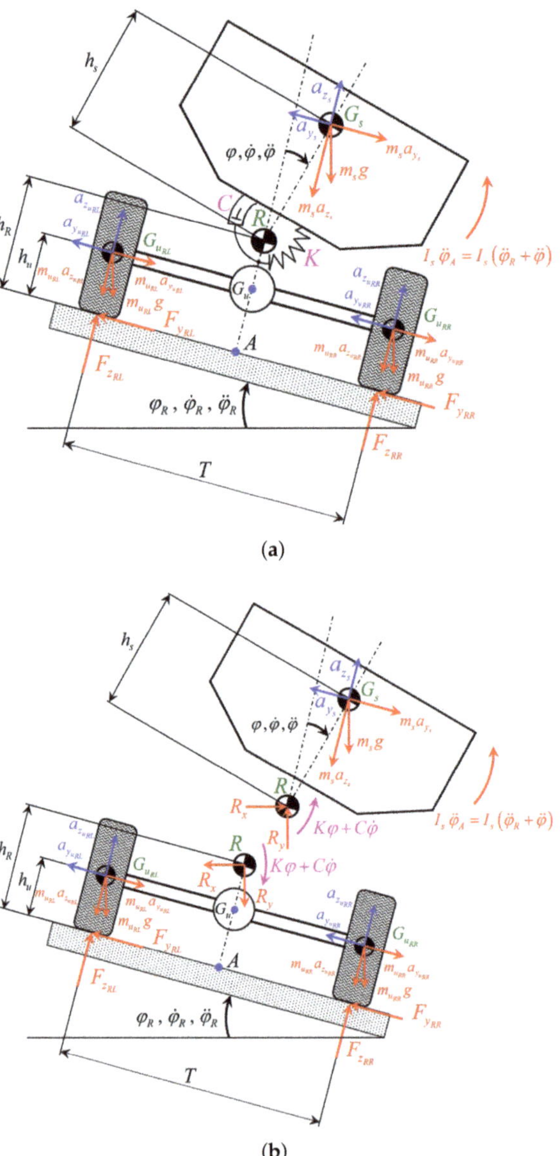

Figure 1. Free body diagrams of: whole vehicle (**a**), sprung and unsprung masses (**b**).

The term G groups together the relative vertical dynamics between the left and right unsprung masses and it is defined as reported in Equation (2):

$$G = \frac{T}{2}\left(m_{u_{FR}}a_{z_{FR}} + m_{u_{FL}}a_{z_{FL}} - m_{u_{RR}}a_{z_{RR}} - m_{u_{RL}}a_{z_{RL}}\right) \tag{2}$$

The roll equilibrium equation for the vehicle unspung mass from Figure 1b is:

$$\left(F_{z_R} - F_{z_L}\right)\frac{T}{2} =$$

$$= \left(F_{y_R} + F_{y_L}\right)h_R - \left[\sum_i \left(m_{u_i} a_{y_{u_i}}\right) + m_u g \sin\phi_R\right](h_R - h_u) + K\phi + C\dot{\phi} - G \quad (3)$$

where K and C are the roll stiffness and damping, respectively. Finally, roll equilibrium equation for the vehicle sprung mass from Figure 1b is:

$$I_s(\ddot{\phi}_R + \ddot{\phi}) + K\phi + C\dot{\phi} = m_s a_{y_s} h_s \cos\phi + m_s a_{z_s} h_s \sin\phi + m_s g h_s \sin\phi \quad (4)$$

The heavy vehicle analysed within the present paper is defined by the parameters listed in Table 1.

Table 1. Vehicle parameters.

Type	Description	Parameter	Value
Springs	Total roll stiffness	K	$209,000$ Nm/rad
Dampers	Total roll damping	C	6122.8 Nms/rad
Masses and inertia	Roll sprung mass moment of inertia	I_s	801.34 kgm^2
	Sprung mass	m_s	1923.9 kg
	FR unsprung mass	$m_{u_{FR}}$	78.715 kg
	FL unsprung mass	$m_{u_{FL}}$	78.715 kg
	RR unsprung mass	$m_{u_{RR}}$	109.314 kg
	RL unsprung mass	$m_{u_{RL}}$	109.314 kg
	Total mass	m	2300 kg
Distances	Front axle from the sprung mass CoG	a	2.119 m
	Rear axle from the sprung mass CoG	b	2.221 m
	Wheelbase	L	4.34 m
	Track width	T	1.674 m
	Unsprung mass CoG height	h_u	0.324 m
	Roll centre height	h_R	0.1998 m
	Sprung mass CoG from the roll centre	h_s	1.0852 m
	Front tyre radius	r_F	0.324 m
	Rear tyre radius	r_R	0.324 m

3. Load Transfer Ratio Estimation

The most accurate, and widely disseminated, index for detecting the incipient rollover risk is the Load Transfer Ratio (LTR) defined as the the relative vertical force on tires between the right and left sides of a vehicle:

$$LTR = \frac{F_{z_R} - F_{z_L}}{F_{z_R} + F_{z_L}} \quad (5)$$

When the LTR is equal to 0, the vertical load is equally distributed between the two sides, by means that the vehicle is far away from the rollover condition. The vehicle is very likely to rollover when the wheels on one side lift off the ground, condition verified when $LTR = \pm 1$.

However, the experimental acquisition of LTR is not feasible to realize since there are no sensors able to provide a direct measurement of the vertical tire load. For this reason, many authors have proposed multiple solutions to provide an accurate estimation of the LTR from the measurements commonly available onboard passenger cars, e.g., accelerations, linear and angular velocities, roll and pitch angles. Simple static [34,35], and dynamic indices have been used, in different works analysed in [24], to detect rollover risk during dynamic manoeuvres. The LTR proposed in [60] only considers the lateral acceleration and the roll angle, while [61,62] neglects the unsprung masses, by placing the roll centre very close to the road plane at $h_R \approx 0$ and assuming the vehicle CoG height as the effective arm for lateral acceleration. Other proposed LTR indices are based on suspension parameters [50,63], by considering the effective torsional stiffness, torsional damping, roll angle and roll rate, and subsequently revised in [64] with the inclusion of the sprung mass lateral acceleration. RIs based on the tyre deflection is proposed in [65,66], also adapted for heavy vehicles. A more complex index, which considers the rolling motion, the lateral acceleration, and the time to wheel lift (TTWL) is proposed in [38], where the TTWL is achieved with a phase plane analysis of roll angle and roll rate. Although a RI for banked roads is already developed in [67], the road bank angle influence receives a lower attention from the literature since its effect is not as relevant as for off-road designed vehicles. Finally the index proposed in [22] tries to to consider both tripped and untripped rollovers, by developing a four dofs vehicle model.

However, most of the above mentioned formulations does not include the influence of road global, i.e., bank angle, and local, i.e., speed bumps or potholes, irregularities on the load transfer among the four wheels, which represents a fundamental aspect for the off-road driving scenario. The present section of the paper describes a model-based and a statistic formulation for the LTR estimation. The first approach aims at providing an analytical methodology to express the LTR as function of the vehicle states and road profile, able to predict their influences on the LTR dynamics. However, the accuracy of the model-based estimation is drastically affected by the model parameters uncertainties, a drawback that is improved by proposing an alternative approach based on the Recurrent Neural-Network (RNN) theory, thus allowing an improved estimation of the LTR directly from conventional measurements.

3.1. Model-Based Estimation

The LTR model-based estimation is obtained from the 3-dofs model presented in the previous section. By combining the roll equilibrium equations of Equations (1) and (4) and by considering the vertical equilibrium equation in Equation (1), the LTR is then calculated as follows:

$$LTR_{est} = \frac{2}{T} \frac{K\phi + C\dot{\phi} + m_s a_{y_s} h_R + \sum_i \left(m_{u_i} a_{y_{u_i}} h_u \right) + (m_s g h_R + m_u g h_u) \sin \phi_R - G}{mg \cos \phi_R + m_s a_{z_s} + \sum_i \left(m_{u_i} a_{z_{u_i}} \right)} \tag{6}$$

For a direct comparison against the LTR formulation commonly available in the literature, the particular cases of a) negligible unsprung dynamics and vertical accelerations ($m_{u_i} = 0$ and $a_{z_s} = a_{z_i} = 0$) and b) negligible unsprung dynamics and vertical accelerations on a flat road ($m_{u_i} = 0$, $a_{z_s} = a_{z_i} = 0$ and $\phi_R = 0$) are also reported in Equation (7) and in Equation (8):

$$LTR_{est,sprung} = \frac{2}{T} \frac{K\phi + C\dot{\phi} + m_s a_{y_s} h_R + m_s g h_R \sin \phi_R}{mg \cos \phi_R} \tag{7}$$

$$LTR_{est,flat} = \frac{2}{T} \frac{K\phi + C\dot{\phi} + m_s a_{y_s} h_R}{mg} \tag{8}$$

Three driving scenarios, shown in Figure 2, are built in IPG CarMaker® to simulate the realistic behavior of a heavy duty vehicle and to compare the efficacy and the reliability of the LTR formulations in Equations (6)–(8):

(a)   A double lane change manouvre (ISO 3888-1) on a flat road at an initial vehicle speed of 100 km/h;

(b)   A straight manoeuvre on a banked road, whose bank angle is smoothly increased from 0 deg to 30 deg, at constant vehicle speed (30 km/h);

(c)   A straight manoeuvre on a road with a flat surface under the left vehicle side and an asymmetrical sinusoidal profile (wavelength equal to the vehicle wheelbase and height amplitude of 10 cm) under the right vehicle side, called asymmetrical waves road in the rest of the paper.

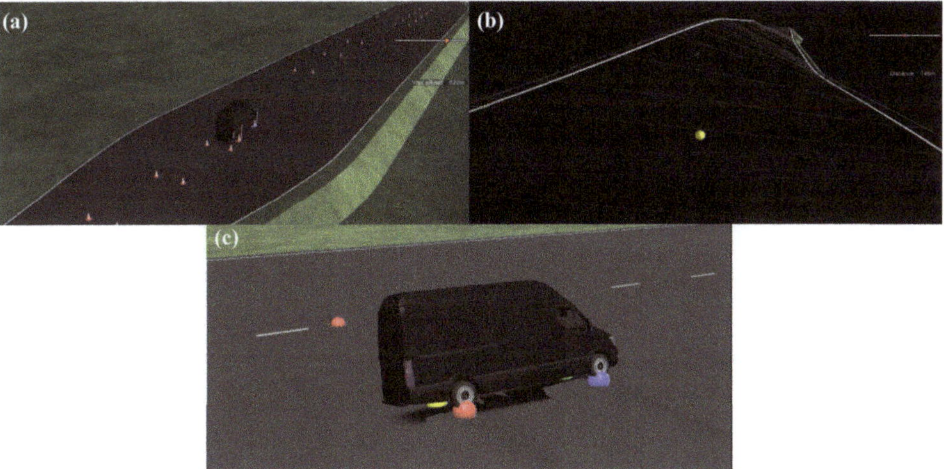

Figure 2. Driving scenarios created in IPG CarMaker® for the *LTR* estimation: double lane change manoeuvre on a flat road (**a**) , straight manoeuvre on a banked road (**b**) and straight manoeuvre on an asymmetrical waves road (**c**).

The first scenario is chosen to evaluate the influence of the maneuver severity, imposed by the driver behavior, on the *LTR* dynamics. The simulation results are reported in Figure 3, in terms of *LTR* estimation, driver commands, vehicle speeds, accelerations and roll angles.

Figure 3 shows that the manoeuvre is aggressive enough to push multiple times the *LTR* towards the rollover condition. It is clear that the lateral acceleration, imposed by the severity of the test and by the dynamic response of the driver, represents the main cause for the marked transfer loads from one vehicle side to the other one. Indeed, the road is flat and does not influence the roll dynamics meanwhile the unsprung mass only slightly affect the *LTR* dynamics, as shown by the comparison of LTR_{est} against $LTR_{est,sprung}$ and $LTR_{est,flat}$, due to the low frequency content of steering wheel angle imposed by the driver.

The second scenario aims at investigating the effect of the road bank angle on the vehicle transfer loads, as shown in the simulation results of Figure 4.

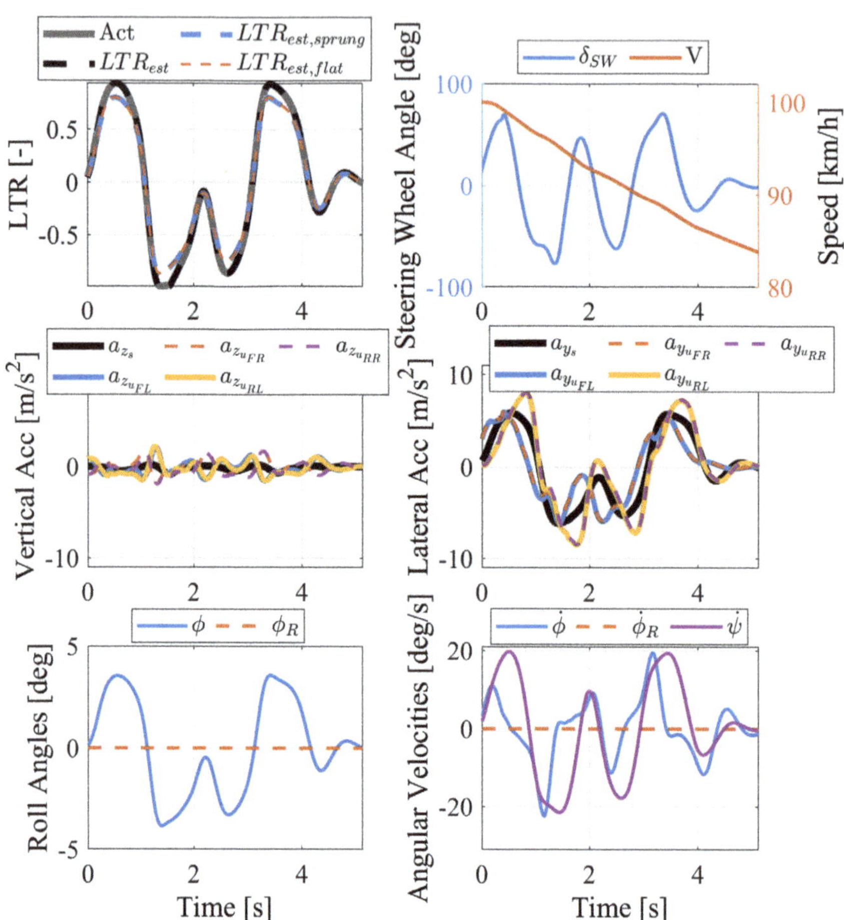

Figure 3. Double lane change manoeuvre at initial speed of 100 km/h on a flat road. From the top left side to the down right side: LTR, steering wheel angle and vehicle speed, vertical accelerations, lateral accelerations, roll angles and angular velocities.

The straight manoeuvre is selected to minimize the influence of lateral accelerations and to focus the attention on the bank angle effect on the load transfer dynamics. The road is banked up to 30 deg that represents a suitable threshold for approaching the rollover limit, identified by the condition $LTR = 1$. The general formulation of LTR_{est} is able to provide a perfect estimation of the actual LTR, meanwhile the basic formulation $LTR_{est,flat}$ completely underestimate the load transfer behavior since it is valid only for flat roads. The estimation $LTR_{est,sprung}$ is closer to the actual LTR behavior than the $LTR_{est,flat}$, but it shows a lower accuracy at high bank angles since the unsprung mass still play an important role in the roll dynamics, although m_u represents only the 16% of the total vehicle mass. Finally, the last driving scenario is designed to evaluate the influence of the unsprung mass on the LTR estimation, by exciting the vehicle roll dynamics with a sinusoidal road profile applied only to the right side of the vehicle. The manoeuvre on the asymmetrical waves road, with a wavelength equal to the vehicle wheelbase, is run at the constant speed of 40 km/h, in order to excite the higher frequency content of the unsprung mass vertical dynamics. The simulation results are shown in Figure 5.

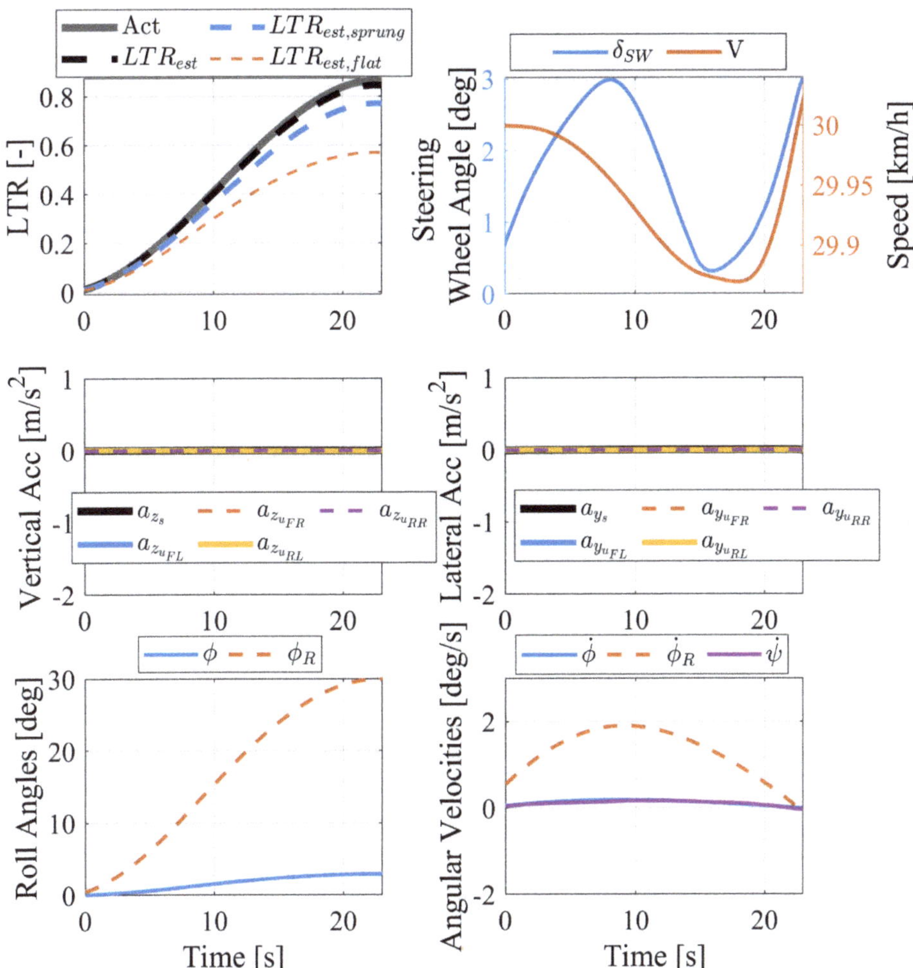

Figure 4. Straight manoeuvre at 30 km/h on a banked road. From the top left side to the down right side: *LTR*, steering wheel angle and vehicle speed, vertical accelerations, lateral accelerations, roll angles and angular velocities.

This manoeuvre can also evaluate the effect of a local road disturbances, i.e., road bump, pothole, etc. ..., which are typical conditions for the off-road applications and it has a different impact on the roll dynamics with respect to a global road geometry such us the presence of a bank angle. Indeed, the roll dynamics derived from the asymmetrical excitation induces a persistent condition of incipient rollover: during the small time frame of 1 s, the tires of the right vehicle side lift off three times. In this contest, the unsprung mass influence becomes much more important with respect the previous driving manoeuvres, and it is appreciable how the generic model-based estimation LTR_{est} is still reliable in detecting the LTR dynamics, if compared against the other two formulations.

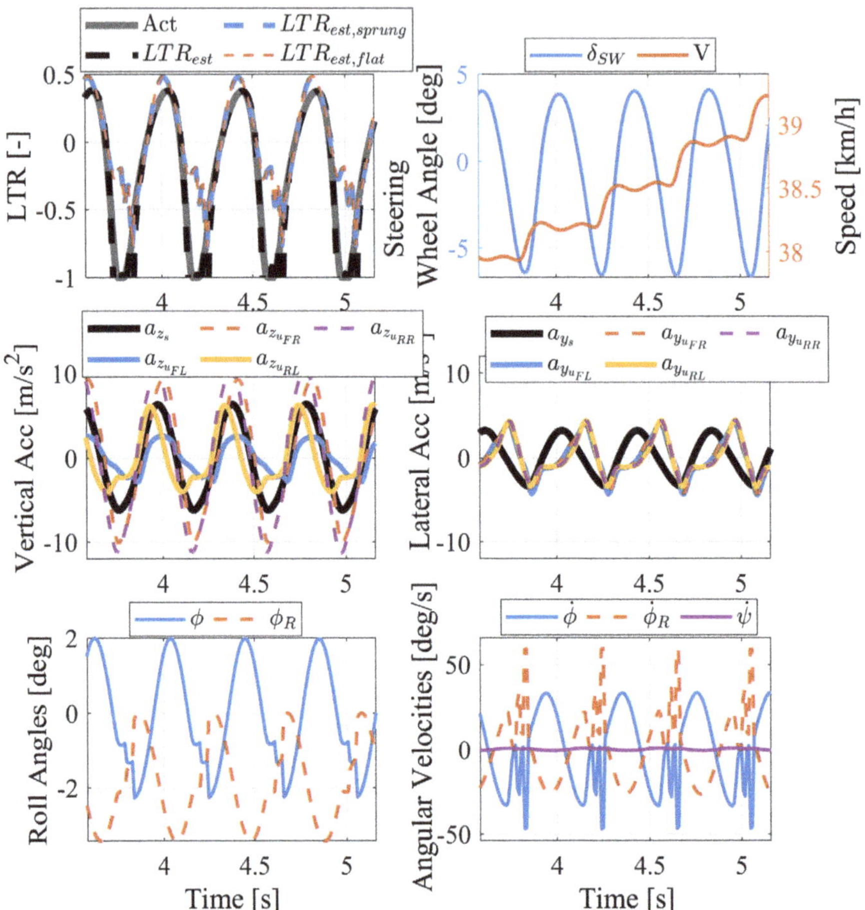

Figure 5. Straight manoeuvre at 40 km/h on an asymmetrical waves road. From the top left side to the down right side: *LTR*, steering wheel angle and vehicle speed, vertical accelerations, lateral accelerations, roll angles and angular velocities.

3.2. Recurrent Neural Network Estimation

As previously anticipated in Section 3, RNNs can be used to improve the estimation of rollover indicators, such as *LTR* and roll angle. The algorithm presented differentiates from the literature in terms of architecture, standardization method, inputs and outputs selection. In particular, the outputs are not defined as discretized quantities (higher or lower rollover risk factors), but they provide directly the estimation of the *LTR*. The RNNs method is chosen by considering the nature of the problem, since they belong to the category of artificial neural network designed to recognize patterns from a set of time histories.

The methodology adopted is characterized by two main phases. The first one consist in the generation of a large time domain dataset which is needed to successfully train the neural networks. Due to the big amount of required data, the time histories are generated by simulating the behaviour of the vehicle in the IPG CarMaker® environment. In particular, a set of 15 maneuvers (and related scenarios), are selected to induce the rollover risk. Each manoeuvre is 500 s long, to impose the same weight for the RNN training. In total, a starting dataset of 750, 000 points is obtained for each variable of interest, sampled at constant frequency of 100 Hz. The variables obtained from IPG CarMaker® used for the neural networks inputs and outputs are shown in Table 2.

Table 2. RNN: inputs and outputs.

Type	Description	Units
Inputs	Longitudinal speed	m/s
	Longitudinal acceleration	m/s^2
	Lateral acceleration	m/s^2
	Vertical acceleration	m/s^2
	roll rate	deg/s
	yaw rate	deg/s
	pitch rate	deg/s
	steer angle	deg
Outputs	Roll angle	deg
	Front load transfer ratio	—
	Rear load transfer ratio	—

The approach developed suggests an integration between the neural network algorithm with the simulation environment (IPG CarMaker®), which represents its data provider. This leads to the following practical implications:

- the RNN algorithm can be developed and tested with a considerable amount of simulated driving scenarios, without requiring an extensive experimental campaign, thus reducing time and costs;
- If the vehicle dynamic behavior is well described by the mathematical model, the neural network designed with a simulated data can be directly deployed on an experimental setup with a lower time and cost effort.

The second phase consisted in the training and testing of the AI algorithms, articulating the process in several steps with the Deep Learning Toolbox available in Matlab®. Firstly, a data pre-process and normalisation elaboration is carried out to obtain dimensionless quantities. The processed data is then adopted to elaborate the architecture of the RNN, by testing multiple solutions, leading to the final solution reported in Table 3.

Table 3. RNN: Architecture.

Type	Description	Characteristics
Layers	Sequence input	Number of features (8)
	LSTM	Number of hidden units (100)
	Fully connected	Number of Responses (3)
	Regression	
Main Hyperparameters	Adam	Adaptive moment estimation
	MaxEpochs	850
	GradientThreshold	1
	InitialLearnRate	0.01
	LearnRateDropPeriod	425
	LearnRateDropFactor	0.2

The features indicated in Table 3, corresponds to the inputs of the RNN, and so the responses for the outputs. In particular, regarding the RNN architecture selection, the number of hidden units (the neurons) is chosen through empirical rules. It has been observed that with more than 100 hidden units, the results do not improve in accuracy and the process becomes excessively time consuming, while a smaller number of neurons inevitably leads to a decline in terms of performance (measured by RMSE and loss function). The number and the typology of layers have been chosen following a similar criteria.

Given the RNN architecture, the algorithm training is ready to start by obtaining a neural network that, provided with the 8 inputs (Table 2), is able to estimate the 3 outputs with good precision and immediate response. This is also validated during the testing phase, where the features are known, and the vehicle responses not. The results related to three different maneuvers, generated in the IPG CarMaker® environment, are presented.

The first maneuver analyzed simulates the vehicle driving over the Bernina Pass. This is a completely unknown road to the RNN, and the results obtained, shown in Figure 6a,b are definitely promising in terms of accuracy (see the resultant Mean Absolute Error and Mean Square Error in Table 4). Moreover, the RNN estimates are collected in the lapse of 0.01–0.05 s. For the sake of brevity, results related to the roll angle have been omitted.

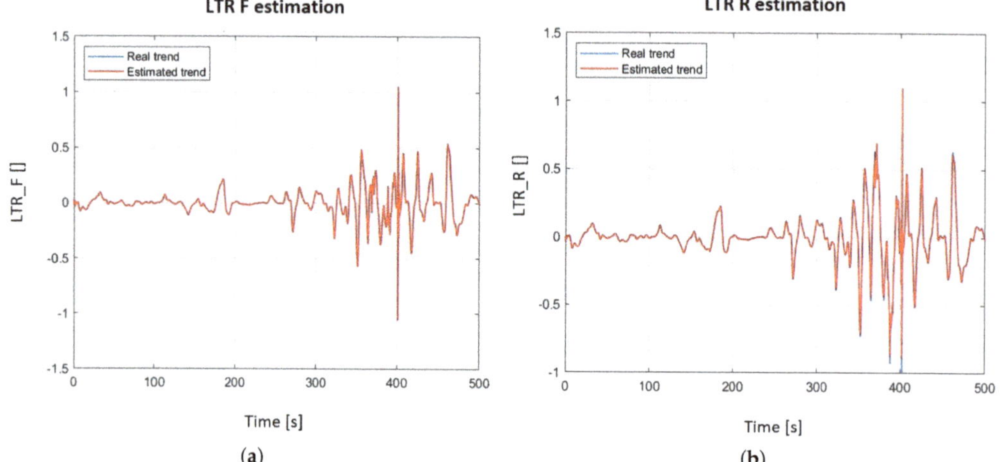

Figure 6. Estimation of the vehicle Load Transfer Ratio at front (**a**) and rear (**b**) axles driving over the Bernina Pass.

The second maneuver presented is obtained from one of the 15 database composing the training. The analysis aims at understanding if similar maneuvers to the ones already adopted for the RNN training, lead to an accuracy worsening, since the weights introduced in the RNN are influenced by the maneuvers set. Results in Figure 7a,b show how the RNN keep their high level of performance. There are only minimal differences that can be considered negligible for the required accuracy.

The Fishook manoeuvre, whose estimating results are shown in Figure 8a,b, produces the less promising results (as confirmed by Table 4). The reasons behind that is that the manoeuvre time length and speed profile are lower w.r.t. to the other manoeuvres introduced in the data set.

To provide a quantitative information related to the accuracy of the neural networks, the Mean Absolute Error (MAE) and the Mean Square Error (MSE) are reported for each manouvre in Table 4.

Table 4. Results: Mean Absolute Error and Mean Square Error related to the tested maneuvers.

Maneuver	Error	Values
	MAE LTR Front	0.0026
	MAE LTR Rear	0.0030
Bernina	MAE Roll	0.0500 deg
	MSE LTR Front	9.7767×10^{-5}
	MSE LTR Rear	1.1046×10^{-4}
	MSE Roll	0.0048

Table 4. *Cont.*

Maneuver	Error	Values
Stelvio	MAE *LTR* Front	0.0066
	MAE *LTR* Rear	0.0063
	MAE Roll	0.0802 deg
	MSE *LTR* Front	1.0434×10^{-4}
	MSE *LTR* Rear	8.8145×10^{-5}
	MSE Roll	0.0141
FishHook	MAE *LTR* Front	0.0146
	MAE *LTR* Rear	0.0138
	MAE Roll	0.106 deg
	MSE *LTR* Front	6.1685×10^{-4}
	MSE *LTR* Rear	5.3661×10^{-4}
	MSE Roll	0.0291

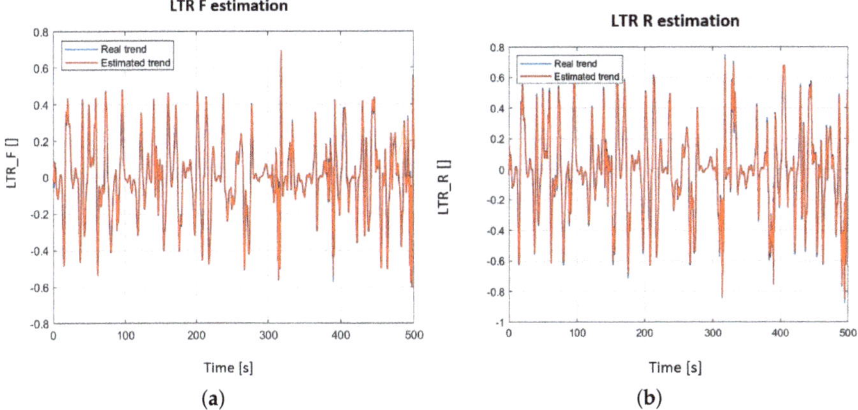

Figure 7. Estimation of the vehicle Load Transfer Ratio at front (**a**) and rear (**b**) axles driving over the Stelvio Pass.

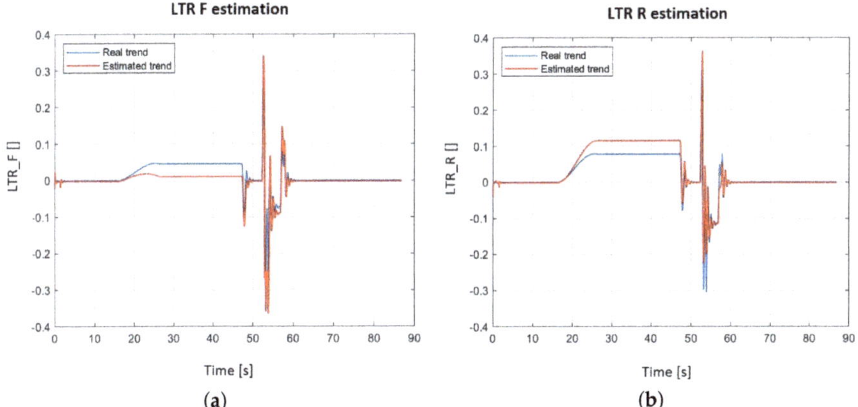

Figure 8. Estimation of the vehicle Load Transfer Ratio at front (**a**) and rear (**b**) axles during a FishHook maneuver.

4. ISO-LTR Phase Plane Portrait

As mentioned in the previous section, the model-based *LTR* formulation, expressed in its more robust and generic form by the Equation (6), is not suitable for a realistic implementation of the LTR estimation but rather it represents a valid tool for the analytical correlation between the vehicle states and the *LTR* dynamics. One of the most adopted tool to evaluate the vehicle states dynamics is the phase plane analysis. By focusing the attention only to the relative roll dynamics between the sprung and the unsprung masses, the vehicle states selected for plotting the phase plane map are the relative roll angle ϕ and roll rate $\dot{\phi}$ between the sprung and the unsprung masses. The phase plane analysis has the fundamental advantage to describe the trajectory of the vehicle states in a unique plot that can also be obtained with complex and non-linear models or directly from experimental measurements. In this paper, the phase-plane plot is drawn by running a fast ramp steering manoeuvre at constant speed in the IPG CarMaker® simulation environment. According the NHTSA's standard (49 CFR Part 575), a steering wheel rate of 720 deg/s is imposed, and the test is repeated at different vehicle speeds with step of 10 km/h. An example of simulation results obtained during a fast ramp steering manoeuvre at 70 km/h on a 10 deg banked road is shown in Figure 9.

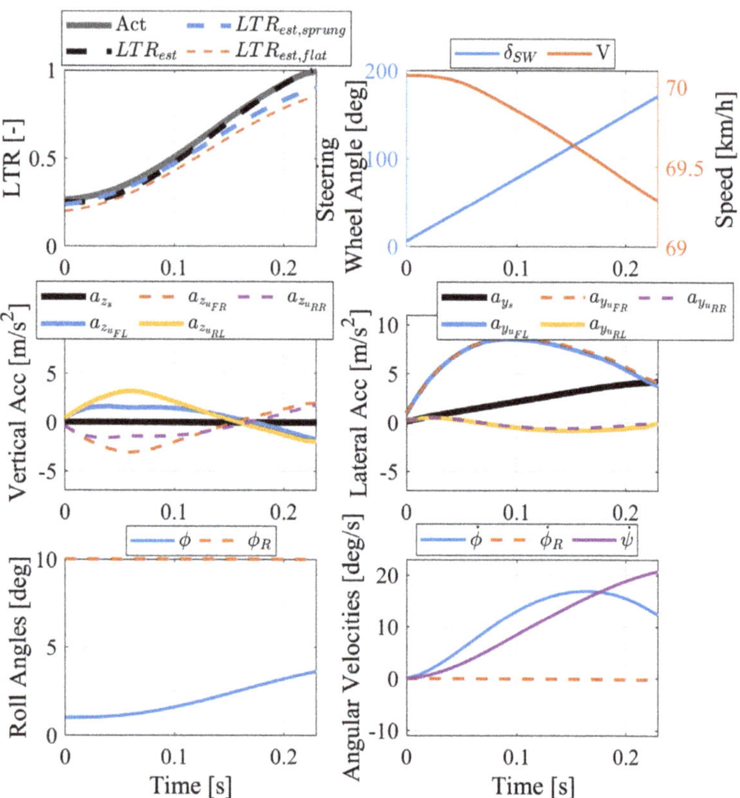

Figure 9. Fast ramp steering manoeuvre at 70 km/h on a 10 deg banked road. From the top left side to the down right side: *LTR*, steering wheel angle and vehicle speed, vertical accelerations, lateral accelerations, roll angles and angular velocities.

This aggressive manoeuvre is able to depict the transient behavior of the vehicle roll dynamics at a constant vehicle speed and road bank angle, through a trajectory line in the phase plane plot. The manoeuvre is repeated for different vehicle speeds, thus obtained the phase plane portrait for a predefined value of the road bank angle. An example of the phase plane portrait for a a 10 deg banked road is reported in Figure 10.

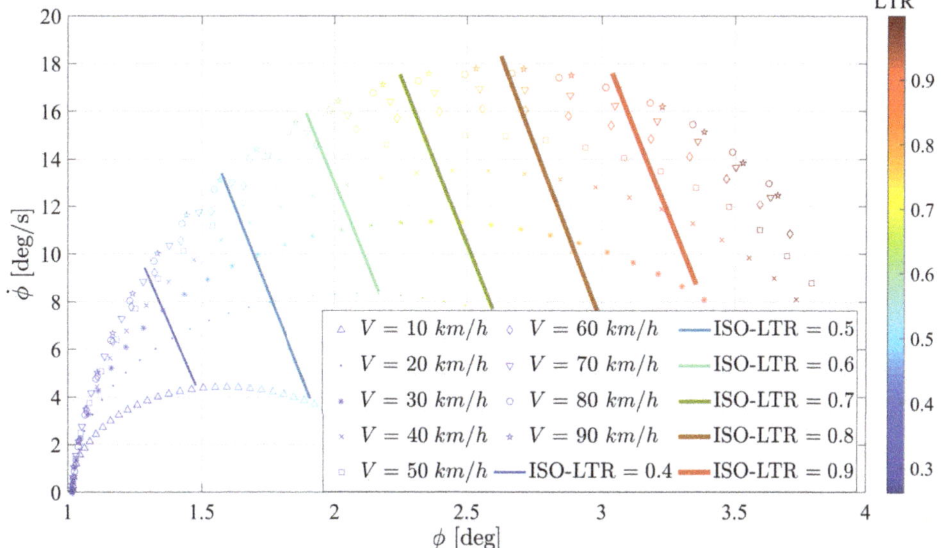

Figure 10. ISO-LTR characteristics obtained during a fast ramp steering manoeuvre on a 10 deg banked road at different speeds.

Each trajectory in the phase plane portrait represents the $\dot{\phi} - \phi$ correlation during a fast ramp steering manoeuvre at constant vehicle speed and include the steady-state and the transient responses of the vehicle roll dynamics. Each trajectory shows a colour variation as function of the corresponding LTR value, passing from a lower rollover risk represented by a blue markers ($LTR = 0$) to a high level of rollover risk with the red markers ($LTR = 1$) indicating that the left wheels lifted off.

When the phase plane portrait $(\dot{\phi}, \phi)$ is built and defined for a specific boundary condition, i.e., the road bank angle, the ISO-LTR characteristics, i.e., the locus of constant vertical load transfers ($LTR = q$, with $q = 0.4, 0.5, \ldots, 0.9$), can be drawn, as shown in Figure 10 for $\phi_R = 10$ deg. The ISO-LTR characteristics map is characterized by parallel lines, each corresponding to a constant LTR level.

By combining the vehicle roll equilibrium equation in Equation (1) with the roll equilibrium equation of the sprung mass in Equation (4), the following analytical expression of the linear ISO-LTR characteristics in the phase-pane is obtained:

$$
\dot{\phi} = -\frac{K}{C}\phi - \left[\frac{m_s a_{y_s} h_R + \sum_i \left(m_{u_i} a_{y u_i} h_u\right)}{C}\right] - \left[\frac{q\frac{T}{2} m_s a_{z_s} + q\frac{T}{2}\sum_i \left(m_{u_i} a_{z u_i}\right) - G}{C}\right] +
$$
$$
+ \left[\frac{q\frac{T}{2} mg \cos\phi_R - (m_s g h_R + m_u g h_u)\sin\phi_R}{C}\right]
\tag{9}
$$

The slope of the ISO-LTR linear characteristics is always negative and it depends on the vehicle suspension system parameters through the total roll stiffness K and roll damping C coefficients. In particular, an increment of the total roll stiffness or, equivalently, a reduction of the total roll damping, would lead a lower ISO-LTR lines sensitivity to the relative roll rate $\dot{\phi}$. On the other hand, the severity of the manoeuvre imposed by the driver, the local road perturbations, e.g., road bumps, potholes, etc. ..., and the road bank angle will affect the last three contributions of the ISO-LTR linear characteristics, respectively, thus substantially producing an horizontal shift of the ISO-LTR characteristics.

It is possible to express the ISO-LTR characteristics in the phase plane through the following linear equation:

$$\dot{\phi}_{LTR=q} = k\phi_{LTR=q} + n \tag{10}$$

where $k = -\frac{K}{C}$ and $n = f\left(a_{y_s}, a_{y_{u_i}}, a_{z_s}, a_{z_{u_i}}, \phi_R\right)$ are the slope and the x intercept of the ISO-LTR corresponding to $LTR = q$, respectively.

To provide a proof of this conclusion, the influence of the the road bank angle on the phase-plane portrait and the distribution of the ISO-LTR lines are shown in Figure 11.

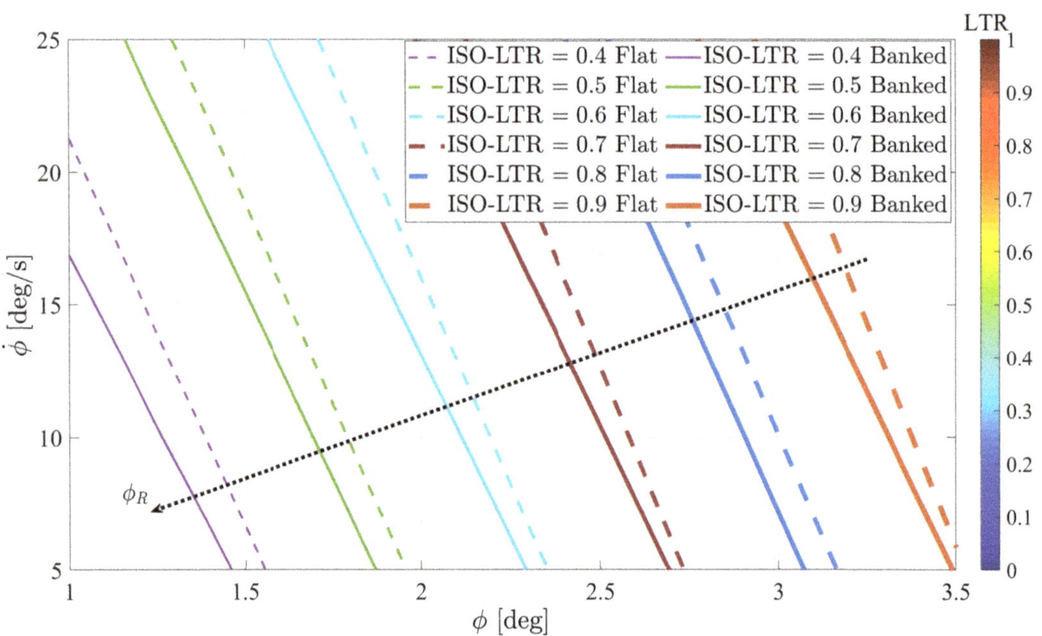

Figure 11. Comparison of the ISO-LTR characteristics for a flat and a 10 deg banked road.

The figure demonstrates that the ISO-LTR lines keep the same negative slope, not influenced by the road bank angle, whose main effect is a shift of the ISO-LTR characteristics towards smaller relative roll angles ϕ.

5. ISO-LTR Predictive Time

The previous sections present the mathematical background to detect the incipient rollover risk through the evaluation of the LTR, its representation on the phase plane plot through the definition of the ISO-LTR characteristics and their perturbations due to the driver behaviour, the local and global road profile geometry. The present section aims at providing a predictive indication of the time the vehicle requires to cross one of the ISO-LTR lines, defined as the ISO-LTR Predictive Time ($ILPT$) in the rest of the paper.

The principle underlying the procedure, based on the formulation proposed by [53], is described by Figure 12 in the phase plane plot.

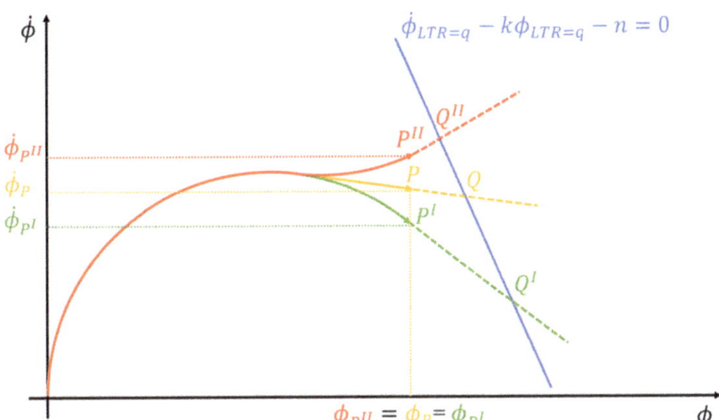

Figure 12. Schematic principle of the *ILPT* definition.

The figure shows three trajectories corresponding to different manoeuvres represented in the phase plane. After a generic time instant, the three trajectories reach the points P, P^{I}, P^{II} approaching to the same roll angle but with different roll rates. The condition of incipient rollover is identified by the blue ISO-LTR line. The *ILPT* is defined as the time required to reach the intersection point between the ISO-LTR line and the tangent line at the current point of the phase plane trajectories. For example, the tangent line in P is expressed by:

$$\dot{\phi} = k_P(\phi - \phi_P) + \dot{\phi}_P \tag{11}$$

where $\ddot{\phi}_P$, $\dot{\phi}_P$ and ϕ_P represent the relative roll acceleration, roll rate and roll angle corresponding to the point P of the phase plane trajectory and $k_P = \frac{\ddot{\phi}_P}{\dot{\phi}_P}$ is the slope of the tangent line. The intersection point between the tangent line and the ISO-LTR line Q is then calculated as follows:

$$\begin{cases} \phi_Q = \frac{k_P\phi_P - \dot{\phi}_P + n}{k_P - k} \\ \dot{\phi}_Q = \frac{kk_P\phi_P - k\dot{\phi}_P + k_P n}{k_P - k} \end{cases} \tag{12}$$

Hence, the *ILPT* represents the time required to reach point Q from point P, under the hypothesis to follow the tangent line expressed by Equation (11), as also described by [53]:

$$ILPT = \frac{[1 - sign(z)]}{2} \sqrt{\frac{(\phi_Q - \phi_P)^2 + (\dot{\phi}_Q - \dot{\phi}_P)^2}{\dot{\phi}_P^2 + \ddot{\phi}_P^2}} \tag{13}$$

with:

$$z = (\dot{\phi}_P - k\phi_P - n)(-\dot{\phi}_P + k\phi_P - n) \tag{14}$$

where z is introduced to saturate the *ILPT* to 0 when the point P crosses the corresponding ISO-LTR line. The calculation of the *ILPT* requires the knowledge of the current relative roll acceleration, roll rate and roll angle and the parameters k and n of the selected ISO-LTR line. The *ILPT* provides a time information only if it exceeds a predefined upper thresholds, i.e., 0.5 s above which the vehicle is far away from the rollover risk.

6. Simulation Results

The predictive capabilities of the *ILPT* are verified in IPG CarMaker® environment through the simulation of the following driving manoeuvres:

- A fast ramp steering on a banked road;
- A double lane change on a flat Road (ISO 3888-1).

The simulation results are analyzed both in the phase plane and in the time domains. The phase-plane analysis provides a graphical interpretation for the *ILPT* calculation. On the other hand, the time domain analysis is carried out to asses the methodology accuracy in predicting the critical condition. Indeed, the reference time T_{Ref}, i.e., the time required to reach a defined *LTR* threshold LTR_{th}, is backward evaluated from the whole post-processed time history and it is compared against the *ILPT* estimation. T_{Ref} is represented in the time domain as -45 deg inclined lines:

$$T_{Ref} = -t + T_{LTR=LTR_{th}} \tag{15}$$

where $T_{LTR=LTR_{th}}$ is the time that $LTR = LTR_{th}$, and it is detected by a backward analysis of the *LTR* time history. The T_{Ref} represents a key quantity to asses the performance of any predictive algorithm to detect a particular event, i.e., the incipient rollover, but it can be only calculated when the whole time history up to the event occurrence is available and it is not suitable for a real-time implementation. The *LTR* threshold, here considered as potentially critical to provoke the vehicle rollover, is assumed to be ± 0.8. The smaller the deviation between *ILPT* and T_{Ref}, the better the predictive capabilities of the proposed methodology.

6.1. Fast Ramp Steering on a Banked Road

The first manoeuvre is one from the simulation set adopted in the previous section to draw the ISO-LTR characteristics of Figure 10.

The resultant phase plane trajectory of the manoeuvre is reported in the topside of the Figure 13, where the graphical solution for the *ILPT* calculation is provided for four time instants. The resultant *ILPT*, elaborated in real-time during the manoeuvre, is plotted in the downside of Figure 13 where the *LTR* time history is also compared against the threshold LTR_{th}.

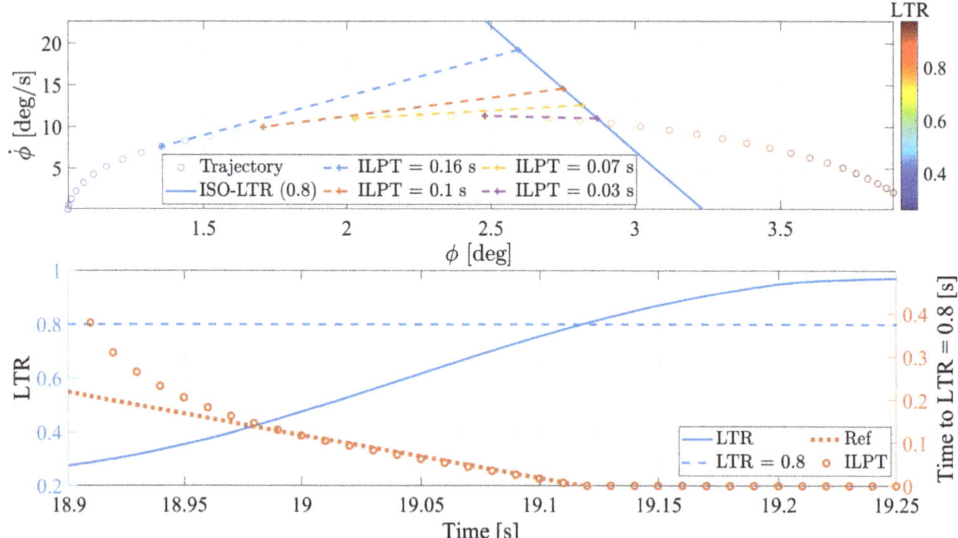

Figure 13. Fast ramp steering manoeuvre at 30 km/h on a 10 deg banked road: phase plane trajectory (top), *LTR*, reference (T_{Ref}) and estimated (*ILTP*) time to reach $LTR = 0.8$ (down).

At the beginning of the manoeuvre, when the *LTR* level indicates a substantial distance from the rollover risk, the *ILPT* (red circles) deviates from the T_{Ref} (red dashed line) to reach the condition of $LTR = 0.8$, event that occurs after 0.24 s from the application of the ramp steering. The time deviations between the estimated *ILPT* and the T_{Ref}

suddenly drops down after 0.05 s, when the *ILPT* starts tracking the reference time until the occurrence of the critical threshold $LTR = 0.8$. This deviation can be graphically explained by the phase-plane portrait in Figure 13. At the beginning of the ramp steering (left side of the trajectory), the trajectory strongly deviates from the tangent line (dashed lines). As soon as the trajectory is approaching the ISO-LTR line (blue continuos line), the tangent line begins to better approximate the future phase-plane trajectory thus allowing the convergence of the *ILPT* towards the T_{Ref}. It is also interesting to note that the *ILPT* slightly underestimates the T_{Ref} when approaching the corresponding ISO-LTR line, thus providing a more conservative estimate.

6.2. Double Lane Change on a Flat Road

The methodology is also applied to a second manoeuvre that is not used to extrapolate the ISO-LTR characteristics or any other results shown in the previous sections. This is the double lane change test which represent a critical manoeuvre from the rollover point of view. The time domain plot of the *LTR*, the *ILPT* and reference time are shown in Figure 14.

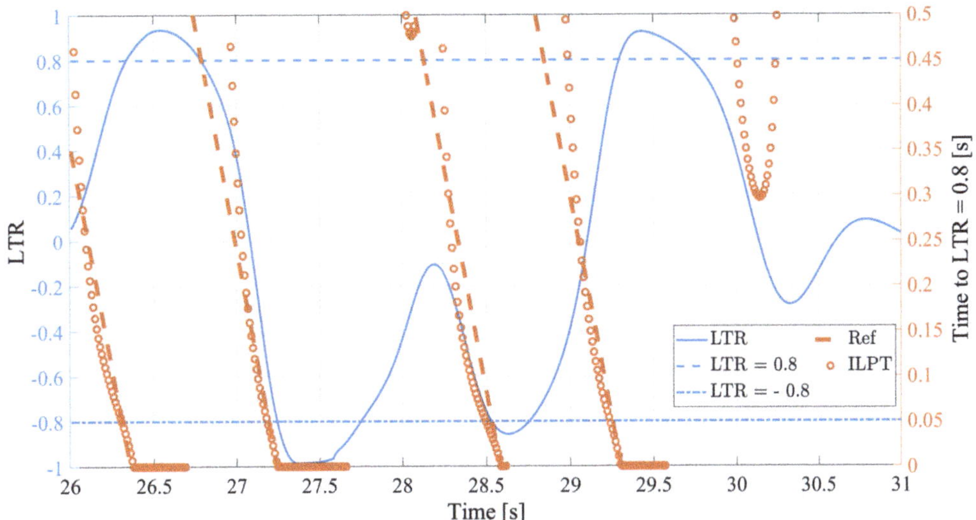

Figure 14. Double lane change manoeuvre at initial speed of 100 km/h on a flat road: *LTR*, reference (T_{Ref}) and estimated (*ILTP*) time to reach $LTR = 0.8$.

During the double change manoeuvre, the vehicle crosses four times the *LTR* threshold. In all cases, the methodology is able to detect in advance the critical point as highlighted by the *ILTP* that starts dropping to zero when the vehicle is approaching the *LTR* threshold. Even during this maneuver, the *ILPT* tends to underestimate the reference time when the vehicle is getting closer to the ISO-LTR boundary, thus providing a conservative and safer prediction. During the last part of the maneuver, the algorithm provides a false positive indication of dangerous situation since the *ILTP* drops to 0.3 s even if the *LTR* did not reach the threshold of 0.8 after this point. This behavior is explained by the sudden reduction of the *LTR* at $t = 30$ s, which is successfully recovered by the stabilizing intervention of the driver.

Even though the vehicle does not rollover during the maneuver, the *ILPT* can predict all the potential dangerous situations in a time frame of a few tenths of seconds, which is suitable for an active control logic to intervene.

7. Conclusions

The activity presented in this paper aims at providing some usefull tools and methodologies to detect and prevent the incipient risk of rollover, applied to heavy vehicles designed for off-road driving applications. The main conclusions drawn from the results and the methodologies presented can be summarized as follows:

- A quantitative indication of the anti rollover risk is represented by the LTR, thus detecting the critical limit above which a wheel lifted off occurs. However, the LTR does not represent a measurable quantity due to the extreme difficulty in estimating the vertical load transfers among the vehicle corners. The paper proposes three LTR model-based formulations, by considering an increased level of complexity. The most generic formulation includes the influence of the road bank angle, unsprung masses and vertical dynamics. Numerical simulations of aggressive manoeuvres on flat roads, banked roads and in presence of asymmetrical speed bump waves, show the significant reliability of the generic model-based formulation when compared to the two simplified versions widespread in the literature;
- The incipient rollover occurs when a defined LTR threshold is approached. The paper proves that the ISO-LTR characteristics, i.e., the combination of vehicle relative roll speeds and angles where the LTR is constant, are linear in the phase plane portrait of the vehicle roll dynamics. This is analytically explained through the LTR model-based formulation, and numerically verified with multiple simulations in IPG CarMaker®. The paper also shows that the LTR model-based formulation provides a qualitative tool to predict how the ISO-LTR lines would change when a road perturbation, i.e., road bank angle or irregularities, is encountered. The ISO-LTR slopes are only influenced by the suspension system configuration and parameters (total roll stiffness and damping), meanwhile the severity of the manoeuvre (lateral and vertical accelerations) and the road global and local perturbations provoke an horizontal shift of the ISO-LTR lines;
- The model-based formulation is essential to analytically evaluate the main influencing factors on the load transfer dynamics between the left and right vehicle sides. However, any mathematical formulation is affected by parameters uncertainties, external disturbances and unmodeled dynamics that compromises its effectiveness for a realistic implementation, especially when noisy experimental measurements are input to the analytical formulation. For this reason, the statistical approach, based on the recurrent neural network principle, is proposed as an alternative methodology to estimate the LTR in a realistic scenario. Indeed, the input of the RNN algorithm are typical measurable quantities available for an experimental implementation, which represent the natural following step the authors are going to explore in the near future. The RNN approach provides excellent results in estimating the front and the rear load transfers even in presence of complex and realistic driving scenarios.
- The detection of current LTR is not sufficient to predict the incipient risk of rollover. A ISO-LTR Predictive Time is then derived to proactively calculate the necessary time to reach a particular LTR threshold. The proposed predictive index is then successfully verified through a fast ramp steering maneuver on a banked road and during an aggressive double lane change manoeuvre. In both cases, the $ILPT$ demonstrates promising predictive capabilities, compatible with the intervention of a control active strategy.

Author Contributions: Conceptualization, A.T., L.D., G.P. and M.V.; methodology, A.T., L.D. and M.V.; software, A.T. and F.V.; validation, A.T., L.D. and F.V.; formal analysis, A.T., L.D. and M.V.; investigation, A.T., L.D., G.P. and M.V.; writing—original draft preparation, A.T., L.D. and F.V.; writing—review and editing, A.T., L.D., F.V., G.P. and M.V.; visualization, L.D.; supervision, A.T., L.D. and M.V. All authors have read and agreed to the published version of the manuscript.

Funding: This research received no external funding.

Institutional Review Board Statement: Not applicable.

Informed Consent Statement: Not applicable.

Data Availability Statement: Not applicable.

Conflicts of Interest: The authors declare no conflict of interest.

References

1. Deutermann, W. *Characteristics of Fatal Rollover Crashes*; NHTSA Tecnical Report DOT HS 809 438; National Center for Statistics and Analysis Research and Development: Washington, DC, USA, 2002; pp. 1–52.
2. Padmanaban, J.; Shields, L.E.; Scheibe, R.R.; Eyges, V.E. A comprehensive review of rollover accidents involving vehicles equipped with Electronic Stability Control (ESC) systems. *Ann. Adv. Automot. Med.* **2008**, *52*, 9–19.
3. Han, I.; Rho, K. Characteristic analysis of vehicle rollover accidents: Rollover scenarios and prediction/warning. *Int. J. Automot. Technol.* **2017**, *18*, 451–461. [CrossRef]
4. *Rollover Resistance*; NHTSA Tecnical Report RIN 2127-AI81; Department of Transportation: Washington, DC, USA, 2001; pp. 1–159.
5. Matolcsy, M. The severity of bus rollover accidents. In Proceedings of the 20th International Technical Conference on the Enhanced Safety of Vehicles (ESV), Lyon, France, 18–21 June 2007; pp. 1–10.
6. Liang, C.C.; Le, G.N. Analysis of bus rollover protection under legislated standards using LS-DYNA software simulation techniques. *Int. J. Automot. Technol.* **2010**, *11*, 495–506. [CrossRef]
7. Winkle, C.B.; Ervin, R.D. Rollover of Heavy Commercial Vehicles. In *Tecnical Report UMTRI-99-19*; University of Michigan-Transportation Research Institute: Ann Arbor, MI, USA, 1999; pp. 1–64.
8. Vella, A.D.; Lisitano, D.; Tota, A.; Wang, B. Analysis of heavy commercial vehicle cornering behaviour through a multibody model. *Int. J. Mech. Control* **2020**, *21*, 39–50.
9. Furleigh, D.; Vanderploeg, M.; Oh, C. Multiple steered axles for reducing the rollover risks of heavy articulated trucks. *SAE Trans.* **1988**, *97*, 837–841.
10. Sampson, D.J.M.; Cebon, D. An investigation of roll control system design for articulated heavy vehicles. In Proceedings of the 4th International Symposium on Advanced Vehicle Control (AVEC'98), Nagoya, Japan, 14–18 September 1998; pp. 1–6.
11. Lin, R.C.; Cebon, D.; Cole, D.J. Active roll control of articulated vehicles. *Veh. Syst. Dyn. Int. J. Veh. Mech. Mobil.* **2007**, *26*, 17–43. [CrossRef]
12. Tota, A.; Velardocchia, M.; Rota, E.; Novara, A. Steering behavior of an articulated amphibious all-terrain tracked vehicle. In Proceedings of the SAE 2020 World Congress & Exhibition (SAE WCX 2020), Online, 21–23 April 2020; pp. 1–11.
13. Tota, A.; Galvagno, E.; Velardocchia, M.; Rota, E.; Novara, A. Articulated Steering Control for an All-Terrain Tracked Vehicle. In Proceedings of the International Federation for the Promotion of Mechanism and Machine Science (IFToMM ITALY 2020), Online, 9–11 September 2020; pp. 823–830.
14. Tota, A.; Galvagno, E.; Velardocchia, M. Analytical Study on the Cornering Behavior of an Articulated Tracked Vehicle. *Machines* **2021**, *9*, 38. [CrossRef]
15. Cossalter, V.; Lot, R.; Massaro, M. The influence of frame compliance and rider mobility on the scooter stability. *Veh. Syst. Dyn.* **2007**, *45*, 313–326. [CrossRef]
16. Vasquez, F.; Lot, R.; Rustighi, E. Optimisation of off-road motorcycle suspensions. In Proceedings of the 15th European Automotive Congress, Madrid, Spain, 2–4 October 2017; pp. 1–10.
17. Bonisoli, E.; Lisitano, D.; Dimauro, L. Detection of critical mode-shapes in flexible multibody system dynamics: The case study of a racing motorcycle. *Mech. Syst. Signal Process.* **2022**, *180*, 109370. [CrossRef]
18. Bonisoli, E.; Lisitano, D.; Dimauro, L. Experimental and numerical mode shape tracing from components to whole motorbike chassis. In Proceedings of the 28th International Conference on Noise and Vibration Engineering (ISMA 2018) and 7th International Conference on Uncertainty in Structural Dynamics (USD 2018), Leuven, Belgium, 17–19 September 2018; pp. 3597–3604.
19. Bonisoli, E.; Lisitano, D.; Dimauro, L.; Peroni, L. A proposal of dynamic behaviour design based on mode shape tracing: Numerical application to a motorbike frame. In Proceedings of the Society for Experimental Mechanics Series (37th IMAC), Orlando, FL, USA, 28–31 January 2019; Springer: New York, NY, USA, 2020; pp. 149–158.
20. Roland Berger. Fuel Cells Hydrogen Trucks. Available online: https://www.fch.europa.eu/publications/study-fuel-cells-hydrogen-trucks (accessed on 27 August 2022).
21. Tota, A.; Galvagno, E.; Dimauro, L.; Vigliani, A.; Velardocchia, M. Energy management strategy for hybrid multimode powertrains: Influence of inertial properties and road inclination. *Appl. Sci.* **2021**, *11*, 11752. [CrossRef]
22. Phanomchoeng, G.; Rajamani, R. New rollover index for the detection of tripped and untripped rollovers. *IEEE Trans. Ind. Electron.* **2013**, *60*, 4726–4736. [CrossRef]
23. Jin, Z.; Zhang, L.; Zhang, J.; Khajepour, A. Stability and optimised H_∞ Control Tripped Untripped Veh. Rollover. *Veh. Syst. Dyn.* **2016**, *54*, 1405–1427. [CrossRef]
24. Ataei, M.; Khajepour, A.; Jeon, S. A general rollover index for tripped and un-tripped rollovers on flat and banked roads. *Proc. Inst. Mech. Eng. Part J. Automob. Eng.* **2019**, *233*, 304–316. [CrossRef]

25. Johansson, B.; Gäfvert, M. Untripped SUV rollover detection and prevention. In Proceedings of the 43rd IEEE Conference on Decision and Control, Atlantis, Bahamas, 14–17 December 2004; pp. 6461–6466.

26. Nalecz, A.G. Influence of vehicle and roadway factors on the dynamics of tripped rollover. *Int. J. Veh. Des.* **2014**, *10*, 321–346.

27. Venturini, S.; Bonisoli, E.; Rosso, C.; Rovarino, D.; Velardocchia, M. Modal analyses and meta-models for fatigue assessment of automotive steel wheels. In Proceedings of the Conference Proceedings of the Society for Experimental Mechanics Series (38th IMAC), Houston, TX, USA, 10–13 February 2020; pp. 155–163.

28. Rovarino, D.; Actis, C.L.; Bonisoli, E.; Rosso, C.; Venturini, S.; Velardocchia, M.; Baecker, M.; Gallrein, A. *A Methodology for Automotive Steel Wheel Life Assessment*; SAE Technical Paper; SAE: St. Joseph, MN, USA, 2020; pp. 1–10. [CrossRef]

29. Vella, A.D.; Tota, A.; Vigliani, A. *On the Road Profile Estimation from Vehicle Dynamics Measurements*; SAE Technical Paper; SAE: St. Joseph, MN, USA, 2021; pp. 1–10. [CrossRef]

30. Yang, H.; Liu, L.Y. A robust active suspension controller with rollover prevention. *SAE Int. J. Passeng. Cars-Mech. Syst.* **2003**, *112*, 992–997.

31. Ataei, M.; Khajepour, A.; Jeon, S. Model predictive rollover prevention for steer-by-wire vehicles with a new rollover index. *Int. J. Control* **2020**, *93*, 140–155. [CrossRef]

32. Fortina, A.; Velardocchia, M.; Sorniotti, A. *Braking System Components Modelling*; SAE Technical Paper; SAE: St. Joseph, MN, USA, 2003; pp. 1–12. [CrossRef]

33. Lapapong, S. Vehicle rollover prevention for banked surfaces. PhD Thesis, The Pennsylvania State University, State College, PA, USA, 2010.

34. Huston, R.L.; Kelly, F.A. Another look at the static stability factor (SSF) in predicting vehicle rollover. *Int. J. Crashworthiness* **2014**, *19*, 567–575. [CrossRef]

35. Pai, J. *Trends and Rollover-Reduction Effectiveness of Static Stability Factor in Passenger Vehicles*; NHTSA Tecnical Report DOT HS 812 444; National Center for Statistics and Analysis Research and Development: Washington, DC, USA, 2017; pp. 1–141.

36. Rajamani, R. *Vehicle Dynamics and Control*, 2nd ed.; Springer: New York, NY, USA, 2011; pp. 1–498.

37. Velardocchia, M.; Vigliani, A. Control systems integration for enhanced vehicle dynamics. *Open Mech. Eng. J.* **2013**, *7*, 58–69. [CrossRef]

38. Yoon, J.; Kim, D.; Yi, K. Design of a rollover index-based vehicle stability control scheme. *Veh. Syst. Dyn. Int. J. Veh. Mech. Mobil.* **2007**, *45*, 459–475.

39. Dahmani, H.; Chadli, M.; Rabhi, A.; El Hajjaji, A. Vehicle dynamic estimation with road bank angle consideration for rollover detection: Theoretical and experimental studies. *Veh. Syst. Dyn.* **2013**, *51*, 1853–1871. [CrossRef]

40. Zhang, X.; Yang, Y.; Guo, K.; Yang, Y.; He, G. Vehicle roll centre estimation with transient dynamics via roll rate. *Veh. Syst. Dyn.* **2020**, *55*, 699–717. [CrossRef]

41. Odenthal, D.; Bünte, T.; Ackermann, J. Nonlinear steering and braking control for vehicle rollover avoidance. In Proceedings of the 1999 European Control Conference (ECC), Karlsruhe, Germany, 31 August–3 September 1999; pp. 1–6.

42. Wielenga, T.J.; Chace, M.A. *A Study in Rollover Prevention Using Anti-Rollover Braking*; SAE Technical Paper; SAE: St. Joseph, MN, USA, 2000; pp. 1–10. [CrossRef]

43. Chen, B.; Peng, H. Differential-braking-based rollover prevention for sport utility vehicles with human-in-the-loop evaluations. *Veh. Syst. Dyn.* **2001**, *36*, 359–389. [CrossRef]

44. Huang, Z.; Nie, W.; Kou, S.; Son, X. Rollover detection and control on the non-driven axles of trucks based on pulsed braking excitation. *Veh. Syst. Dyn.* **2018**, *56*, 1864–1882. [CrossRef]

45. Guizhen, Y.; Honggang, L.; Pengcheng, W.; Xinkai, W.; Yunpeng, W. Real-time bus rollover prediction algorithm with road bank angle estimation. *Chaos, Solitons Fractals* **2016**, *89*, 270–283. [CrossRef]

46. Cao, J.; Jing, L.; Guo, K.; Yu, F. Study on integrated control of vehicle yaw and rollover stability using nonlinear prediction model. *Math. Probl. Eng.* **2013**, *2013*, 643548. [CrossRef]

47. Ricco, M.; Zanchetta, M.; Rizzo, G.C.; Tavernini, D.; Sorniotti, A.; Chatzikomis, C.; Velardocchia, M.; Geraerts, M.; Dhaens, M. On the design of yaw rate control via variable front-to-total anti-roll moment distribution. *IEEE Trans. Veh. Technol.* **2020**, *69*, 1388–1403. [CrossRef]

48. Ricco, M.; Percolla, A.; Rizzo, G.C.; Zanchetta, M.; Tavernini, D.; Dhaens, M.; Geraerts, M.; Vigliani, A.; Tota, A.; Sorniotti, A. On the model-based design of front-to-total anti-roll moment distribution controllers for yaw rate tracking. *Veh. Syst. Dyn.* **2022**, *60*, 569–596. [CrossRef]

49. Goldman, R.W.; El-Gindy, M.; Kulakowski, B.T. Rollover dynamics of road vehicles: Literature survey. *Int. J. Heavy Veh. Syst.* **2001**, *8*, 103–141. [CrossRef]

50. Solmaz, S.; Corless, M.; Shorten, R. A methodology for the design of robust rollover prevention controllers for automotive vehicles with active steering. *Int. J. Control* **2007**, *80*, 1763–1779. [CrossRef]

51. Imine, H.; Benallegue, A.; Madani, T.; Srairi, S. Rollover risk prediction of heavy vehicle using high-order sliding-mode observer: Experimental results. *IEEE Trans. Veh. Technol.* **2014**, *63*, 2533–2543. [CrossRef]

52. Larish, C.; Piyabongkarn, D.; Tsourapas, V.; Rajamani, R. A new predictive lateral load transfer ratio for rollover prevention systems. *IEEE Trans. Veh. Technol.* **2013**, *62*, 2928–2936. [CrossRef]

53. Zhang, X.; Yang, Y.; Guo, K.; Lv, J.; Peng, T. Contour line of load transfer ratio for vehicle rollover prediction. *Veh. Syst. Dyn.* **2017**, *55*, 1748–1763. [CrossRef]

54. Yoon, J.; Cho, W.; Koo, B.; Yi, K. Unified chassis control for rollover prevention and lateral stability. *IEEE Trans. Veh. Technol.* **2009**, *58*, 596–609. [CrossRef]

55. Tota, A.; Velardocchia, M.; Güvenç, L. Path Tracking Control for Autonomous Driving Applications. In Proceedings of the International Conference on Robotics in Alpe-Adria-Danube Region (RAAD 2017), Turin, Italy, 21–23 June 2017; pp. 456–467

56. Sellami, Y.; Imine, H.; Boubezoul, A.; Cadiou, J.C. Rollover risk prediction of heavy vehicles by reliability index and empirical modelling. *Veh. Syst. Dyn.* **2018**, *56*, 385–405. [CrossRef]

57. Zhu, T.; Yin, X.; Na, X.; Li, B. Research on a Novel Vehicle Rollover Risk Warning Algorithm Based on Support Vector Machine Model. *IEEE Access* **2020**, *8*, 108324–108334. [CrossRef]

58. Chen, X.; Chen, W.; Hou, L.; Hu, H.; Bu, X.; Zhu, Q. A novel data-driven rollover risk assessment for articulated steering vehicles using RNN. *J. Mech. Sci. Technol.* **2020**, *34*, 2161–2170. [CrossRef]

59. Baldi, M.M.; Perboli, G.; Tadei, R. Driver maneuvers inference through machine learning. In *Lecture Notes in Computer Science (including Subseries Lecture Notes in Artificial Intelligence and Lecture Notes in Bioinformatics) 10122 LNCS*; Springer: Cham, Switzerland, 2016; pp. 182–192. [CrossRef]

60. Ackermann, J.; Odenthal, D. Damping of vehicle roll dynamics by gain scheduled active steering. In Proceedings of the 1999 European Control Conference (ECC), Karlsruhe, Germany, 31 August–3 September 1999; pp. 4100–4106.

61. Rajamani, R.; Piyabongkarn, D.N.; Tsourapas, V.; Lew, J.Y. Real-time estimation of roll angle and CG height for active rollover prevention applications. In Proceedings of the IEEE 2009 American Control Conference, St. Louis, MO, USA, 10–12 June 2009.

62. Rajamani, R.; Piyabongkarn, D.N. New paradigms for the integration of yaw stability and rollover prevention functions in vehicle stability control. *IEEE Trans. Intell. Transp. Syst.* **2013**, *14*, 249–261. [CrossRef]

63. Lee, S.; Yakubl, F.; Kasahara, M.; Mori, Y. Rollover prevention with predictive control of differential braking and rear wheel steering. In Proceedings of the 6th IEEE Conference on Robotics, Automation and Mechatronics (RAM), 12–15 November 2013; pp. 144–149. [CrossRef]

64. Huang, H.H.; Yedavalli, R.K.; Guenther, D.A. Active roll control for rollover prevention of heavy articulated vehicles with multiple-rollover-index minimisation. *Veh. Syst. Dyn.* **2012**, *50*, 471–493. [CrossRef]

65. Gaspar, P.; Szaszi, I.; Bokor, J. Reconfigurable control structure to prevent the rollover of heavy vehicles. *Control Eng. Pract.* **2005**, *13*, 699–711. [CrossRef]

66. Boada, M.J.L.; Boada, B.L.; Gauchia Babe, A.; Calvo Ramos, J.A.; Lopez, V.D. Active roll control using reinforcement learning for a single unit heavy vehicle. *Int. J. Heavy Veh. Syst.* **2009**, *16*, 596–609. [CrossRef]

67. Dahmani, H.; Chadli, M.; Rabhi, A.; El Hajjaji, A. Fuzzy observer for detection of impending vehicle rollover with road bank angle considerations. In Proceedings of the IEEE 18th Mediterranean Conference on Control and Automation (MED'10), Marrakech, Marocco, 23–25 June 2010; pp. 1497–1502. [CrossRef]

Article

Discriminant Analysis of the Vibrational Behavior of a Gas Micro-Turbine as a Function of Fuel [†]

Vincenzo Niola *, Sergio Savino, Giuseppe Quaremba, Chiara Cosenza, Armando Nicolella and Mario Spirto

Department of Industrial Engineering, University of Naples "Federico II", Via Claudio 21, 80125 Napoli, Italy
* Correspondence: vniola@unina.it
† This paper is an extended version of "Spirto, M.; Savino, S.; Quaremba, G.; Cosenza, C.; Nicolella, A.; Niola, V. Discriminant analysis of the vibrational dynamics of a gas microturbine with different fuels. In Proceedings of the 5th Jc-IFToMM International Symposium, the 28th Jc-IFToMM Symposium on Theory of Mechanism and Machines, Kyoto, Japan, 16 July 2022".

Abstract: Several studies were conducted previously on fuel and biofuel performance of micro-turbines. The present paper combines experimental and statistical approaches to study the vibrational behavior of a gas micro-turbine supplied with different pure fuels and admixed with rapeseed oils. Experimental tests carried out at different operating conditions have allowed us to build a classification model through using discriminant analysis. The classification model can distinguish the vibrational behavior occurring when the turbine is fueled with kerosene, or pure and admixed diesel with rapeseed oil. Moreover, the methodology has even allowed us to highlight differences in vibrational behavior caused by small amounts of rapeseed oil admixed in the fuel. The model reliability, in terms of Cohen's kappa, results in optimal data classification.

Keywords: gas micro-turbine; discriminant analysis; vibrational analysis; diesel blends; kerosene; biofuel; statistical index; accelerometer; confusion matrix; Cohen's kappa

Citation: Niola, V.; Savino, S.; Quaremba, G.; Cosenza, C.; Nicolella, A.; Spirto, M. Discriminant Analysis of the Vibrational Behavior of a Gas Micro-Turbine as a Function of Fuel. *Machines* **2022**, *10*, 925. https://doi.org/10.3390/machines10100925

Academic Editors: Marco Ceccarelli, Giuseppe Carbone and Alessandro Gasparetto

Received: 27 September 2022
Accepted: 8 October 2022
Published: 12 October 2022

Publisher's Note: MDPI stays neutral with regard to jurisdictional claims in published maps and institutional affiliations.

1. Introduction

In recent years, significant effort has been made to lessen the environmental impact of operating gas turbines [1–3].

Micro gas turbines have the potential source to be an alternative power unit (APU) in range-extended electric vehicles (REEVs) [4].

Due to their improved thermal efficiency and reduced fuel consumption, diesel engines make up a significant portion of the passenger car industry in Europe. However, due to increasingly strict exhaust emission laws, particularly for NOx and PM, a more complex and expensive gas exhaust treatment system is required for most diesel vehicles, which also results in increased fuel usage. By 2030, several EU regulations and incentives would call for up to 45% bio-components in fossil fuel. It has been demonstrated that biofuels in the form of pure plant oils cannot be used for automotive applications in diesel engines efficiently because of technical issues brought on by their considerably higher viscosity, corrosive character, and increased exhaust smoke emissions [5].

Biofuels are, broadly speaking, any fuel source made from organic material, such as firewood, charcoal, animal fats and oils, dung, and vegetable oils. To mimic the performance and physical properties of fossil fuels, biofuels must be modified. The development of new power and propulsion technologies will necessitate a better control of the harmful chemicals generated by combustion sources. As energy demand increases, concerns about global warming could result in restrictions on greenhouse gas emissions from fossil fuel facilities. This has prompted substantial study into carbon capture and storage, as well as a growth in renewable energy sources. Many investigations of fuel preparation and emission characterizations of systems powered entirely or partially by pure vegetable oils are available [6–13].

A crucial aspect is the relation between spray quality and combustion performance in micro gas turbines burned with biofuels and biomasses [14,15].

Pure vegetable oils might cause unwanted vibrations by damaging the injection system or combustion device. The ideal answer would be to have a tool that could foresee and/or track the vibrational state of the combustion device in real time. By using appropriate quantifiers generated from microphone and accelerometer inputs, investigations based on acoustic and vibrational measurements seem to offer an intriguing diagnostic and predictive answer [16–20]. Other authors have proposed a neural-network-based tool that allows them to ensure protection and the safety measures against the instability phenomena in a gas turbine based on the modeling of its dynamic behavior [21].

In this context, Allouis et al. [22] have evaluated the impact of biofuel properties on emissions and performances of a micro gas turbine using combustion vibrations detection. Other methodologies described in [23–25] are employed for detecting anomalies in mechanical systems such as gas micro-turbines.

The present paper aims to study the vibrational behavior of a low-emission gas micro-turbine for power generation, fueled with different liquid fuels, including commercial diesel oil and its blend with pure rapeseed oil. Kerosene is the design fuel for this turbine. The authors of this current paper have presented a preliminary work on this topic [26].

In the first part, the study describes the experimental phase in which vibrational signals have been acquired through accelerometers properly mounted on the injector. Afterward, signal processing has been performed through discriminant analysis. A classification model has been developed to distinguish the vibrational behavior associated to the turbine fueled with kerosene, pure diesel, and admixed diesel with rapeseed oil.

2. Materials

The study system (the same as of [22]) is a Capstone 30 model 18-blade gas micro-turbine (Figure 1) with a maximum power output of 30 kW. Since its output power is less than 100 kW, it can be included in the micro-turbine category.

Figure 1. Turbine side view.

The maximum fuel flow rate is 10 kg/h, and the exhaust gas temperature is approximately 590 °C, while at the system discharge it is 276 °C (as nominal temperature). The test bench of the system is reported in Figure 2.

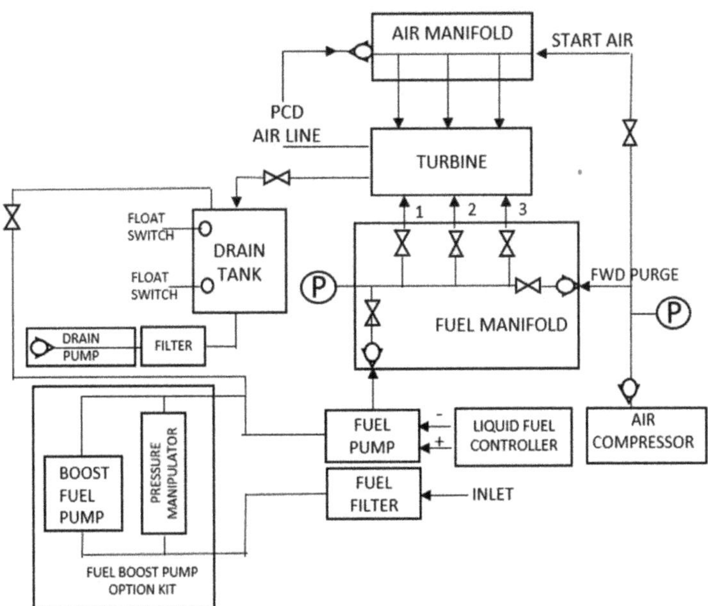

Figure 2. System test bench.

Two different blends of diesel with rapeseed oil (respectively at 1 and 3% by volume) and kerosene (the turbine's designed fuel) are used as turbine fuels. The choice of such low percentages of rapeseed oil in the two additive fuels is necessary to characterize the machine's behavior from a vibrational point of view and to grasp the minimal variations resulting from the use of fuels with a very similar composition.

As for the choice of rotation speeds, speeds lower than the maximum one (90,000 rpm) were chosen to study the machine's behavior under conditions that gradually deviate from the maximum. Tests were conducted for turbine rotation speeds of 75,000, 80,000, and 85,000 rpm. A piezoelectric uniaxial accelerometer (PCB 352C22) and a data acquisition system (LMS SCADAS Mobile SM01) were used to sense and acquire the turbine vibrations. The accelerometer was mounted on a rigid bracket in-built with the injector. The need to use the bracket is due to the high temperature reached by the turbine: in the injector, the temperature can reach a high value to damage the accelerometer.

The accelerometer signals were acquired for a duration of 10 s with a sampling frequency of 102,400 Hz. As a preliminary step, the accelerometer was calibrated using a Bruel & Kjaer Calibration Exciter Type 4294. Below is the procedure followed to carry out the experimental tests at 85,000 rpm (the procedure is similar for the other two speeds):

1. Calibration of the accelerometer;
2. Implementation of the accelerometer to the brackets and anchoring them to the injector;
3. Preparation of fuel mixtures with additives using a graduated cylinder;
4. Starting the turbine with the design fuel (kerosene) and the required power of 10 kW;
5. Waiting for the turbine operating conditions by checking that the exhaust gas temperature is 590 °C as declared by the name plate;
6. Adjustment of turbine power (approx. 19 kW) to reach the preset speed of 85,000 rpm;
7. Accelerometer acquisition for kerosene supply;
8. Adjustment of power output to 10 kW;
9. Gradual switchover from kerosene fueling to pure diesel;
10. Adjustment of turbine power (approx. 19 kW) to reach the preset speed of 85,000 rpm;

11. Accelerometer acquisition for pure diesel fuel supply;
12. Adjustment of power output to 10 kW;
13. Gradual switchover from pure diesel fuel to 1% rapeseed oil diesel;
14. Adjustment of turbine power (approx. 19 kW) to reach the preset speed of 85,000 rpm;
15. Accelerometer acquisition for feeding 3% rapeseed oil diesel;
16. Adjustment of power output to 10 kW;
17. Gradual switchover from 1% rapeseed oil to 3% rapeseed oil admixed fuel;
18. Adjustment of turbine power (approx. 18.5 kW) to reach the preset speed of 85,000 rpm;
19. Accelerometer acquisition for 3% rapeseed oil diesel;
20. Adjustment of power output to 10 kW;
21. Gradual switchover from 3% rapeseed oil diesel supply to pure kerosene;
22. Turbine shutdown with design fuel (kerosene).

Table 1 shows a summary of the acquisition data.

Table 1. Tests summary.

Test Number	Fuel	RPM
1	97% Diesel and 3% rapeseed oil	75,000
2	99% Diesel and 1% rapeseed oil	75,000
3	Kerosene	75,000
4	100% Diesel	75,000
5	97% Diesel and 3% rapeseed oil	80,000
6	99% Diesel and 1% rapeseed oil	80,000
7	Kerosene	80,000
8	100% Diesel	80,000
9	97% Diesel and 3% rapeseed oil	85,000
10	99% Diesel and 1% rapeseed oil	85,000
11	Kerosene	85,000
12	100% Diesel	85,000

3. Methods

The acquired vibrational signals were analyzed by means of discriminant analysis of statistical indices directly calculated on the raw signals.

The method used for the creation of the classificatory model obtained through discriminant analysis is stepwise, i.e., the variables are not inserted into the model at the same time but in steps, i.e., only those variables with the greater discriminating weight are inserted. At each step, the variable with the lowest Wilks lambda value and, in turn, the highest F coefficient value is entered, i.e., the variable that contributes to better differentiate the groups. More details on this type of analysis can be found in the bibliographical references [26,27].

The chosen statistical indices for the analysis are asymmetry, kurtosis, shape factor, quadratic oscillation index, root mean square value, crest factor, non-normalized Shannon entropy, logarithmic entropy, synchrony index, Pearson's correlation index, Kendall's correlation index, and Spearman's correlation index [27–30]. The statistical indices used to create the three discriminating models (Section 4) will be briefly described. In the description, n is the length of the signal and x_i is the i-th component of the signal. The quadratic oscillation index (O^2) that provides indications of the oscillating phenomenon is described by Equation (1):

$$O^2 = \sqrt{\frac{1}{n-1}\sum_{i=1}^{n-1}(x_i - x_{i+1})^2} \tag{1}$$

The non-normalized Shannon entropy (SE) that provides a measure of the degree of order or disorder of a signal is described by Equation (2):

$$SE = -\sum_{i=1}^{n} x_i^2 \log\left(x_i^2\right) \tag{2}$$

The asymmetry (γ), which is described in Equation (3), gives an indication of the distance between the mean value and the mode of the signal:

$$\gamma = \frac{m_3}{\sigma^3} \tag{3}$$

where m_3 is the third-order central moment of the signal, while σ is the standard deviation.

The root mean square (RMS) value, which is described in Equation (4), gives an indication of the loads acting on the system and the system speed:

$$RMS = \sqrt{\frac{1}{n}\sum_{i=1}^{n} x_i^2} \tag{4}$$

Finally, to assess the degree of accuracy and reliability of the classification, Cohen's kappa (κ) was calculated [31,32]. It is a statistical coefficient representing the degree of accuracy and reliability in a statistical classification defined from a confusion matrix, as reported in Equation (5):

$$\kappa = \frac{P_0 - P_e}{1 - P_e} \tag{5}$$

P_0 is the sum of the probabilities along the main diagonal of the confusion matrix, as given in Equation (6):

$$P_0 = \sum_{i=1}^{n} P_{i,i} \tag{6}$$

where $P_{i,i}$ is the probability of the generic element along the main diagonal of the confusion matrix. The P_e formula is given in Equation (7):

$$P_e = \sum_{i=1}^{n} P_{i,TOT} \cdot P_{TOT,i} \tag{7}$$

where $P_{i,TOT}$ is the total probability along the i-th row, and $P_{TOT,i}$ is the total probability along the i-th column of the confusion matrix.

4. Results and Discussion

The first operation carried out on the sampled signals was the calculation of the statistical indices indicated in paragraph 3. From the single sampled signal at a given speed of the micro-turbine powered with a given fuel, 125 values of the generic statistical index were obtained, i.e., one value was calculated every 100 revolutions of the turbine: this means that a total of 500 values of the generic index were obtained for each speed considered. An example-blocks diagram of the algorithm adopted to calculate the generic index I for the four signals (for each speed) was reported in Figure 3.

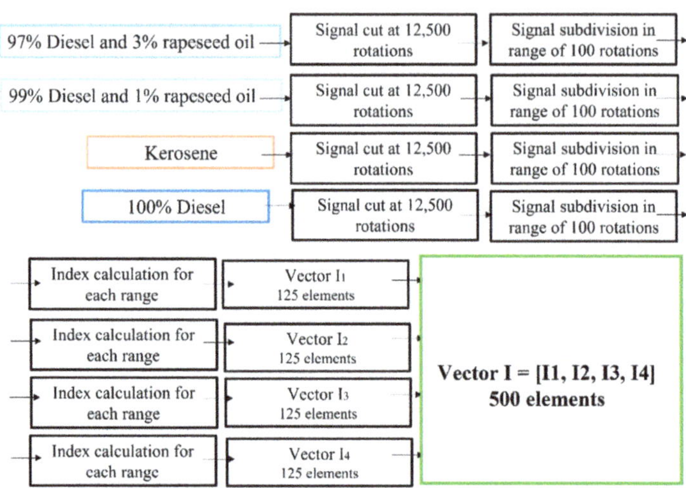

Figure 3. Algorithms implemented for calculating indices.

Subsequently, each of the calculated indices was used as a potential variable for the creation of the classification model for each speed. The software used to create the classification model returned only three variables (indices) out of all those inserted in the model. Table 2 summarizes the variables entered into the model, i.e., those variables that were able to explain 100% of the cumulative variance for each speed analyzed.

Table 2. Variables chosen for the model.

Step	75,000 rpm	80,000 rpm	85,000 rpm
1	Quadratic Oscillation Index (O^2)	Non-Normalized Shannon Entropy (SE)	Root Mean Square Value (RMS)
2	Non-Normalized Shannon Entropy (SE)	Asymmetry (γ)	Quadratic Oscillation Index (O^2)
3	Asymmetry (γ)	Root Mean Square Value (RMS)	Asymmetry (γ)

Starting from these variables, it is possible to write three discriminant functions, called F_i, for each analyzed speed:

$$F_1 = a_1 X_1 \tag{8}$$

$$F_2 = b_1 X_1 + b_2 X_2 \tag{9}$$

$$F_3 = c_1 X_1 + c_2 X_2 + c_3 X_3 \tag{10}$$

where X_i are the variables (indices) chosen by the model and shown in Table 2, while a_i, b_i, and c_i are the standardized coefficients associated with each function returned by the software.

The three discriminant functions (Table 3) are different for each turbine speed. The coefficients a, b, and c are different for each case study.

Table 4 displays the variance percentages contributed by the individual functions and the cumulative variance. For all three turbine speeds, it is possible to exceed 95% of the cumulative variance with only two functions. The third added function contributes a value of less than 5% to the cumulative variance, which is necessary to attain 100% of the total cumulative variance.

Table 3. Discriminant function.

Function	75,000 rpm	80,000 rpm	85,000 rpm
1	$F_1 = a_1 O^2$	$F_1 = a_1 SE$	$F_1 = a_1 RMS$
2	$F_2 = b_1 O^2 + b_2 SE$	$F_2 = b_1 SE + b_2 \gamma$	$F_2 = b_1 RMS + b_2 O^2$
3	$F_3 = c_1 O^2 + c_2 SE + c_3 \gamma$	$F_3 = c_1 SE + c_2 \gamma + c_3 RMS$	$F_3 = c_1 RMS + c_2 O^2 + c_3 \gamma$

Table 4. Variance percentage.

RPM	75,000	75,000	80,000	80,000	85,000	85,000
Function	% Function Variance	% Cumulative Variance	% Function Variance	% Cumulative Variance	% Function Variance	% Cumulative Variance
1	86.9	86.9	74.6	74.6	73.2	73.2
2	9.3	96.2	22.1	96.7	22.0	95.2
3	3.8	100	3.3	100	4.8	100

4.1. Variables Included in the Classification Model

In the following, the obtained indices for the fulfillment of the classification model at each speed (Table 2) are reported. All the following diagrams are shown as a function of the number of periods where the single period corresponds to 100 turbine revolutions.

Figure 4 shows the trends of the three indices (variables) inserted in the discriminant model for the speed of 75,000 rpm.

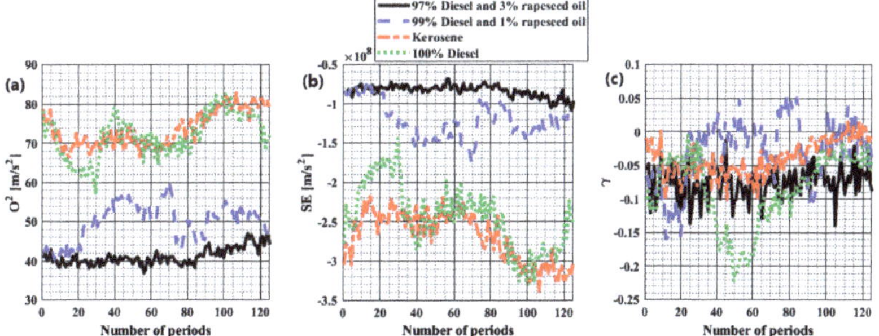

Figure 4. Indices for 75,000 rpm: (**a**) quadratic oscillation index, (**b**) non-normalized Shannon entropy, and (**c**) asymmetry.

In Figure 4a, which shows the trend of the quadratic oscillation index, the clear distinction between the four fuels can be seen immediately. Indeed, while kerosene and pure diesel are in the upper part of the diagram, the two admixed diesels clearly present lower values, with the diesel with 1% rapeseed oil averagely higher than the diesel with 3% rapeseed oil: this phenomenon can be seen in the greater tendency of diesel with 1% rapeseed oil towards the behavior of pure diesel. Since high values of this index emphasize that the system dynamic energy is better utilized, it can be said that using pure rather than admixed fuels makes the system better in usage use of dynamic energy. This can be explained by the fact that admixed diesel has an explosive power lower than the pure, reducing its energy output. This makes it possible to see that, by using biofuels, CO_2 emissions and other pollutants can certainly be reduced at the expense of efficiency.

The distinction between pure and admixed fuels can also be seen in Figure 4b, in which the trend of non-normalized Shannon entropy is shown, by observing that admixed fuels exhibit higher values than pure fuels. This trend highlights how admixed fuels show little stationary and periodic vibrational cycles in contrast to pure fuels: this phenomenon is due to the reduction in explosive power caused by the presence of the oil and the resulting non-homogeneity of the mixture.

The asymmetry index is the last one adopted for modeling the 75,000 rpm case, shown in Figure 4c. By simply plotting the trend, it is not possible to show a difference between the four fuels used because of the overlapping of the curves and their highly chaotic course. The only possible observation is the main trend to negative values for all four fuels.

Figure 5 shows the trends of the three indices (variables) inserted in the discriminant model for the speed of 80,000 rpm.

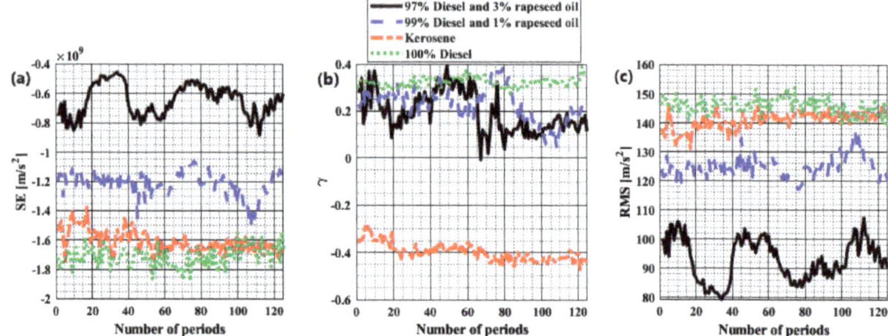

Figure 5. Indices for 80,000 rpm: (**a**) non-normalized Shannon entropy, (**b**) asymmetry, and (**c**) root mean square value.

In Figure 5a, in which the trend of non-normalized Shannon entropy is shown, a clear distinction between pure and additive fuels can be seen, as in the case for 75,000 rpm. Pure diesel and kerosene present the lowest values for this index; diesel with 1% rapeseed oil gives higher values than the latter and diesel with 3% rapeseed oil gives the highest values among the four curves. This indicates how the vibrational dynamics of additive fuels are more irregular and unsteady than those of pure fuels and how the most irregular dynamics are in the case of diesel with 3% rapeseed oil, as expected, the latter being characterized by the higher percentage of rapeseed oil.

In Figure 5b, the asymmetry trend is reported, which tends to form two distinct clusters: the diesel group, including both pure and additive fuels, and the kerosene. The kerosene assumes negative values, which is not the case for diesel, where values are generally higher. This makes it possible to say that from a dynamic point of view, the use of diesel, pure or with rapeseed oil admixed, because it is characterized by a positive asymmetry, results in stronger vibrations than the use of kerosene, which, on the other hand, results in a much lower vibration. This was desirable since kerosene is the design fuel of the turbine.

Finally, Figure 5c shows the trend of the last of the three indices for the 80,000 rpm case, the RMS. The trend of the four curves allows a good distinction between pure and admixed fuels (note the separation of the red and the green curve from the other two). Furthermore, this graph emphasizes how the energy involved is greater in the case of pure fuels and, therefore, the explosive power decreases as the percentage of rapeseed oil admixed to diesel increases.

Figure 6 shows the trends of the three indices (variables) entered into the discriminant model for the speed of 85,000 rpm.

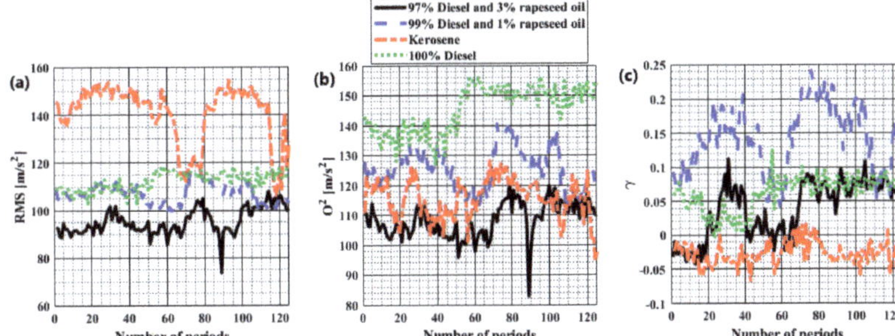

Figure 6. Indices for 85,000 rpm: (**a**) root mean square value, (**b**) quadratic oscillation index, and (**c**) asymmetry.

In Figure 6a, the RMS trend is shown. In this graph, the kerosene values are generally higher than those of the other fuels: as previously stated, this phenomenon was desired because kerosene is the turbine's design fuel, and the turbine works at a speed of only 5000 rpm lower than the maximum speed (90,000 rpm). Kerosene is the fuel with the best possible working conditions in terms of energy compared to all the others; on the other hand, the admixed diesel with the highest percentage of rapeseed oil has the worst working conditions in terms of energy.

In Figure 6b, a good distinction between pure diesel and admixed diesel can be seen, emphasizing how the turbine rotates smoothly and periodically using a pure fuel, as expected, compared to additive ones.

In the last diagram, i.e., the one in Figure 6c, the curves relating to admixed diesels are much more chaotic and assume a much wider range of values; those relating to pure fuels are much more stable and assume a much narrower range of values.

4.2. Cluster Analysis

As follows, the diagrams of the discriminant functions (Table 3) have been displayed for each speed. All functions are normalized in the range $[0, 1]$.

Figure 7 shows the clusters of the discriminant function for the speed of 75,000 rpm.

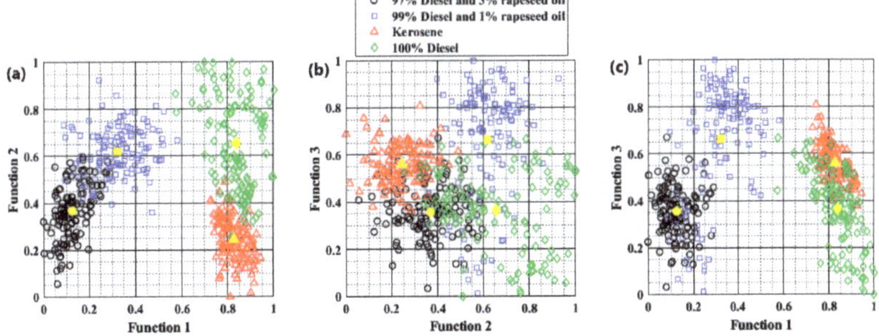

Figure 7. Clusters for 75,000 rpm: (**a**) functions 1 and 2, (**b**) functions 2 and 3, and (**c**) functions 1 and 3.

Functions 1 and 2 (Figure 7a) explain most of the total variance, showing a clear distinction between pure diesel and kerosene, which occupy the right-hand side of the diagram, while on the other side there are the two additive diesel fuels, thus creating two families: pure and admixed fuels. It is possible to observe different behavior depending on whether a pure or non-pure fuel is used. Function 1 is associated with the quadratic oscillation index: the higher this index takes on values, the more the system is characterized by a better use of dynamic power. Indeed, it can be seen from the diagram that dynamic power is better used when using pure rather than admixed fuels: kerosene and diesel assume a very high value of function 1 and, therefore, of the quadratic oscillation index compared to the admixed fuels. This is in full agreement with what was previously stated for the index diagram in Figure 4a, namely that using biofuels reduces emissions of pollutants at the expense of system efficiency. Function 2, which is obtained by adding to the term associated with the quadratic oscillation index a term relating to non-normalized Shannon entropy, the values are spread with respect to the centroids (yellow in the diagrams): as previously stated for this index, it highlights the presence of vibrational cycles that are not stationary and not periodic. It follows that the discriminant analysis at a speed of 75,000 rpm does not allow any clear distinction, since pure diesel describes power conditions like those of kerosene (function 1), while the admixed fuels assume high entropy values, which could compromise the integrity of the system.

Functions 2 and 3 (Figure 7b) do not allow a good distinction between the different types of fuels. Indeed, they explain only 13.1% of the total variance (Table 4) so that the different fuels make a single cluster, preventing any differentiation between the fuels.

Functions 1 and 3 (Figure 7c) allow a distinction between admixed fuels like in Figure 7a but less sharply: this is due to the lower discriminating power of function 3 compared to function 2 (9.3% of function 2 against 3.8% of function 3). Indeed, the overlap between the kerosene and pure diesel clusters is greater than the case in Figure 7a. Once again, the non-admixed fuels occupy the right-hand of the diagram, i.e., the dynamic power of the turbine is fully exploited (function 1 being composed of only the quadratic oscillation index), while the admixed fuels occupy the left-hand of the diagram, emphasizing an opposite behavior to the other two fuels. Finally, the cluster is much more dispersed with respect to the *y*-axis, particularly for the case of admixed diesel with 1% rapeseed oil.

Figure 8 shows the clusters of the discriminant function for the speed of 80,000 rpm.

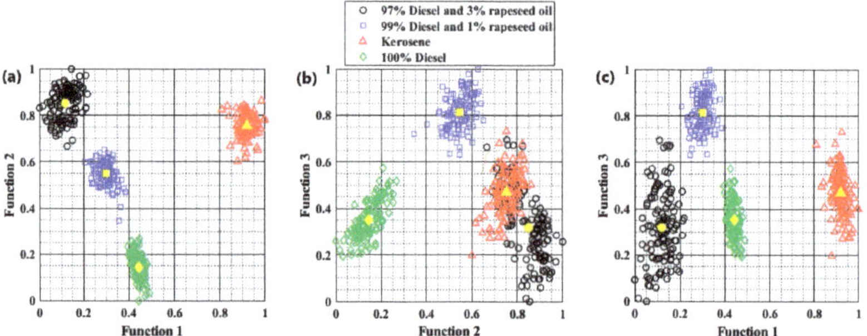

Figure 8. Clusters for 80,000 rpm: (**a**) functions 1 and 2, (**b**) functions 2 and 3, and (**c**) functions 1 and 3.

The highest discriminating power is associated with functions 1 and 2 (Figure 8a), where it is possible to notice a pattern that clearly distinguishes the four fuels: particularly, the kerosene has values of function 1 different from the remaining fuels. Observing the non-normalized Shannon entropy (function 1), the points are really close to their own centroids: the system vibrates more contained than the 75,000 rpm case. In function 2, the entropy term is more prevalent than the asymmetry: from the diagram in Figure 5b, the kerosene values are clearly lower than the other three fuels, an event that is not repeated in the plot of the clusters (Figure 8a). Finally, it shows the tendency of admixed diesel with 1% rapeseed oil to pure diesel to be greater than diesel with 3% rapeseed oil: this is due to the different percentages of rapeseed oil added to diesel. In general, it can be seen in the diagram that the four clusters are distinct, separated from each other and closer to their own centroids: this allows us to say that it is the best classification obtained.

Since functions 2 and 3 (Figure 8b) have the least discriminating power, they do not allow a good distinction between the different types of fuels as in the case of Figure 8a. Although pure diesel and admixed diesel with 1% rapeseed oil are well-distinguished and separated from the others, kerosene and admixed diesel with 3% rapeseed oil tend to overlap and form a single cluster.

Functions 1 and 3 (Figure 8c) allow a good distinction between the various fuels, although the cumulative variance percentage of 77.9% is lower than in functions 1 and 2 (96.7%). Even though they are slightly less sharply defined than in Figure 8a (this is due to the lower discriminating power of function 3 compared to function 2), the four clusters are still separated so that the diesel and the kerosene family stand out. Therefore, the considerations made previously on function 1 for Figure 8a are the same for the same function but in Figure 8c, i.e., the clusters are concentrated to their centroids and the system vibrates smoother than in the case of 75,000 rpm. Function 3 does not allow a good distinction between the fuels: the four clusters have an elongated shape along y-axis and a same range of values, except for admixed diesel with 1% rapeseed oil.

Figure 9 shows the clusters of the discriminant function for the speed of 85,000 rpm.

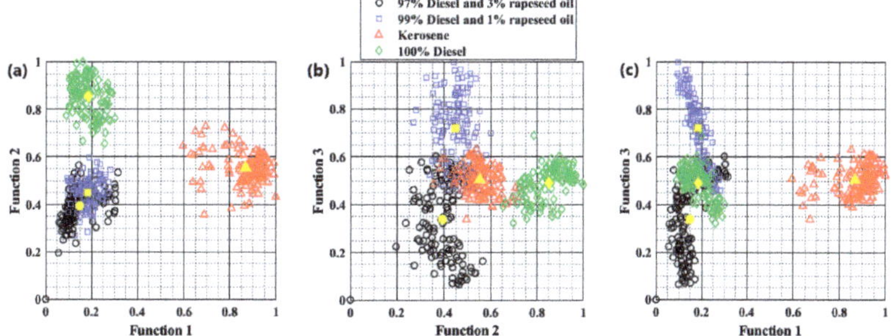

Figure 9. Clusters for 85,000 rpm: (**a**) functions 1 and 2, (**b**) functions 2 and 3, and (**c**) functions 1 and 3.

In the diagram of Figure 9a, knowing that function 1 is associated with the RMS, function 1 indicates how much energy is involved in terms of signal power where function 1, relatively to the kerosene, assumes greater values than the other fuels, as expected since kerosene is the design fuel, and the turbine is at a speed of 85,000 rpm (closer to the maximum speed). Function 2 allows a clear distinction between pure and admixed diesel: the first is clearly distinguished from the latter with a well-defined cluster, while the admixed diesel with 1% and 3% rapeseed oil overlap and there is not a big difference between the two. Kerosene has the highest power because it has the highest RMS. Since kerosene is the most expensive fuel among the proposed ones, a good alternative is using the pure diesel, which is far cheaper than the kerosene but less powerful (lower RMS). The pure and admixed diesel fuels have about the same function 1 range values: it means that it is possible to use admixed diesel keeping a similar RMS. Increasing the amount of rapeseed oil, the power output of the turbine and the pollutants decrease (this is also confirmed by the diagram in Figure 6a).

Functions 2 and 3 (Figure 9b) do not allow a good distinction between the various clusters: the four fuels tend to overlap and mix. As said for the previous speed rates, this event is due to the low percentage of cumulative variance provided by functions 2 and 3 (26.8%).

As in the case with functions 1 and 2, in the case of functions 1 and 3 (Figure 9c) it is possible to distinguish, on the left hand, the diesel family and, on the right hand, the kerosene, regarding function 1; furthermore, there is no distinction between pure diesel and admixed fuels. As regards function 3, it can be observed that the assumed values by pure fuels is much more concentrated around their centroid, unlike admixed fuels, which present much wider and dispersed values.

4.3. Classification Summary

Finally, the confusion matrices and the Cohen's kappa for each of the analyzed speeds are reported to highlight the goodness of the classification performed. In all confusion matrices, there are target classes along the first row and output classes along the first columns.

In the 75,000 rpm case (Table 5), it can be seen that 90.4% of the samples are correctly classified, which is an optimal result. In the kerosene case, 100% of the samples are correctly classified, while the lowest percentage is in the case of admixed diesel with 3% rapeseed oil where 19.2% of the samples are included in the case of admixed diesel with 1% rapeseed oil. Finally in the pure diesel case, 84% of the samples are correctly classified, while the remaining are confused with kerosene. It is interesting to note that it is never confused with admixed diesel, except for one sample confused with admixed diesel with 3% rapeseed oil: this underlines the clear separation between the two families of fuels shown in the clusters of Figure 7a.

Table 5. Confusion matrix for 75,000 rpm.

Fuel	97% Diesel and 3% Rapeseed Oil	99% Diesel and 1% Rapeseed Oil	Kerosene	100% Diesel	Total %
97% Diesel and 3% rapeseed oil	101	24	0	0	80.8%
99% Diesel and 1% rapeseed oil	4	121	0	0	96.8%
Kerosene	0	0	125	0	100%
100% Diesel	0	1	19	105	84%
Total %	83.4%	95.3%	86.8%	100%	90.4%

In the case at 80,000 rpm (Table 6), 100% of the samples is correctly classified: this is the best possible case of classification since no value of the chosen variables will be associated with a different predicted class. This result was desirable from the simple observation of the clusters in Figure 8a, which are all distinct and separate from the others, and this confirmed that this is the best possible classification.

Table 6. Confusion matrix for 80,000 rpm.

Fuel	97% Diesel and 3% Rapeseed Oil	99% Diesel and 1% Rapeseed Oil	Kerosene	100% Diesel	Total %
97% Diesel and 3% rapeseed oil	125	0	0	0	100%
99% Diesel and 1% rapeseed oil	0	125	0	0	100%
Kerosene	0	0	125	0	100%
100% Diesel	0	0	0	125	100%
Total %	100%	100%	100%	100%	100%

In the case at 85,000 rpm (Table 7), 93.4% of samples are correctly classified. Both kerosene and pure diesel fuel contribute to this high percentage as both report a 100% correct classification. The lowest percentage of correct classification is for admixed diesel with 1% rapeseed oil: 80% is for the 1% rapeseed oil while the remaining 20% is assigned to 3% rapeseed oil. Moreover, for the admixed diesel with 3% rapeseed oil, 93.6% of samples are correctly classified, while the remaining are confused with admixed diesel with 1% rapeseed oil. This was also evident from the diagram in Figure 9a in which the clusters of the two admixed diesels tend to overlap and form a single cluster. It is possible to say that the case analyzed at 85,000 rpm constitutes a good distinction between pure fuels (the latter being correctly classified as 100%) and admixed fuels.

Table 7. Confusion matrix for 85,000 rpm.

Fuel	97% Diesel and 3% Rapeseed Oil	99% Diesel and 1% Rapeseed Oil	Kerosene	100% Diesel	Total %
97% Diesel and 3% rapeseed oil	100	25	0	0	80%
99% Diesel and 1% rapeseed oil	8	117	0	0	93.6%
Kerosene	0	0	125	0	100%
100% Diesel	0	0	0	125	100%
Total %	92.6%	82.4%	100%	100%	93.4%

To confirm the optimal results obtained with the proposed classification method, Table 8 shows the Cohen's kappa values for each turbine speed. The values of this index are always greater than 0.8: this implies an optimal classification, especially for the case at 80,000 rpm.

Table 8. Cohen's kappa.

RPM	Lower Extreme	Mean Value	Upper Extreme
75,000	0.8376	0.8720	0.9064
80,000	1	1	1
85,000	0.8830	0.9120	0.9410

5. Conclusions

From the simple observation of the selected indices of the discriminant model, the following results could be deduced:

- For the 75,000 rpm case, a strong distinction emerges between pure and admixed fuels: for the pure fuels, the quadratic oscillation index assumes the greatest values suggesting that the explosive power of the fuel is fully exploited while there is a decrease in the explosive power and pollutants for the admixed fuels. Non-normalized Shannon entropy show a no-stationary and no-periodic vibrational cycle, in particular for the admixed fuels;
- For the 80,000 rpm case, just as for the previous case, it is possible to well-distinguish pure and admixed fuels (by non-normalized Shannon entropy and RMS), but at the same time, thanks to the asymmetry, an excellent distinction emerges between the turbine design fuel (kerosene) and the others (diesel family);
- For the 85,000 rpm case, the distinction is evident of how dynamic energy is best exploited using kerosene (by RMS), and how vibrations tend to be more regular with pure fuels (by asymmetry).

For all three analyzed speeds, the diagrams of discriminant functions 2 and 3 do not allow a good differentiation of the fuels as expected given the low percentage of cumulative variance (Table 4).

As regards the diagrams of functions 1 and 3, they certainly allow better differentiation than the case of functions 2 and 3 because of the higher percentage of cumulative variance, but they still do not allow optimal differentiation for each speed.

The best differentiation is provided by the diagrams obtained from functions 1 and 2, which provide the highest percentage of cumulative variance. Indeed:

- The 75,000 rpm case showed the distinction between the pure and admixed family: they are in two different zones of the diagram in Figure 7a. The admixed diesel with 1% rapeseed oil has a greater tendency toward pure diesel than the diesel with 3% rapeseed oil;
- The 80,000 rpm case provided the best possible differentiation among all the fuels because of the formation of four distinct, separate and non-overlapping clusters. In this diagram, it is possible to distinguish between the diesel and the kerosene family;

- As for the previous speed, the case at 85,000 rpm led to the distinction between the diesel and the kerosene family but not an optimal distinction between the two admixed fuels because of overlapping clusters.

The results show that the best possible classification occurs in the case at 80,000 rpm where the clusters are well-defined and spaced out. This classification model can be used for quality check of the purchased fuel, especially diesel. It could happen that the purchased fuel has been admixed with crude oils, which can negatively affect its quality because of different molecular structures. Another application can be the quantity check of rapeseed oil admixed with diesel. Comparing the vibrational behavior of the turbine powered by the supplied fuel with the vibrational response of the same with the desired fuel, it is possible to say whether the fuel purchased from the supplier meets the required standards or not.

Finally, it is possible to make some final observations on the individual velocities with respect to the diagrams obtained by plotting functions 1 and 2:

- In the case at 75,000 rpm, the clusters with respect to function 2 were always very dispersed around their centroids, underlining how the turbine presents vibrational cycles that are not very stationary and not very periodic, while the clusters with respect to function 1 show that the dynamic power of the system is better exploited with pure fuels. The best performance is with the pure fuels, but the lower emissions and costs are with the admixed fuels.
- In the case at 80,000 rpm, the clusters with respect to function 1 are always grouped around their own centroids, highlighting more stationary and periodic vibrations than the other cases. This is the best working condition for the turbine.
- In the case of 85,000 rpm, referring to function 1, the kerosene has the highest values and better exploits dynamic energy, while the diesel family presents a similar range of values. Regarding function 2, there is a clear distinction between admixed and pure diesels: the pure diesel presents higher explosive power. Thus, kerosene has the best possible performance but is the most expensive. The three diesels are cheaper but make the turbine less efficient. Lastly, there is not a great difference between using an admixed diesel with 1 or 3% rapeseed oil.

From the confusion matrices and the calculation of Cohen's kappa, it was possible to establish the worth of the classification method studied and to see that, for all three velocities studied, optimal results were obtained.

The operating conditions of a gas micro-turbine powered with different types of fuel was discussed in this work. As discussed in the current literature, vibrational analysis has already been addressed to study the behavior of gas turbines through different approaches. The innovation of this work is the application of discriminant analysis to accelerometric signals, highlighting how it can provide indications regarding the differences in the system vibrations when powered with pure fuels or biofuels.

Author Contributions: Conceptualization, C.C., M.S. and A.N.; methodology, V.N., S.S. and G.Q.; software, C.C., M.S. and A.N.; validation, V.N., S.S. and G.Q.; investigation, C.C., M.S. and A.N.; data curation, C.C., M.S. and A.N.; writing—original draft preparation, C.C., M.S. and A.N.; writing—review and editing, V.N., S.S. and G.Q.; visualization, C.C., M.S. and A.N.; supervision, V.N., S.S., G.Q. All authors have read and agreed to the published version of the manuscript.

Funding: This paper received no external funding.

Data Availability Statement: Not available.

Conflicts of Interest: The authors declare no conflict of interest.

References

1. Rochelle, D.; Najafi, H. A review of the effect of biodiesel on gas turbine emissions and performance. *Renew. Sustain. Energy Rev.* **2019**, *105*, 129–137. [CrossRef]
2. Bazooyar, B.; Darabkhani, H.G. Design, manufacture and test of a micro-turbine renewable energy combustor. *Energy Convers. Manag.* **2020**, *213*, 112782. [CrossRef]

3.   Ameli, S.M.; Agnew, B.; Potts, I. Integrated distributed energy evaluation software (IDEAS) Simulation of a micro-turbine based CHP system. *Appl. Therm. Eng.* **2007**, *27*, 2161–2165. [CrossRef]
4.   Karvountzis-Kontakiotis, A.; Andwari, A.M.; Pesyridis, A.; Russo, S.; Tuccillo, R.; Esfahanian, V. Application of Micro Gas Turbine in Range-Extended Electric Vehicles. *Energy* **2018**, *147*, 351–361. [CrossRef]
5.   Menkiel, B.; Donkerbroek, A.; Uitz, R.; Cracknell, R.; Ganippa, L. Combustion and soot processes of diesel and rapeseed methyl ester in an optical diesel engine. *Fuel* **2014**, *118*, 406–415. [CrossRef]
6.   Daho, T.; Vaitilingom, G.; Sanogo, O.; Ouiminga, S.K.; Zongo, A.S.; Pirioud, B.; Koulidiati, J. Combustion of vegetable oils under optimized conditions of atomization and granulometry in a modified fuel oil burner. *Fuel* **2014**, *118*, 329–334. [CrossRef]
7.   Bayındır, H.; Zerrakki Isık, M.; Argunhan, Z.; Lütfü Yücel, H.; Aydın, H. Combustion, performance and emissions of a diesel power generator fueled with biodiesel-kerosene and biodiesel-kerosene-diesel blends. *Energy* **2017**, *123*, 241–251. [CrossRef]
8.   Nascimento, M.A.; Lora, E.S.; Corrêa, P.S.; Andrade, R.V.; Rendon, M.A.; Venturini, O.J.; Ramirez, G.A. Biodiesel fuel in diesel micro-turbine engines: Modelling and experimental evaluation. *Energy* **2008**, *33*, 233–240. [CrossRef]
9.   Rakopoulos, D.C.; Rakopoulos, C.D.; Giakoumis, E.G. Impact of properties of vegetable oil, bio-diesel, ethanol and n-butanol on the combustion and emissions of turbocharged HDDI diesel engine operating under steady and transient conditions. *Fuel* **2015**, *156*, 1–19. [CrossRef]
10.   Chiaramonti, D.; Rizzo, A.M.; Spadi, A.; Prussi, M.; Riccio, G.; Martelli, F. Exhaust emissions from liquid fuel micro gas turbine fed with diesel oil, biodiesel and vegetable oil. *Appl. Energy* **2013**, *101*, 349–356. [CrossRef]
11.   Boomadevi, P.; Paulson, V.; Samlal, S.; Varatharajan, M.; Sekar, M.; Alsehli, M.; Tola, S. Impact of microalgae biofuel on microgas turbine aviation engine: A combustion and emission study. *Fuel* **2021**, *302*, 121–155. [CrossRef]
12.   Chiong, M.C.; Chong, C.T.; Ng, J.H.; Lam, S.S.; Tran, M.V.; Chong, W.W.F.; Valera-Medina, A. Liquid biofuels production and emissions performance in gas turbines: A review. *Energy Convers. Manag.* **2018**, *173*, 640–658. [CrossRef]
13.   Enagi, I.I.; Al-Attab, K.A.; Zainal, Z.A. Liquid biofuels utilization for gas turbines: A review. *Renew. Sustain. Energy Rev.* **2018**, *90*, 43–55. [CrossRef]
14.   Sallevelt, J.L.H.P.; Gudde, J.E.P.; Pozarlik, A.K.; Brem, G. The impact of spray quality on the combustion of a viscous biofuel in a micro gas turbine. *Appl. Energy* **2014**, *132*, 575–585. [CrossRef]
15.   Lv, X.; Liu, X.; Gu, C.; Weng, Y. Determination of safe operation zone for an intermediate temperature solid oxide fuel cell and gas turbine hybrid system. *Energy* **2016**, *99*, 91–102. [CrossRef]
16.   Reggio, F.; Ferrari, M.L.; Silvestri, P.; Massardo, A.F. Vibrational analysis for surge precursor definition in gas turbines. *Meccanica* **2019**, *54*, 1257–1278. [CrossRef]
17.   Kaczmarczyk, T.Z.; Żywica, G.; Ihnatowicz, E. Measurements and vibration analysis of a five-stage axial-flow microturbine operating in an ORC cycle. *Diagnostyka* **2017**, *18*, 51–58.
18.   Waumans, T.; Waumans, T.; Vleugels, P.; Peirs, J.; Al-Bender, F.; Reynaerts, D. Rotordynamic behaviour of a micro-turbine rotor on air bearings: Modelling techniques and experimental verification. In Proceedings of the International Conference on Noise and Vibration Engineering, Heverlee, Belgium, 18–20 September 2006.
19.   Talebi, S.S.; Tousi, A.M. The effects of compressor blade roughness on the steady state performance of micro-turbines. *Appl. Therm. Eng.* **2017**, *115*, 517–527. [CrossRef]
20.   Cafaro, S.; Traverso, A.; Ferrari, M.L.; Massardo, A.F. Performance Monitoring of Gas Turbine Components: A Real Case Study Using a Micro Gas Turbine Test Rig. In Proceedings of the ASME Turbo Expo, Genova, Italy, 8–12 June 2009.
21.   Rahmounea, M.B.; Hafaifaa, A.; Kouzoua, A.; Chenc, X.; Chaibetd, A. Gas turbine monitoring using neural network dynamic nonlinear autoregressive with external exogenous input modelling. *Math. Comput. Simul.* **2021**, *179*, 23–47. [CrossRef]
22.   Allouis, C.; Amoresano, A.; Capasso, R.; Langella, G.; Niola, V.; Quaremba, G. The impact of biofuel properties on emissions and performances of a micro gas turbine using combustion vibrations detection. *Fuel Process. Technol.* **2018**, *179*, 10–16. [CrossRef]
23.   Amoresano, A.; Avagliano, V.; Niola, V.; Quaremba, G. The assessment of the in-cylinder pressure by means of the morpho-dynamical vibration analysis—Methodology and application. *Int. Rev. Mech. Eng.* **2013**, *7*, 999–1006.
24.   Niola, V.; Quaremba, G.; Forcelli, A. The detection of gear noise computed by integrating the Fourier and Wavelet methods. *WSEAS Trans. Signal Process.* **2008**, *4*, 60–67.
25.   Niola, V.; Quaremba, G. The Gear Whine Noise: The influence of manufacturing process on vibro-acoustic emission of gear-box. In Proceedings of the 10th WSEAS International Conference, Recent Researches in Communications, Automation, Signal Processing, Nanotechnology, Astronomy and Nuclear Physics, Athens, Greece, 20 February 2011.
26.   Spirto, M.; Savino, S.; Quaremba, G.; Cosenza, C.; Nicolella, A.; Niola, V. Discriminant analysis of the vibrational dynamics of a gas microturbine with different fuels. In Proceedings of the 5th Jc-IFToMM International Symposium, the 28th Jc-IFToMM Symposium on Theory of Mechanism and Machines, Kyoto, Japan, 16 July 2022.
27.   Niola, V.; Quaremba, G. *Sistemi Vibrazionali Complessi Teoria, Applicazioni e Metodologie Innovative di Analisi*, 1st ed.; Nane Edizioni: Italy, 2015.
28.   Niola, V.; Spirto, M.; Savino, S.; Cosenza, C. Vibrational analysis to detect cavitation phenomena in a directional spool valve. *Int. J. Mech. Control* **2021**, *22*, 11–16.

29. Cosenza, C.; Nicolella, A.; Genovese, A.; Niola, V.; Savino, S.; Spirto, M. A Vision Based Approach to Study Lubrication Conditions in Gearwheels. In Proceedings of the 4th International Conference of the IFToMM Italy, Naples, Italy, 7 September 2022.
30. Niola, V.; Savino, S.; Quaremba, G.; Cosenza, C.; Spirto, M.; Nicolella, A. Study on the Dispersion of Lubricant Film from a Cylindrical Gearwheels with Helical Teeth by Vibrational Analysis. *WSEAS Trans. Appl. Theor. Mech.* **2021**, *16*, 274–282. [CrossRef]
31. Warrens, M.J. Five Ways to Look at Cohen's Kappa. *J. Psychol. Psychother.* **2015**, *5*, 1–4. [CrossRef]
32. Sun, S. Meta-analysis of Cohen's kappa. *Health Serv. Outcomes Res. Methodol.* **2011**, *11*, 145–163. [CrossRef]

Article

Tools and Methods for Human Robot Collaboration: Case Studies at i-LABS

Massimo Callegari *, Luca Carbonari, Daniele Costa, Giacomo Palmieri, Matteo-Claudio Palpacelli, Alessandra Papetti and Cecilia Scoccia

Department of Industrial Engineering and Mathematical Sciences, Università Politecnica delle Marche, 60024 Ancona, Italy
* Correspondence: m.callegari@univpm.it

Abstract: The collaboration among humans and machines is one of the most relevant topics in the Industry 4.0 paradigm. Collaborative robotics owes part of the enormous impact it has had in small and medium size enterprises to its innate vocation for close cooperation between human operators and robots. The i-Labs laboratory, which is introduced in this paper, developed some case studies in this sense involving different technologies at different abstraction levels to analyse the feasibility of human-robot interaction in common, yet challenging, application scenarios. The ergonomics of the processes, safety of operators, as well as effectiveness of the cooperation are some of the aspects under investigation with the main objective of drawing to these issues the attention from industries who could benefit from them.

Keywords: human-robot collaboration; human-in-the-loop; virtual reality; obstacle avoidance

Citation: Callegari, M.; Carbonari, L.; Costa, D.; Palmieri, G.; Palpacelli, M.-C.; Papetti, A.; Scoccia, C. Tools and Methods for Human Robot Collaboration: Case Studies at i-LABS. *Machines* **2022**, *10*, 997. https://doi.org/10.3390/machines10110997

Academic Editors: Huosheng Hu, Marco Ceccarelli, Giuseppe Carbone and Alessandro Gasparetto

Received: 22 July 2022
Accepted: 7 September 2022
Published: 30 October 2022

Publisher's Note: MDPI stays neutral with regard to jurisdictional claims in published maps and institutional affiliations.

1. Introduction

Until the introduction of Industry 4.0 principles, robotics in factories was mostly about machines replacing laborers who were tasked with non-ergonomic duties. The exploitation of robots was almost limited to manipulation of heavy loads or in uncomfortable positions, execution of dangerous tasks due to toxic payloads or environment, and execution of monotonous repetitive operations. Nowadays, thanks to the wide spread of *collaborative robots* (or *cobots*), the trend is shifting, especially in small and medium enterprises (SME's) [1]. In fact, small batch production and high level of product customization make these industrial entities still based on the versatility of human labor. Cobots, for their part, had a chance to easily insert themselves in this productive paradigm for they have been specifically developed for coexistence with people.

In this scenario, the collaboration of the researchers of Università Politecnica delle Marche together with five companies operating in the technology sector resulted in the creation of a laboratory which aims to disseminate the principles of Industry 4.0 in the local industrial fabric. Among them, collaborative robotics plays a crucial role, alongside with the development of tools and strategies for *Human Robot Collaboration* [2,3]. The aim is that of achieving the seamless team dynamics of an all-human team, and to this goal the last decade of research focused on several aspects [4,5] going from the machines themselves (exteroception [6,7], collision avoidance [8,9], intrinsic safety of the mechanics [10]), to their relation capabilities with people (interaction modalities [11,12], control oriented perception such as gesture recognition [13] and gaze tracking [14,15]).

Many relevant studies of the recent past can be mentioned trying to assess some of the HRC related aspects. Rusch et al. [16] quantified the beneficial impact, under both ergonomic and economic point of view, of preliminary simulated design of HRC scenarios. Papetti et al. [17] proposed a quantitative approach to evaluate a HRC simulated application. Similar topics were analysed in [18] where authors considered such opposing effects on collaborative production lines. However, besides the economic impact, ergonomics analyses surely play a fundamental role for injury hazard management [19,20]. This becomes even more important considering the conclusions drawn by Dafflon et al. [21]: in light of their claims, completely unmanned factories are not reasonably possible mainly due to the un-feasibility of such complex control systems. Therefore, processes involving humans (*Human In The Loop*, HITL) will be implemented more and more frequently, justifying the effort of developing dedicated simulation environments and evaluation metrics for anthropocentric approaches [22–25].

However, the perception of operator commands in HITL tasks remains a widely investigated topic [26–28], since it still represents a critical point of the HRC process. Robustness, effectiveness, and above all safety are the keywords to be kept in mind. Thus, the interest aroused by strategies not involving contact among machines and workers is well understendable. Dinges et al. [29] considered the use of facial expressions to assess aggravated HRC scenarios; Yu et al. [30] used a multi-sensor approach to sense both postures and gestures to be interpreted as commands. In all of these examples, the key role played by artificial vision is evident, as also featured by authors of [31].

In the following, some of these topics are exemplified by case studies approached by the researchers at i-Labs. The aim is that of providing an impression of the ongoing research developed to pursue the principles of Industry 4.0 for what concerns the cooperation among humans and machines. In particular, a simulated environment is used to evaluate the ergonomics of an industrial work-cell in the first example. The second case considers the use of virtual reality for testing of obstacle avoidance control strategies applied to a redundant collaborative industrial manipulator. The third case study shows how a simple yet effective gesture recognition strategy can be developed to control tasks executed by a dual arm cobot. For the last case study, an obstacle avoidance control algorithm has been implemented on a collaborative robot to validate the effectiveness of the proposed law.

2. Simulation Methods in HRC Design

The method proposed in this section aims to support the human-oriented HRC design by merging the enabling technologies of industry 4.0 with the concepts of human ergonomics, safety, and performance. As shown in Figure 1, it starts with the human work analysis according to objective ergonomic assessments. The XSens™MVN inertial motion capture system is used to measure joint angles and detect awkward postures. The analysis results drive the collaborative robotic cell design; for example, allocating the non-ergonomic tasks to the robot. The preliminary concept is then virtually simulated to identify critical issues from different perspectives (technical constraints, ergonomics, safety, etc.) and improve the interaction modalities. In these two phases, the NX (https://www.plm.automation.siemens.com/global/uk/products/nx/ (accessed on 1 January 2022)) and Tecnomatix (https://www.plm.automation.siemens.com/global/en/products/tecnomatix/ (accessed on 1 January 2022)) Process Simulate by Siemens are respectively used. The simulation contributes to the design optimization so that all the requirements are satisfied. The realization of the physical prototype allows the HRC experimentation, optimization, and validation before being implemented in the real production line. The experimentation phase includes a new ergonomic assessment to quantitatively estimate the potential benefits for the operator.

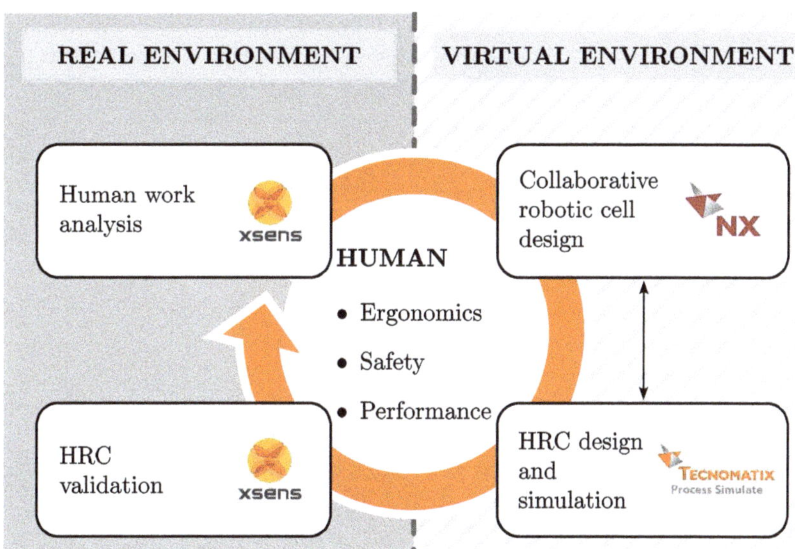

Figure 1. Simulation based method for human-oriented HRC design.

The industrial case study refers to the drawers' assembly line of LUBE Industries, the major kitchen manufacturer in Italy. Currently, a traditional robot and a CNC machine make the first part of the production line automated. Then, three manual stations complete the drawer assembly. Figure 2a shows one of the involved operators, equipped with 18 Xsens MTw (Wireless Motion Tracker), performing the tasks. The analysis involved all the labourers qualified for that specific task for a total of 6 operators (3 males, 3 females) all having average anthropometric characteristics. All the participants were informed about the goal of the study and the procedure. Also they were asked to read and sign the consent form.

The human work analysis highlighted a medium-high ergonomic risk (asymmetrical posture and stereotypy) for the operator dedicated to screwing. Before the objective evaluation, this workstation was mistakenly considered the most ergonomic of the three. Accordingly, the HRC design was based on the need to assign the screwing task to the robot to preserve the operator's health. Figure 2b shows the preliminary concept of the workstation that includes the following elements: conveyor belt; Universal Robot UR10e; collaborative screwdriver with a flexible extension; an aluminum structure where the robot is fixed by four M9 holes; L-shaped squaring system; clamping system; dispenser for feeding the screwdriver, easily and safely accessible by the operator to reload it; two plexiglass fences to reduce collisions risk.

The simulation highlighted several critical issues to be solved. To overcome some safety and technical problems, a new type of screwdriver was designed and implemented, which provides a maximum torque of 12 Nm and weighs 2.3 kg. To improve the line balancing the first two manual stations were aggregated in the cooperative robotic cell. However, the ergonomics simulation (Figure 2c) highlighted potential ergonomics risks for the operator during the drawer rotation.

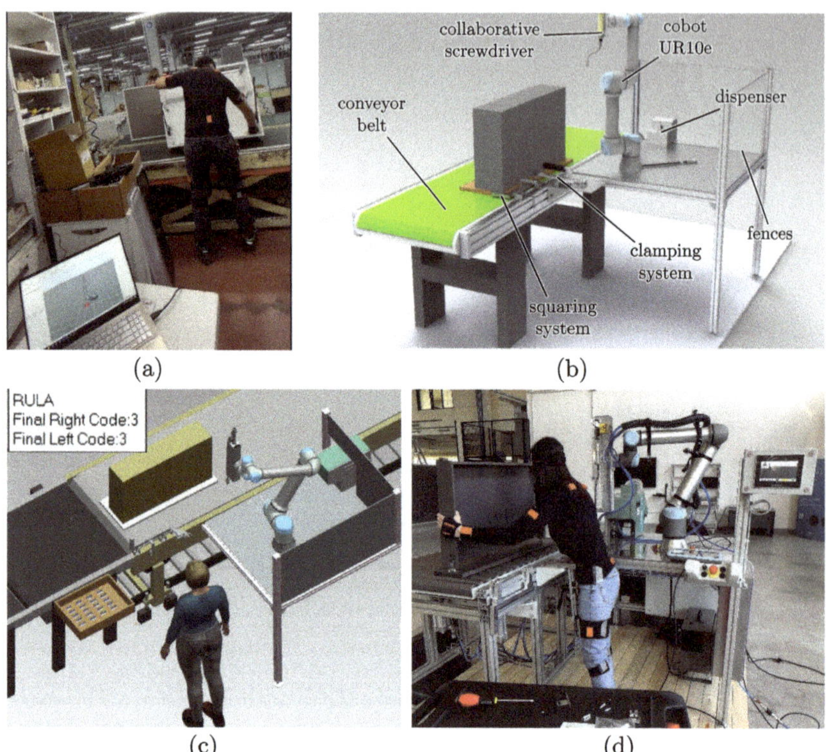

Figure 2. (**a**) human work analysis; (**b**) preliminary concept of cooperative robotic cell; (**c**) HRC simulation; (**d**) physical prototype at i-Labs.

Then, an idle rotating roller was inserted so that the operator can manually turn the conveyor belt comfortably and safely. The simulation also showed that the new cooperative cell increased the current takt time. Consequently, an autoloader (FM-503H) was introduced to avoid unnecessary cobot movements. It consists of a tilting blade (suction) that loads the screws from a collection basket. Finally, the human-robot interaction has been improved by introducing the Smart Robots (http://smartrobots.it/product/ (accessed on 1 January 2022)) vision system, which monitors the area of cooperation and enables the robot program according to human gestural commands. It allows the management of a high number of product variants.

In the new HRC workstation (Figure 2d), installed at i-Labs, negligible ergonomics risks, according to RULA (Rapid Upper Limb Assessment) and OCRA methods, were observed. As summarized in Table 1, also the Pilz Hazard Rating (PHR), which is calculated by (1), was reduced to acceptable values.

Table 1. Performance comparison between the "as-is" workstation, the preliminary concept of cooperative robotic cell and the final prototype.

	AS-IS	Pre-Simulation HRC	Validated HRC
RULA	Right = 5 🟠 Left = 3 🟡	Right = 3 🟡 Left = 3 🟡	Right = 2 🟢 Left = 2 🟢
OCRA	Right = 13.3 🟠 Left = 1.9 🟢	Right = 5.7 🟢 Left = 1.9 🟢	Right = 5.7 🟢 Left = 1.9 🟢
PHR		Event 1 = 14.06 ⚠ Event 2 = 15.62 ⚠ Event 3 = 20.62 ⚠	Event 1 = 5.62 ✓ Event 2 = 0.94 ✓ Event 3 = 0.47 ✓
Takt time **Balance**	57 s ~ 70%	< 57 s ~ 74%	< 57 s ~ 80%

$$PHR = DPH \times PO \times PA \times FE \tag{1}$$

where:

- DPH is Degree of Possible Harm
- PO is Probability of Occurrence
- PA is Possibility of Avoidance
- FE is Frequency and/or Duration of Exposure

As shown, the new production line respects the cycle time (57 s) imposed by the CNC machine and the introduction of a collaborative robot led to better line balancing (+10%). The lab test was aimed at recreating the actual working conditions, therefore both a male and a female volunteer (actually two researchers of the i-Labs group) have been asked to take part in the investigation. Also, in this case, their anthropometric characteristics can be considered average.

3. HITL: Gesture Recognition Examples

In the search for effective ways to increase collaboration between workers and robots, new types of communication can be considered. Voice commands, for example, are typical in many systems today, such as vehicles, home automation systems, assistive robotics, etc., but they are not suitable for the industrial environment, where noise and the coexistence of multiple workers in a shared space make their implementation impossible. On the other hand, image sensors, such as standard or RGBD cameras, can be exploited to capture workers' gestures, which can then be interpreted by artificial intelligence algorithms to generate a command to be sent to the robot.

A distinction should be made between applications that require the use of wearable devices and those that rely on direct hand and body recognition. Regarding the former type, the most common technology for hand gesture recognition are glove devices. As part of their work at i-Labs, the authors developed the Robotely software suite for the specific purpose of overcoming the need for wearing gloves by adopting a hand recognition system using a machine learning approach [32]. As shown in Figure 3, the software exploits MediaPipe libraries in order to recognize 20 landmarks of the hand and identify up to three different hand configurations that can be converted to commands which can be sent to whatever robot or system. A specific test case was carried out at i-Labs: a PC with a standard webcam was used to run Robotely and connected to a Universal Robots UR10e cobot in order to sequence some operations between the operator and the robot.

Figure 3. Example of gesture recognition using a standard webcam and Robotely software [32].

A second example is shown in Figure 3 where the commercial Smart Robots system is integrated with the ABB YuMi cobot. Smart Robots is a programmable system based on RGBD sensors directly integrated into the robot control software. Again, three different hand gestures can be recognized and converted into commands, such as stop in Figure 4a, calling a program variation such as a quality inspection (Figure 4b), and opening the gripper for part rejection (Figure 4c). In addition, it is possible to define areas of the workspace corresponding to the activation of different program sequences: when the operator's hand enters a predefined area of the scene (once, or typically twice to avoid inadvertent activation) a program state is changed. In Figure 4d), for example, the operator covers a letter drawn on the table with his hand to order the robot to compose that letter with LEGO bricks.

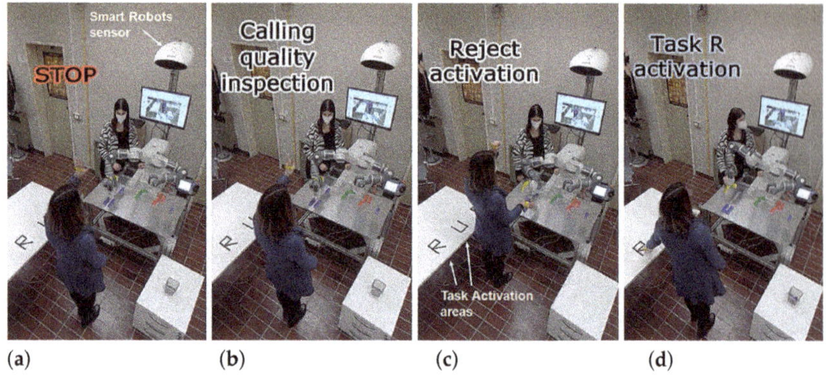

Figure 4. (a–d): example of gesture recognition phases using Smart Robots on YuMi Robot.

4. Safe Autonomous Motions: Collision Avoidance Strategies

This section recalls the obstacle avoidance algorithm introduced by authors in [8,9] and presents some experimental results. The research in this field is presented hereby as an advancement in the direction of safe human-robot coexistence. Also, the resulting control algorithm is at the base of the test case presented in the subsequent section, which makes use of virtual reality tools to test and calibrate the empirical parameters introduced by the algorithm.

For a start, it is possible to consider the velocity kinematics of a generic manipulator owning a number of actuators greater or equal to 6 (depending on its degree of redundancy). In matrix form, the velocity kinematics of such a system can be written as $\dot{x} = J\dot{q}$, where J is the $6 \times n$ arm Jacobian (with n number of the joints composing the kinematic chain of the manipulator, $n \geq 6$) and $\dot{q}$ is the vector of joint rates. In such terms, the vector of variables is $q = \begin{bmatrix} q_1 & \cdots & q_n \end{bmatrix}^T$ where $q_1 \ldots q_n$ are the n joint variables of the arm kinematic chain. With such notation, the Jacobian J depends on the structure of the manipulator. The inverse of the Jacobian J of the redundant system can be obtained as a damped inverse:

$$J^* = J^T(JJ^T + \lambda^2 I)^{-1} \tag{2}$$

where λ is the damping factor, modulated as a function of the smaller singular value of the Jacobian matrix (the interested reader is addressed to [8,9] for further details). The inverse $\mathbf{J}^*$ can be used to compute the joint velocities needed to perform a given trajectory with a Closed-Loop Inverse Kinematic (CLIK) approach:

$$\dot{\mathbf{q}} = \mathbf{J}^*(\dot{\mathbf{x}} + \mathbf{K}\mathbf{e}) \tag{3}$$

where $\dot{\mathbf{x}}$ is the vector of planned velocities, $\mathbf{K}$ is a gain matrix (usually diagonal) to be tuned on the application, and $\mathbf{e}$ is a vector of orientation and position errors ($\mathbf{e}_r$ and $\mathbf{e}_p$), defined as:

$$\mathbf{e} = \begin{bmatrix} \mathbf{e}_r \\ \mathbf{e}_p \end{bmatrix} = \begin{bmatrix} \frac{1}{2}(\mathbf{i} \times \mathbf{i}_d + \mathbf{j} \times \mathbf{j}_d + \mathbf{k} \times \mathbf{k}_d) \\ \mathbf{P} - \mathbf{P}_d \end{bmatrix} \tag{4}$$

In (4) the subscript d stands for desired planned variable, $\mathbf{P}$ is the position of the end-effector, while $\mathbf{i}$, $\mathbf{j}$ and $\mathbf{k}$ are the unit vectors of the end-effector reference frame.

The collision avoidance strategy is then implemented as a further velocity component (to be added to the trajectory joint velocities) capable of distancing the end-effector and the other parts of the robot from a given obstacle. Such contribution is a function of the distance among the obstacle and every body of the robotic system. To this purpose, the bodies have been represented by means of two control points, $\mathbf{A}$ and $\mathbf{B}$ referring to Figure 5. Called $\mathbf{C}$ the centre of a generic obstacle, the distance among it and the segment $\overline{AB}$ can be differently computed in three different scenarios:

- $(\mathbf{C} - \mathbf{A})^T(\mathbf{B} - \mathbf{A}) \leq 0$, Figure 5a: in this case the obstacle is closer to the $\mathbf{A}$ tip than to any other point of $\overline{AB}$; the distance d among the obstacle center and the line $\overline{AB}$ coincides with the length of $\overline{AC}$.
- $0 < (\mathbf{C} - \mathbf{A})^T(\mathbf{B} - \mathbf{A}) < |\overline{AB}|^2$, Figure 5b: the minimum distance d lies within points $\mathbf{A}$ and $\mathbf{B}$. In this case, it is:

$$d = \frac{|(\mathbf{C} - \mathbf{A}) \times (\mathbf{B} - \mathbf{A})|}{|\overline{AB}|} \tag{5}$$

- $(\mathbf{C} - \mathbf{A})^T(\mathbf{B} - \mathbf{A}) \geq |\overline{AB}|^2$, Figure 5c: point $\mathbf{C}$ is closer to any other point, therefore $d = |\overline{BC}|$.

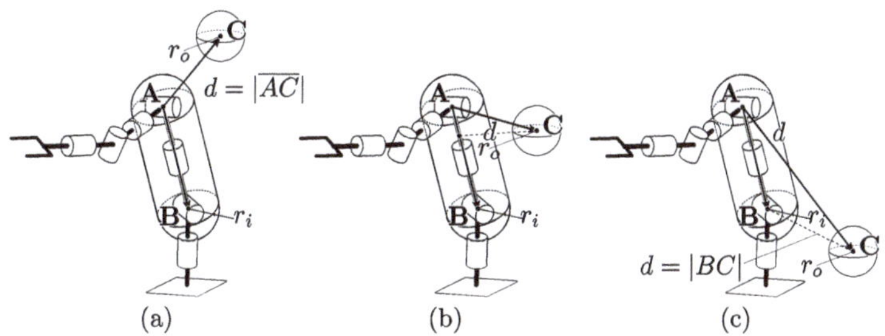

Figure 5. (**a–c**): obstacle to robot body distance in the three considered cases.

It is worth remarking that for this paper only spherical obstacles were considered, although similar approaches can be developed for objects of any shape starting from the distance primitives here defined.

Now the set of repulsive velocities for the ith body of the system can be introduced as:

$$\dot{\mathbf{q}}_{r,i} = \psi v_r \left(k \mathbf{J}^*_{\mathbf{A},i} \frac{(\mathbf{A} - \mathbf{C})}{|\overline{AC}|} + (1 - k)\mathbf{J}^*_{\mathbf{B},i} \frac{(\mathbf{B} - \mathbf{C})}{|\overline{BC}|} \right) \tag{6}$$

where:

- ψ is an activation parameter, function of d, of the obstacle dimension r_o and the length r_i which characterize the body ith (as represented in Figure 5, the dimension r_i defines a region around the line $\overline{AB}$ given by the intersection of two spheres centred in **A** and **C**, and a cylinder aligned with $\overline{AB}$, of radius r_i). The activation parameter can be whatever function such that $\psi = 0$ if $i > r_i + r_o$, and $\psi = 1$ if $i \leq r_i + r_o$. Actually, such a transition can be made smoother by the adoption of feasible functions (polynomials, logarithmic, etc.).
- v_r is a customized scalar representing the module of the repulsive velocity provided by obstacle to the body.
- k is a parameter which depends on the three cases of Figure 5: in the first case $k = 1$ so that only the point **A** influences $\dot{\mathbf{q}}_{r,i}$; in the second case both **A** and **B** are considered proportionally to their distance from **C**, thus $k = 1 - (\mathbf{C} - \mathbf{A})^T (\mathbf{B} - \mathbf{A})/|\overline{AB}|^2$; at last, in the third case $k = 0$ so that only the point **B** is relevant to $\dot{\mathbf{q}}_{r,i}$.
- $\mathbf{J}^*_{\mathbf{A},i}$ and $\mathbf{J}^*_{\mathbf{B},i}$ are the damped inverse of the Jacobian matrices of points **A** and **B**.

At this point, the CLIK control law (3) can be completed as:

$$\dot{\mathbf{q}} = \mathbf{J}^*(\dot{\mathbf{x}} + \mathbf{Ke}) + \sum_{i=1}^{m} \dot{\mathbf{q}}_{r,i} \tag{7}$$

being m the number of segments used to describe the manipulator.

At last, some experimental results are shown as demonstration of the obstacle avoiding strategy consistency. Going a little into details, the CLIK control law of the KUKA KMR iiwa (a 7 axes redundant collaborative manipulator) was built using 6 different segments, while the two obstacles were modelled as spheres. As shown by the test rig image (Figure 6), the robot carries a collaborative gripper by Schunk which is also considered for the definition of the robot segments (cfr Figure 7). The results are presented in terms of joints positions and rates, and pose and velocity of the robot end-effector (EE). Each variable is referred, for comparison, to the values collected during the execution of the same trajectory without obstacles in the robot workspace. The robot was controlled via Matlab with a cycle frequency of ~ 150 Hz. It is worth remarking that the visualization provided in Figure 7 is a plot built on experimental data. The trajectory under investigation is a simple linear motion with constant orientation of the EE. Two obstacles have been put in the robot workspace, one directly on the planned trajectory (O_1 in Figure 7) and the other in the space occupied by the robot non-terminal bodies (O_2 in Figure 7). The results show how the control law is able to follow the given trajectory even in presence of the two obstacles. As expected, obstacle O_1 (met approximately in the time span 2–4.5 s, gray area in graphs of Figure 7) prevents the robot EE from maintaining the desired pose: the control law permits the dodging of O_1 deviating the bare minimum from the planned motion. Regarding obstacle O_2 (met in the span 5.5–7.5 s, purple areas), the control law exploits the robot redundancy to avoid the contact among the robot elbow and the obstacle while maintaining the EE on the right trajectory and orientation.

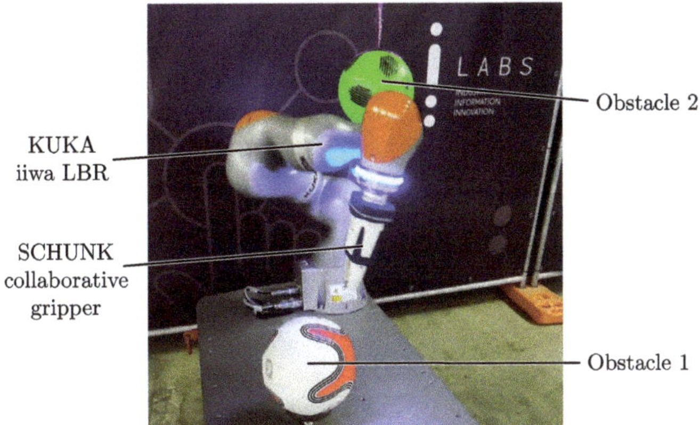

Figure 6. Experimental test rig: the KUKA iiwa LBR avoiding two different obstacles in its workspace.

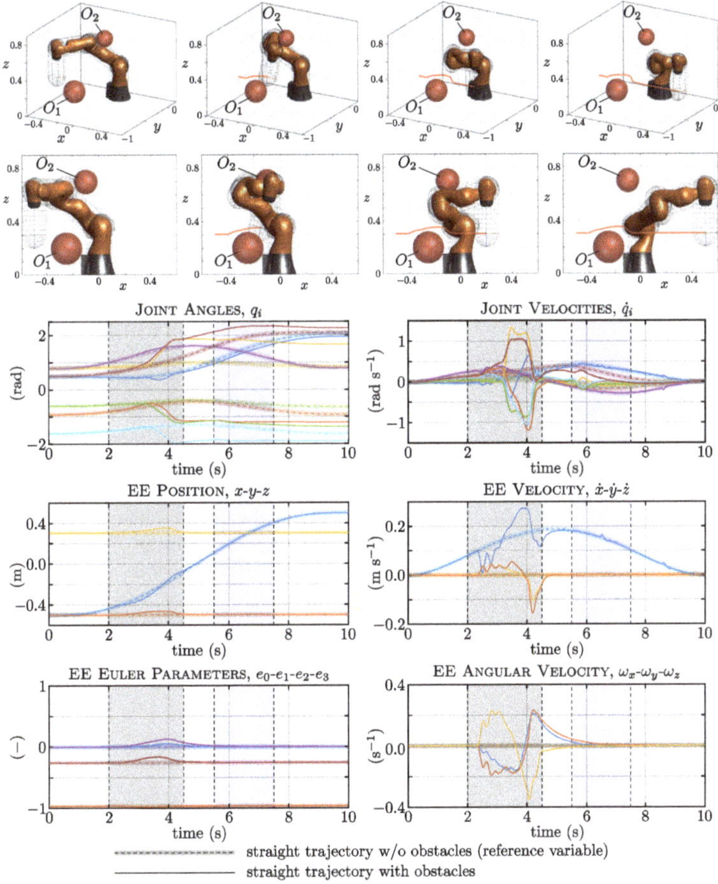

Figure 7. Experimental results: the KUKA iiwa LBR avoiding two different obstacles in its workspace.

5. Virtual Reality Based Design Methods

In this section a virtual implementation of the obstacle avoidance control strategy is presented. Aside the experimentation previously shown, the algorithm has been tested in advance in a simulated interactive scenario to demonstrate the feasibility of the control law, and to tune the parameters used to optimize the robot response to a dynamic obstacle. Such an approach, schematically shown in Figure 8, represents an efficient paradigm for design and virtual testing of HITL applications. Starting in a simulated environment, the interaction workflow can be designed considering also the presence of labourer and their impact on the production. A further optimization step can be added to optimize the interaction among humans and robots exploiting virtual reality. The VR tools permit one to disregard the modeling of labourer behaviour since their presence is played by actual humans who bring in the design process not only the actual expertise of the production line final users, but also the randomness of human actions and movements with further advantages in terms of security features design.

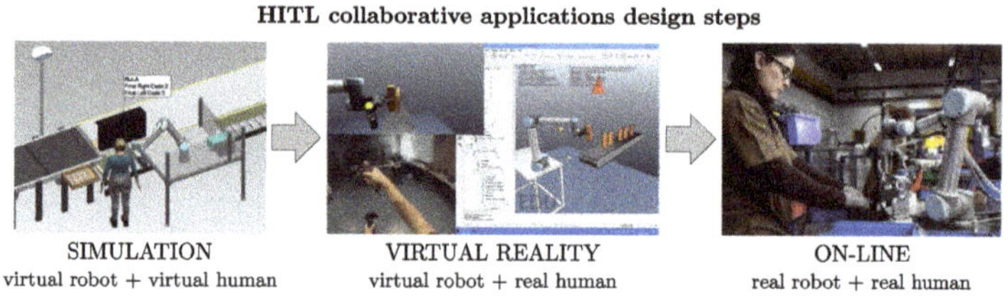

HITL collaborative applications design steps

SIMULATION
virtual robot + virtual human

VIRTUAL REALITY
virtual robot + real human

ON-LINE
real robot + real human

Figure 8. Design steps of a human in the loop (HITL) applications.

For this purpose, the idea was to create a *virtual reality* (VR) application in which a real obstacle is inserted in the simulation loop, while the robot is still completely virtual. The advantages of this approach are many: firstly, the algorithms can be modified and fine-tuned ensuring safety for the operators, secondly these changes can be made easily and quickly, reducing the development time of the real application.

The control architecture used for the VR application is shown in Figure 9, while Figure 10 is the experimental rig. The core of the system is the VR engine, running on a standard PC, developed by the SteamVR development suite with Unreal Engine 4 used to model the kinematics of the KUKA robot. A classic HTC VR set is used to equip the operator: the HMD provides an immersive three-dimensional representation of the workspace, which in this case is simply the robot mounted on a stand; the controller, manually managed by the operator, is rigidly connected to a virtual obstacle, shaped like a sphere, which can be moved in the virtual space to interfere with the robot's motion. The control of the robot is executed by the same PC in a parallel thread developed in Matlab, with a frame rate set to a typical value for communication protocols with robots (e.g., 125 Hz or 250 Hz, depending on the manufacturer). The frame rate of the VR engine is set to 60 Hz, which is the standard display refresh rate in VR applications. The communication between the two executed threads is realized by TCP/IP protocol. A classic pick and place task is simulated, thus the robot controller sends joint rotations to the VR engine at each time step, similarly to what is done in real robotic systems. The trajectory of the robot is updated in real time by the collision avoidance algorithm if the obstacle enters in a safety region of the manipulator. The position of the obstacle is known once the VR engine reads the coordinates of the controller held by the operator and sends this information to the robot controller.

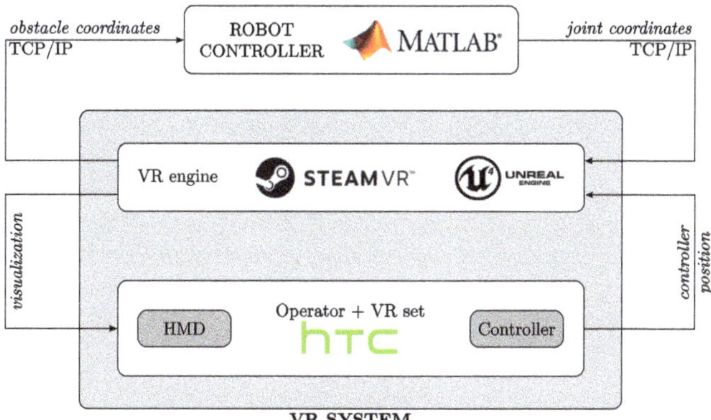

Figure 9. Architecture of the control loop for the VR application.

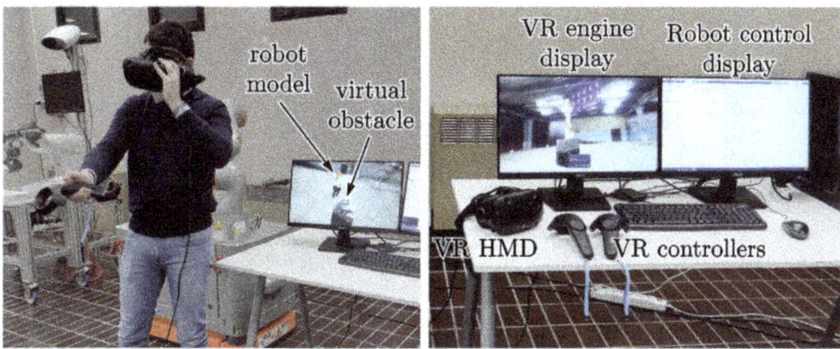

Figure 10. VR system components.

6. Concluding Remarks

The paper showed the results obtained by the i-Labs laboratory in the wide field of research on human-robot interaction. Many technologies have been investigated and experimented to horizontally approach the issue of close cooperation among labourers and machines, going from off-line design, to on-line safety oriented control strategies.

The first case study showed a possible approach to the design and optimization of an assembly work-cell. The optimization, oriented at enhancing the labourer ergonomics and safety, allowed the realization of the prototype work-cell for the cobot assisted assembly of furniture pieces. The outcome quality has been quantitatively estimated.

Similar aspects of Human In The Loop operations have also been investigated in the second case study, which was about the realization of a visual based interaction paradigm among humans and robot. In this case, a depth camera was used to recognize the labourer commands which allowed the execution of simple tasks by the robot. This kind of contactless interaction pushed a little bit further in direction of security, which have been the core focus of the third case study.

The third case introduced an innovative control algorithm for active obstacle avoidance, applicable to any cobot. The strategy, based on the method of repulsive fields of velocity, was tested on a redundant industrial cobot to evaluate the possibilities offered in terms of both safety and task execution ability. At last, a Virtual Reality approach for the detail design an exploitation of the obstacle avoidance strategy has been described in the fourth case study.

The technologies which have been separately presented in this manuscript, actually represent some of the enabling technologies in the field of Industry 4.0 and 5.0 which place the human operator at the very center of the production process. Such a knowledge base can be exploited for the enhancement of industrial processes, especially in SMEs, which represent the main matter of the local industrial tissue.

Author Contributions: Conceptualization, G.P. and A.P.; methodology, G.P. and A.P.; software, L.C.; validation, D.C. and C.S.; formal analysis, L.C.; investigation, M.-C.P.; resources, M.C.; data curation, C.S.; writing—original draft preparation, L.C.; writing—review and editing, M.C.; visualization, C.S.; supervision, G.P. and A.P.; project administration, M.-C.P. All authors have read and agreed to the published version of the manuscript.

Funding: This work was partly funded by the project URRA, "Usability of robots and reconfigurability of processes: enabling technologies and use cases", on the topics of User-Centered Manufacturing and Industry 4.0, which is part of the project EU ERDF, POR MARCHE Region FESR 2014/2020–AXIS 1–Specific Objective 2–ACTION 2.1, "HD3Flab-Human Digital Flexible Factory of the Future Laboratory", coordinated by the Polytechnic University of Marche.

Data Availability Statement: No publicly archived dataset has been used or generated during the research.

Conflicts of Interest: The authors declare no conflict of interest.

References

1. Galin, R.; Meshcheryakov, R. Automation and robotics in the context of Industry 4.0: The shift to collaborative robots. In *IOP Conference Series: Materials Science and Engineering*; IOP Publishing: Bristol, UK, 2019; Volume 537, p. 032073.
2. Vicentini, F. Terminology in safety of collaborative robotics. *Robot. Comput. Integr. Manuf.* **2020**, *63*, 101921. [CrossRef]
3. Costa, G.d.M.; Petry, M.R.; Moreira, A.P. Augmented reality for human–robot collaboration and cooperation in industrial applications: A systematic literature review. *Sensors* **2022**, *22*, 2725. [CrossRef] [PubMed]
4. Ajoudani, A.; Zanchettin, A.M.; Ivaldi, S.; Albu-Schäffer, A.; Kosuge, K.; Khatib, O. Progress and prospects of the human–robot collaboration. *Auton. Robot.* **2018**, *42*, 957–975. [CrossRef]
5. Mukherjee, D.; Gupta, K.; Chang, L.H.; Najjaran, H. A survey of robot learning strategies for human-robot collaboration in industrial settings. *Robot. Comput. Integr. Manuf.* **2022**, *73*, 102231. [CrossRef]
6. Morato, C.; Kaipa, K.; Zhao, B.; Gupta, S.K. Safe human robot interaction by using exteroceptive sensing based human modeling. In *International Design Engineering Technical Conferences and Computers and Information in Engineering Conference*; American Society of Mechanical Engineers: New York, NY, USA, 2013; Volume 55850, p. V02AT02A073.
7. Dawood, A.B.; Godaba, H.; Ataka, A.; Althoefer, K. Silicone-based capacitive e-skin for exteroception and proprioception. In Proceedings of the 2020 IEEE/RSJ International Conference on Intelligent Robots and Systems (IROS), Las Vegas, NV, USA, 25–29 October 2020; pp. 8951–8956.
8. Chiriatti, G.; Palmieri, G.; Scoccia, C.; Palpacelli, M.C.; Callegari, M. Adaptive obstacle avoidance for a class of collaborative robots. *Machines* **2021**, *9*, 113. [CrossRef]
9. Palmieri, G.; Scoccia, C. Motion planning and control of redundant manipulators for dynamical obstacle avoidance. *Machines* **2021**, *9*, 121. [CrossRef]
10. Lee, W.B.; Lee, S.D.; Song, J.B. Design of a 6-DOF collaborative robot arm with counterbalance mechanisms. In Proceedings of the 2017 IEEE International Conference on Robotics and Automation (ICRA), Marina Bay Sands, Singapore, 29 May–3 June 2017; pp. 3696–3701.
11. Katyara, S.; Ficuciello, F.; Teng, T.; Chen, F.; Siciliano, B.; Caldwell, D.G. Intuitive tasks planning using visuo-tactile perception for human robot cooperation. *arXiv* **2021**, arXiv:2104.00342.
12. Saunderson, S.P.; Nejat, G. Persuasive robots should avoid authority: The effects of formal and real authority on persuasion in human-robot interaction. *Sci. Robot.* **2021**, *6*, eabd5186. [CrossRef] [PubMed]
13. Kollakidou, A.; Haarslev, F.; Odabasi, C.; Bodenhagen, L.; Krüger, N. HRI-Gestures: Gesture Recognition for Human-Robot Interaction. In Proceedings of the 17th International Joint Conference on Computer Vision, Imaging and Computer Graphics Theory and Applications–Volume 5 VISAPP, online, 5–7 February 2022; pp. 559–566. [CrossRef]
14. Sharma, V.K.; Biswas, P. Gaze Controlled Safe HRI for Users with SSMI. In Proceedings of the 20th International Conference on Advanced Robotics (ICAR), Ljubljana, Slovenia, 6–10 December 2021; pp. 913–918.
15. Li, W.; Yi, P.; Zhou, D.; Zhang, Q.; Wei, X.; Liu, R.; Dong, J. A novel gaze-point-driven HRI framework for single-person. In *International Conference on Collaborative Computing: Networking, Applications and Worksharing*; Springer: Berlin/Heidelberg, Germany, 2021; pp. 661–677.
16. Rusch, T.; Spitzhirn, M.; Sen, S.; Komenda, T. Quantifying the economic and ergonomic potential of simulated HRC systems in the focus of demographic change and skilled labor shortage. In Proceedings of the 2021 IEEE International Conference on Industrial Engineering and Engineering Management (IEEM), Marina Bay Sands, Singapore, 13–16 December 2021; pp. 1377–1381.
17. Papetti, A.; Ciccarelli, M.; Scoccia, C.; Germani, M. A multi-criteria method to design the collaboration between humans and robots. *Procedia CIRP* **2021**, *104*, 939–944. [CrossRef]

18. Stecke, K.E.; Mokhtarzadeh, M. Balancing collaborative human–robot assembly lines to optimise cycle time and ergonomic risk. *Int. J. Prod. Res.* **2021**, *60*, 25–47. [CrossRef]
19. Ranavolo, A.; Chini, G.; Draicchio, F.; Silvetti, A.; Varrecchia, T.; Fiori, L.; Tatarelli, A.; Rosen, P.H.; Wischniewski, S.; Albrecht, P.; et al. Human-Robot Collaboration (HRC) Technologies for reducing work-related musculoskeletal diseases in Industry 4.0. In *Congress of the International Ergonomics Association*; Springer: Berlin/Heidelberg, Germany, 2021; pp. 335–342.
20. Advincula, B. User experience survey of innovative softwares in evaluation of industrial-related ergonomic hazards: A focus on 3D motion capture assessment. In Proceedings of the SPE Annual Technical Conference and Exhibition, Dubai, UAE, 23 September 2021.
21. Dafflon, B.; Moalla, N.; Ouzrout, Y. The challenges, approaches, and used techniques of CPS for manufacturing in Industry 4.0: A literature review. *Int. J. Adv. Manuf. Technol.* **2021**, *113*, 2395–2412. [CrossRef]
22. Maurice, P.; Schlehuber, P.; Padois, V.; Measson, Y.; Bidaud, P. Automatic selection of ergonomie indicators for the design of collaborative robots: A virtual-human in the loop approach. In Proceedings of the 2014 IEEE-RAS International Conference on Humanoid Robots, Madrid, Spain, 18–20 November 2014; pp. 801–808.
23. Garcia, M.A.R.; Rojas, R.; Gualtieri, L.; Rauch, E.; Matt, D. A human-in-the-loop cyber-physical system for collaborative assembly in smart manufacturing. *Procedia CIRP* **2019**, *81*, 600–605. [CrossRef]
24. Brunzini, A.; Papetti, A.; Messi, D.; Germani, M. A comprehensive method to design and assess mixed reality simulations. *Virtual Real.* **2022**, 1–19. [CrossRef]
25. Ciccarelli, M.; Papetti, A.; Cappelletti, F.; Brunzini, A.; Germani, M. Combining World Class Manufacturing system and Industry 4.0 technologies to design ergonomic manufacturing equipment. *Int. J. Interact. Des. Manuf.* **2022**, *16*, 63–279. [CrossRef]
26. Liu, H.; Wang, L. Latest developments of gesture recognition for human–robot collaboration. In *Advanced Human-Robot Collaboration in Manufacturing*; Springer: Berlin/Heidelberg, Germany, 2021; pp. 43–68.
27. Askarpour, M.; Lestingi, L.; Longoni, S.; Iannacci, N.; Rossi, M.; Vicentini, F. Formally-based model-driven development of collaborative robotic applications. *J. Intell. Robot. Syst.* **2021**, *102*, 1–26. [CrossRef]
28. Scoccia, C.; Ciccarelli, M.; Palmieri, G.; Callegari, M. Design of a Human-Robot Collaborative System: Methodology and Case Study. International Design Engineering Technical Conferences and Computers and Information in Engineering Conference. *Am. Soc. Mech. Eng.* **2021**, *85437*, V007T07A045.
29. Dinges, L.; Al-Hamadi, A.; Hempel, T.; Al Aghbari, Z. Using facial action recognition to evaluate user perception in aggravated HRC scenarios. In Proceedings of the 12th International Symposium on Image and Signal Processing and Analysis (ISPA), Zagreb, Croatia, 13–15 September 2021; pp. 195–199.
30. Yu, J.; Li, M.; Zhang, X.; Zhang, T.; Zhou, X. A multi-sensor gesture interaction system for human-robot cooperation. In Proceedings of the 2021 IEEE International Conference on Networking, Sensing and Control (ICNSC), Nanjing, China, 30 October–2 November 2021; Volume 1, pp. 1–6.
31. Pandya, J.G.; Maniar, N.P. Computer Vision-Guided Human–Robot Collaboration for Industry 4.0: A Review. *Recent Adv. Mech. Infrastruct.* **2022**, 147–155.
32. Scoccia, C.; Menchi, G.; Ciccarelli, M.; Forlini, M.; Papetti, A. Adaptive real-time gesture recognition in a dynamic scenario for human-robot collaborative applications. In Proceedings of the 4th IFToMM ITALY Conference, Napoli, Italy, 7–9 September 2022; pp. 637–644.

Article

Configuration-Dependent Substructuring as a Tool to Predict the Vibrational Response of Mechanisms

Jacopo Brunetti [1], Walter D'Ambrogio [1,*] and Annalisa Fregolent [2]

[1] Dipartimento di Ingegneria Industriale e dell'Informazione e di Economia, Università dell'Aquila, Piazzale Pontieri, Monteluco di Roio, I-67100 L'Aquila, Italy
[2] Dipartimento di Ingegneria Meccanica e Aerospaziale, Università di Roma La Sapienza, Via Eudossiana, 18, I-00184 Roma, Italy
* Correspondence: walter.dambrogio@univaq.it

Abstract: Dynamic substructuring allows us to predict the dynamic behavior of mechanical systems built by linking together several subsystems, whose dynamic behavior is known. The classical formulation, originally conceived for invariant systems, was extended by the authors to include mechanical systems made by invariant subsystems that may be coupled in different configurations. A mechanism is a typical example of a mechanical system built by coupling together invariant subsystems; during its motion, it can take several configurations that significantly affect its vibrational behavior. Therefore, the configuration-dependent substructuring approach can provide meaningful insights into the dynamic behavior of the mechanism. In this paper, the proposed approach is exploited to evaluate the vibrational behavior of a three-point linkage, a widely used mechanism to connect agricultural tractors to operating machines, considering a significant range of operative configurations. The proposed substructuring approach is able to predict the frequency response functions, the natural frequencies and the mode shapes of the mechanism in a wide range of configurations.

Keywords: configuration-dependent substructuring; dynamics of linkages; three-points linkage; vibrations

Citation: Brunetti, J.; D'Ambrogio, W.; Fregolent, A. Configuration-Dependent Substructuring as a Tool to Predict the Vibrational Response of Mechanisms. *Machines* **2022**, *10*, 1146. https://doi.org/10.3390/machines10121146

Academic Editors: Marco Ceccarelli, Giuseppe Carbone and Alessandro Gasparetto

Received: 2 August 2022
Accepted: 28 November 2022
Published: 1 Decmber 2022

Publisher's Note: MDPI stays neutral with regard to jurisdictional claims in published maps and institutional affiliations.

1. Introduction

Dynamic substructuring allows us to predict the dynamic behavior of mechanical systems built by coupling together several subsystems, whose dynamic behavior is known. The classical formulation [1] deals with invariant systems. Applications on configuration-dependent systems are presented in [2,3] without exploiting dynamic substructuring. Dynamic substructuring was extended in recent years to include mechanical systems made by invariant subsystems that may assume different configurations during motion. Specifically, the problem of frictional sliding contact and of rolling contact are respectively considered in [4,5], while friction-induced vibrations are tackled in [6,7]. More generally, configuration-dependent substructuring can deal with mechanical systems composed of subsystems in relative motion with respect to each other, such as mechanisms and linkages as outlined in [8]. In fact, a linkage is a typical example of a mechanical system built by coupling together invariant subsystems; during its motion it can take several configurations that significantly affect its vibrational behavior. A classical approach to deal with the dynamics of mechanisms with deformable links is provided by the framework of multibody systems dynamics [9,10]. The multibody approaches are typically able to perform the kinematic and dynamic analysis of each link in the time domain by considering at the same time rigid body motion and vibrations. This is particularly useful in simulations aimed to verify the behavior of the mechanism under specified loads. However, a frequency domain analysis of such transient responses might not provide significant results for other design purposes mainly because the input forces might not properly excite the mechanism.

On the contrary, configuration-dependent substructuring can be directly formulated in the frequency domain. On one side, this allows the description of a complex link, that would be difficult to model properly, in terms of an experimentally determined Frequency Response Function. So far, multibody approaches do not allow the use of experimentally identified components. On the other side, the vibrational behavior of the mechanism can be analyzed at every position using the configuration-dependent frequency response function from which configuration-dependent natural frequencies and vibration modes can be identified. This quasi-static information set is very important for design purposes since it highlights most of the potential dynamic problems that can occur during the mechanism's operation.

Since the proposed approach is quasi-static and a linearization of the system is performed on each configuration, the effects of friction and of nonlinearities in general, are not accounted. Friction contact problems can be accounted in the subsructuring framework as shown in [6,7]. However, in this case, the problem must be solved in the time domain and no experimentally derived models can be used. For these reasons, the configuration-dependent substructuring approach cannot be seen as an alternative to multibody analysis but rather as a technique providing different and complementary information.

The original contribution of this paper lies in envisaging the application of substructuring techniques, typically devised for structures, to mechanisms and linkages. Specifically, as a proof of concept, configuration-dependent substructuring is applied on a three-point linkage, a widely used mechanism to connect agricultural tractors to operating machines [11]. The vibrational behavior of this system is evaluated throughout a significant range of operative configurations. Frequency response functions, natural frequencies and mode shapes of the mechanism are predicted for each analyzed configuration.

2. Substructure Coupling in the Frequency Domain

A dynamic system made up of n coupled subsystems is considered. Each subsystem r can be described using the mass, stiffness and damping matrices $\mathbf{M}^{(r)}$, $\mathbf{K}^{(r)}$ and $\mathbf{C}^{(r)}$, from which the dynamic stiffness matrix can be computed as $\mathbf{Z}^{(r)}(\omega) = \mathbf{K}^{(r)} - \omega^2\mathbf{M}^{(r)} + i\omega\mathbf{C}^{(r)}$.

For a given linear and invariant subsystem r, the equation of motion can be expressed in the frequency domain as:

$$\mathbf{Z}^{(r)}(\omega)\boldsymbol{u}^{(r)}(\omega) = \boldsymbol{f}^{(r)}(\omega) + \boldsymbol{g}^{(r)}(\omega) \tag{1}$$

where:

$\mathbf{Z}^{(r)}$: dynamic stiffness matrix of subsystem r;

$\boldsymbol{u}^{(r)}$: vector of displacements of subsystem r;

$\boldsymbol{f}^{(r)}$: vector of external forces acting on subsystem r;

$\boldsymbol{g}^{(r)}$: vector of connecting forces with other subsystems (internal constraints).

By writing the equation of motion of the n subsystems in a block diagonal format, it is obtained, after leaving out the frequency dependence:

$$\mathbf{Z}\boldsymbol{u} = \boldsymbol{f} + \boldsymbol{g} \tag{2}$$

with

$$\mathbf{Z} = \begin{bmatrix} \mathbf{Z}^{(1)} & & \\ & \ddots & \\ & & \mathbf{Z}^{(n)} \end{bmatrix}, \quad \boldsymbol{u} = \begin{Bmatrix} \boldsymbol{u}^{(1)} \\ \vdots \\ \boldsymbol{u}^{(n)} \end{Bmatrix}, \quad \boldsymbol{f} = \begin{Bmatrix} \boldsymbol{f}^{(1)} \\ \vdots \\ \boldsymbol{f}^{(n)} \end{Bmatrix}, \quad \boldsymbol{g} = \begin{Bmatrix} \boldsymbol{g}^{(1)} \\ \vdots \\ \boldsymbol{g}^{(n)} \end{Bmatrix}$$

To couple together the n subsystems, compatibility and equilibrium conditions must be enforced. Compatibility at the interface DoFs means that any pair of corresponding DoFs $u_l^{(r)}$ and $u_m^{(s)}$, i.e., DoF l on subsystem r and DoF m on subsystem s must share the same displacement, that is $u_l^{(r)} - u_m^{(s)} = 0$.

Generally, this condition can be written as:

$$\mathbf{B}u = 0 \tag{3}$$

where each row of **B** refers to a pair of corresponding DoFs.

Equilibrium implies that internal constraint forces must be balanced. They arise when two subsystems are connected together at a pair of corresponding DoFs. Therefore, the sum of internal constraint forces must be zero for any pair of corresponding DoFs, i.e., $g_l^{(r)} + g_m^{(s)} = 0$.

Moreover, if DoF k on subsystem q is not a coupling DoF, $g_k^{(q)} = 0$.

Generally, the equilibrium conditions can be written as:

$$\mathbf{L}^T g = 0 \tag{4}$$

where **L** is a localization matrix.

The system of Equations (2)–(4) provides the so-called three-field formulation, defining the coupling between any number of subsystems:

$$\begin{cases} \mathbf{Z}u = f + g \\ \mathbf{B}u = 0 \\ \mathbf{L}^T g = 0 \end{cases} \tag{5}$$

In the dual assembly [1,12], all DoFs are retained, i.e., each coupling DoF among two substructures appears twice. The equilibrium condition $g_l^{(r)} + g_m^{(s)} = 0$ at a pair of coupling DoFs is ensured by selecting $g_l^{(r)} = -\lambda$ and $g_m^{(s)} = \lambda$. Therefore, the connecting forces can be written in the form:

$$g = -\mathbf{B}^T \lambda \tag{6}$$

where the Lagrance multipliers λ represent the intensities of connecting forces.

The equilibrium condition (4) can be rewritten as:

$$\mathbf{L}^T g = -\mathbf{L}^T \mathbf{B}^T \lambda = 0 \qquad \forall \lambda \tag{7}$$

The Equation (7) is always satisfied, thus the three-field formulation (5) reduces to:

$$\begin{cases} \mathbf{Z}u + \mathbf{B}^T \lambda = f \\ \mathbf{B}u = 0 \end{cases} \tag{8}$$

To eliminate λ, the following steps can be performed. The first of Equation (8) becomes:

$$u = -\mathbf{Z}^{-1}\mathbf{B}^T \lambda + \mathbf{Z}^{-1} f \tag{9}$$

Equation (9) can be substituted in the second of Equation (8) giving:

$$\mathbf{B}\mathbf{Z}^{-1}\mathbf{B}^T \lambda = \mathbf{B}\mathbf{Z}^{-1} f \quad \Rightarrow \quad \lambda = \left(\mathbf{B}\mathbf{Z}^{-1}\mathbf{B}^T \right)^{-1} \mathbf{B}\mathbf{Z}^{-1} f \tag{10}$$

By substituting λ in the first of Equation (8), one finally gets:

$$\mathbf{Z}u + \mathbf{B}^T \left(\mathbf{B}\mathbf{Z}^{-1}\mathbf{B}^T \right)^{-1} \mathbf{B}\mathbf{Z}^{-1} f = f$$

$$\Rightarrow \quad u = \left(\mathbf{Z}^{-1} - \mathbf{Z}^{-1}\mathbf{B}^T \left(\mathbf{B}\mathbf{Z}^{-1}\mathbf{B}^T \right)^{-1} \mathbf{B}\mathbf{Z}^{-1} \right) f \tag{11}$$

Since $\mathbf{H}^{(r)} = [\mathbf{Z}^{(r)}]^{-1}$ is the Frequency Response Function (FRF) matrix of the r-th subsystem, it can be written:

$$\mathbf{Z}^{-1} = \mathbf{H} = \begin{bmatrix} \mathbf{H}^{(1)} & & \\ & \ddots & \\ & & \mathbf{H}^{(n)} \end{bmatrix} \tag{12}$$

Therefore, Equation (11) becomes:

$$u = \left(\mathbf{H} - \mathbf{H}\mathbf{B}^T \left(\mathbf{B}\mathbf{H}\mathbf{B}^T \right)^{-1} \mathbf{B}\mathbf{H} \right) f \tag{13}$$

The FRF matrix of the coupled system $\mathbf{H}_c$ satisfies a relation of the kind $u = \mathbf{H}_c f$, thus:

$$\mathbf{H}_c = \mathbf{H} - \mathbf{H}\mathbf{B}^T \left(\mathbf{B}\mathbf{H}\mathbf{B}^T \right)^{-1} \mathbf{B}\mathbf{H} \tag{14}$$

Because of dual assembly, $\mathbf{H}_c$ contains twice the rows and columns corresponding to the coupling DoFs. Consequently, one row and one column for each coupling DoF can be eliminated.

2.1. Configuration Dependent Interface

When a relative motion exists between two coupled bodies, systems built from time-invariant component subsystems subjected to configuration-dependent coupling conditions can be considered.

In this case, configuration-dependent compatibility and equilibrium conditions are found. For a given configuration χ, compatibility can be expressed as:

$$\mathbf{B}_\mathbf{C}(\chi)u(\chi) = 0 \tag{15}$$

where each row of $\mathbf{B}_\mathbf{C}(\chi)$ refers to a pair of corresponding DoFs at configuration χ.

The equilibrium condition $g_l^{(r)}(\chi) + g_m^{(s)}(\chi) = 0$ at a pair of corresponding DoFs is again ensured by selecting $g_l^{(r)}(\chi) = -\lambda(\chi)$ and $g_m^{(s)}(\chi) = \lambda(\chi)$. Therefore, the connecting forces can be written in the form:

$$g(\chi) = -\mathbf{B}_\mathbf{E}^T(\chi)\lambda(\chi) \tag{16}$$

where $\lambda(\chi)$ are configuration-dependent Lagrange multipliers corresponding to connecting force intensities. Moreover, $\mathbf{B}_\mathbf{E}(\chi)$ is generally different from the matrix $\mathbf{B}_\mathbf{C}(\chi)$ used to enforce the compatibility condition, because $\mathbf{B}_\mathbf{E}(\chi)$ should also account for possible friction forces arising at the interface. If friction forces are neglected:

$$\mathbf{B}_\mathbf{C}(\chi) = \mathbf{B}_\mathbf{E}(\chi) = \mathbf{B}(\chi) \tag{17}$$

2.2. Configuration Dependent Frequency Response Function

The frequency response function of the coupled system is configuration-dependent and can be computed as follows. Equation (8) can be rewritten by considering a configuration-dependent interface:

$$\begin{cases} \mathbf{Z}u(\chi) + \mathbf{B}^T(\chi)\lambda(\chi) = f \\ \mathbf{B}(\chi)u(\chi) = 0 \end{cases} \tag{18}$$

For each configuration, the same procedure outlined in Equations (9)–(13) can be followed to eliminate $\lambda(\chi)$ from the first of Equation (18). Finally, the FRF matrix of the coupled system with configuration-dependent interface is obtained as:

$$\mathbf{H}_c(\chi) = \mathbf{H} - \mathbf{H}\mathbf{B}^T(\chi)\left(\mathbf{B}(\chi)\mathbf{H}\mathbf{B}^T(\chi)\right)^{-1}\mathbf{B}(\chi)\mathbf{H} \tag{19}$$

3. Mechanisms Description in the Substructuring Framework

Mechanisms are composed by bodies connected together by kinematic constraints. In the substructuring framework, each link can be considered as an invariant subsystem, while kinematic constraints can be expressed as configuration-dependent compatibility conditions. By considering for instance two bodies connected by a revolute joint (Figure 1), the compatibility conditions can be written as:

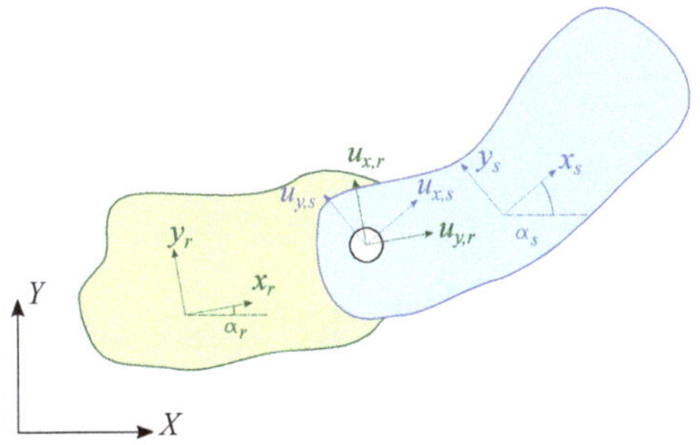

Figure 1. Two bodies r and s connected by a revolute joint in the plane orthogonal to the revolute joint axis: local ($x_r y_r$ and $x_s y_s$) and global (XY) reference frames.

$$\begin{cases} u_{x,r}\cos(\alpha_r) - u_{y,r}\sin(\alpha_r) - u_{x,s}\cos(\alpha_s) + u_{y,s}\sin(\alpha_s) = 0 \\ u_{x,r}\sin(\alpha_r) + u_{y,r}\cos(\alpha_r) - u_{x,s}\sin(\alpha_s) - u_{y,s}\cos(\alpha_s) = 0 \\ u_{z,r} - u_{z,s} = 0; \end{cases} \tag{20}$$

where α_r and α_s are the angles between two corresponding axes of the local and global reference frame, and u are the displacements in the local reference frames. In matrix form:

$$\begin{bmatrix} \cos(\alpha_r) & -\sin(\alpha_r) & 0 & -\cos(\alpha_s) & \sin(\alpha_s) & 0 \\ \sin(\alpha_r) & \cos(\alpha_r) & 0 & -\sin(\alpha_s) & -\cos(\alpha_s) & 0 \\ 0 & 0 & 1 & 0 & 0 & -1 \end{bmatrix} \begin{Bmatrix} u_{x,r} \\ u_{y,r} \\ u_{z,r} \\ u_{x,s} \\ u_{y,s} \\ u_{z,s} \end{Bmatrix} = 0 \tag{21}$$

Therefore, the portion $\tilde{\mathbf{B}}$ of the matrix $\mathbf{B}$, that enforces the compatibility between two bodies connected by a revolute joint, depends on the angles α_r and α_s between the local and global reference frames. It can be expressed as:

$$\tilde{\mathbf{B}}(\alpha_r, \alpha_s) = \begin{bmatrix} \cos(\alpha_r) & -\sin(\alpha_r) & 0 & -\cos(\alpha_s) & \sin(\alpha_s) & 0 \\ \sin(\alpha_r) & \cos(\alpha_r) & 0 & -\sin(\alpha_s) & -\cos(\alpha_s) & 0 \\ 0 & 0 & 1 & 0 & 0 & -1 \end{bmatrix} \tag{22}$$

It accounts for the dependence on the system configuration.

In frequency-based substructuring, the dynamics of each link can be expressed using the FRF matrix. The FRF matrix can be either measured experimentally or evaluated using a numerical model. It must be defined on all the degrees-of-freedom (DoFs) necessary to define the kinematic constraints with the other links.

Note that, with respect to the classical multi-body approach, frequency-based substructuring allows not only to use experimental models of components, but provides results in the frequency domain as well.

4. Application

The configuration-dependent substructuring is exploited to evaluate the configuration-dependent dynamics of a 1 DoF planar mechanism, the three-points linkage (Figure 2), that is typically used to connect operating machines to agricultural tractors.

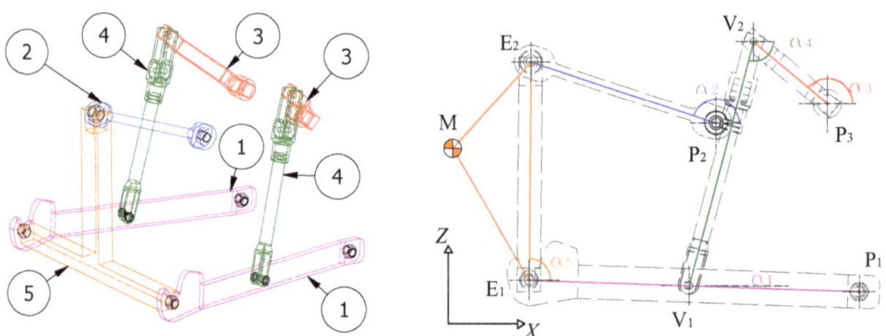

Figure 2. Three points linkage. (**Left**) 3D scheme. (**Right**) kinematic scheme.

It is in fact a Watt six bar linkage. The schematic in Figure 2 shows a typical rear three-point linkage and highlights its main components: ① lower links; ② upper link; ③ input cranks; ④ lift rods; ⑤ implement. The implement is part of the operating machine and is considered as a rigid body, whilst all the other bodies are assumed to be deformable. In Figure 2, the boundary nodes of the different components are highlighted; P_i and E_i indicate the boundary nodes connecting the three-point linkage to the tractor and the operating machine, respectively; V_i are the boundary nodes connecting the components of the linkage to each other; M is the center of gravity of the operating machine. It is based on a four-bar linkage composed of the lower links, the implement and the upper link, whose configurations are controlled by the kinematic chain composed by the input cranks and the lift rods. Note that the mechanism is planar, since all trajectories are parallel to the plane XZ under the assumption of rigid links. However, when considering deformable links, vibrations can also occur along the Y direction. The main dimensions of the linkage, the inertial properties of the attached operating machine and the material properties are listed in Table 1. All the elements of the linkage are represented in scale in Figure 2.

Table 1. Principal dimensions of the three-points linkage and inertial properties of the operating machine.

Quantity	Value	
P_1	(0.000, 0.000) m	Position of point P_1
P_2	(−0.334, 0.403) m	Position of point P_2
P_3	(−0.064, 0.385) m	Position of point P_3
$\overline{P_3V_2}$	0.203 m	Length of the input crank
$\overline{V_2V_1}$	0.534 m	Lenght of the lift rods
$\overline{P_1E_1}$	0.810 m	Lenght of the lower link
$\overline{P_1V_1}$	0.366 m	Distance between P_1 and V_1
$\overline{P_2E_2}$	0.420 m	Lenght of the upper link
$\overline{E_1E_2}$	0.438 m	Lenght of the implement
$\overline{E_1M}$	1.020 m	Distance between E_1 and M
$\overline{E_2M}$	1.095 m	Distance between E_2 and M
M_{om}	800 kg	Mass of the operating machine
I_{om}	200 kg·m^2	Moment of inertia of the operating machine
E	210 GPa	Young's modulus
ρ	7800 kg/m^3	Density
ν	0.3	Poisson's ratio

Since the three-points linkage is a 1 DoF mechanism, its configuration is defined by a single input coordinate, i.e., the angular position of the input cranks. Figure 3 shows three configurations of the linkage obtained for different angular position α_3 of the input cranks, i.e., 90, 140 and 180 degrees. The angular positions α_i of all the other links can be obtained as the outcome of the position analysis of the mechanism as shown in Figure 4 for α_3 varying from 90 to 180 degrees.

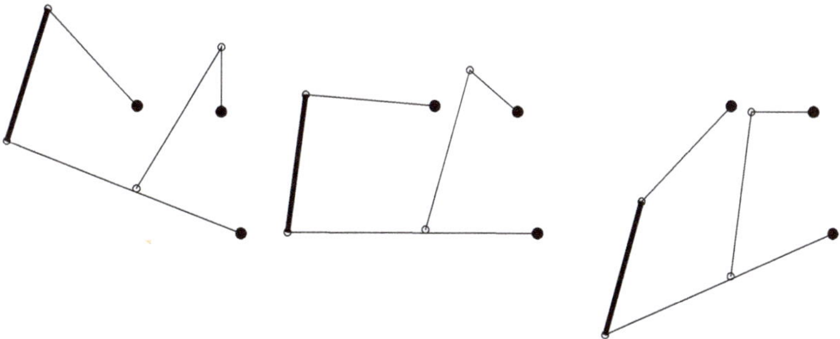

Figure 3. Three configurations of the linkage. (**Left**) $\alpha_3 = 90°$. (**Middle**) $\alpha_3 = 140°$. (**Right**) $\alpha_3 = 180°$.

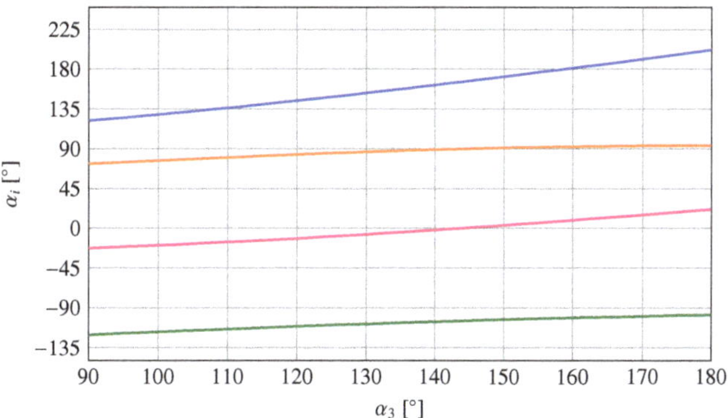

Figure 4. Angular position of the mechanism's links. α_1 (——); α_2 (——); α_4 (——); α_5 (——).

5. Results

For each joint, the angular positions of the connected links are used to express the compatibility condition, according to Equation (22). Furthermore, a rigid transformation is used to express constraint about the motion of the implement. All constraint equations are gathered in a overall compatibility matrix **B**.

Each component of the three-points linkage is modeled using a commercial FE software and a Craig-Bampton modal reduction [13] is performed retaining only the physical connecting DoFs with other components together with an appropriate number of fixed interface modes (See Table 2). Therefore for each subsystem, the mass, damping and stiffness reduced order matrices are used to obtain the FRF matrix.

Table 2. Number of physical and modal DoFs retained for each subsystem.

Link	Physical DoFs	Modal DoFs
1 lower links	9	20
2 upper link	6	20
3 input cranks	6	20
4 lift rods	6	20
5 implement	3	0

Finally, Equation (19) can be used to obtain the configuration-dependent FRFs, evaluated for α_3 spanning the angular range 90–180 degrees with an angular step of 2 degrees.

Figure 5 show the configuration-dependent drive point FRFs of the end point of the lower link (Node E_1) along the z direction in the frequency band 0–200 Hz. It can be noticed that the first natural frequency increases from 2.9 to 6.4 Hz, the second natural frequency decreases from 105.9 to 57.8 Hz and the third natural frequency lies in the interval from 152.8 to 191.0 Hz.

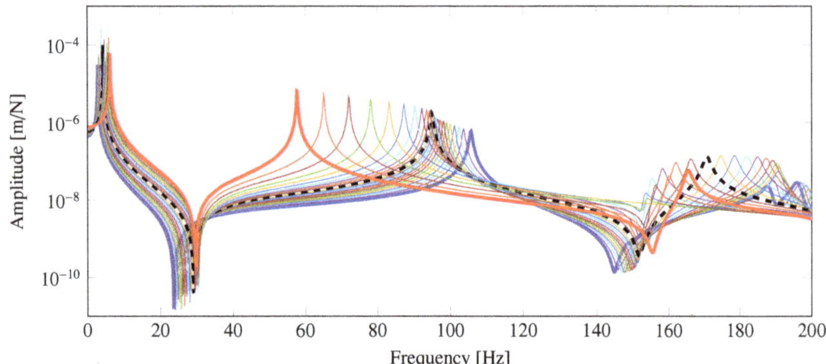

Figure 5. Configuration dependent drive point FRF of node E_1 in the y direction. (——) $\alpha_3 = 90°$; (——) $\alpha_3 = 180°$; (- - -) $\alpha_3 = 140°$.

For each configuration, the first three natural frequencies and the mode shapes of the whole mechanism in the plane XZ are identified. Figure 6 shows the configuration-dependent natural frequencies of the identified vibration modes of the three-point linkage.

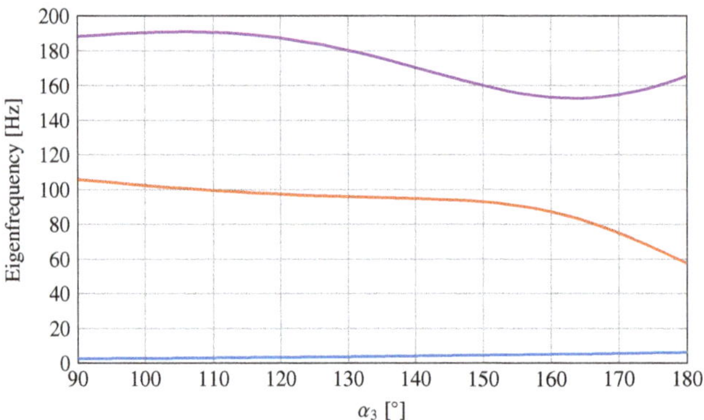

Figure 6. Configuration dependent natural frequencies of vibration modes from one to three of the three-point linkage, as function of the angle α_3 of the input crank. First natural frequency (——); Second natural frequency (——); Third natural frequency (——).

The results highlight the dependency of the natural frequencies on the configuration. Moreover, Figures 7–9 show the displacements of nodes E_1, E_2, V_1 and V_2 for the first three mode shapes of the linkage in three different configurations.

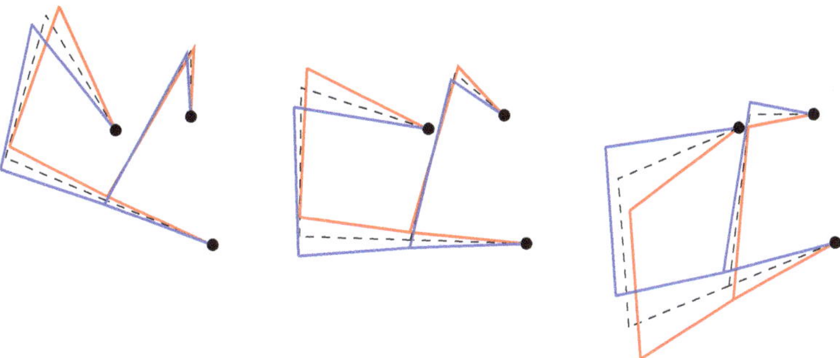

Figure 7. Mode 1 in three different configurations. (**Left**) $\alpha_3 = 90°$. (**Middle**) $\alpha_3 = 140°$. (**Right**) $\alpha_3 = 180°$. (- - -) Undeformed model; (——) and (——) extreme deformed configurations during oscillation.

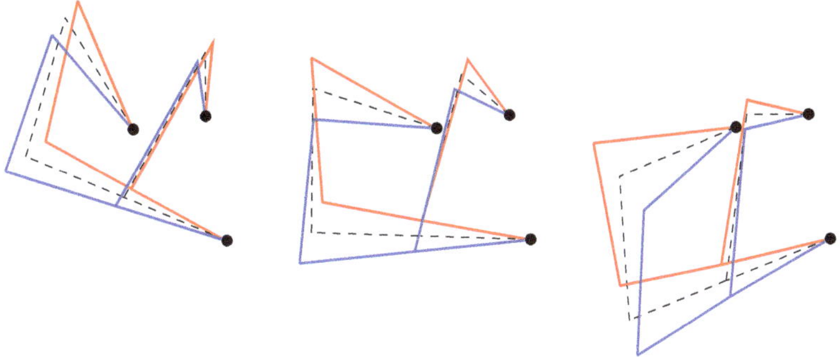

Figure 8. Mode 2 in three different configurations. (**Left**) $\alpha_3 = 90°$. (**Middle**) $\alpha_3 = 140°$. (**Right**) $\alpha_3 = 180°$. (- - -) Undeformed model; (——) and (——) extreme deformed configurations during oscillation.

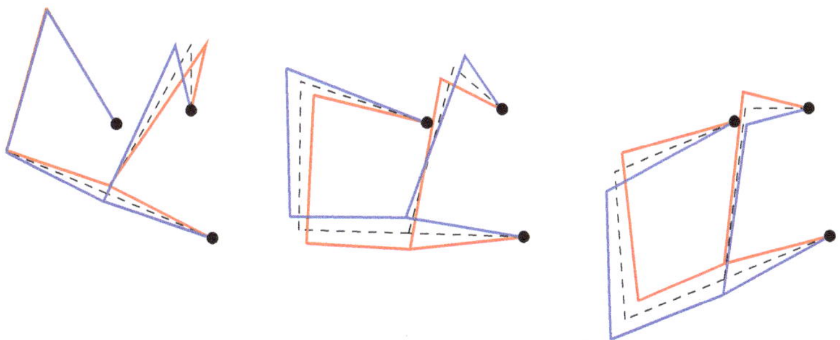

Figure 9. Mode 3 in three different configurations. (**Left**) $\alpha_3 = 90°$. (**Middle**) $\alpha_3 = 140°$. (**Right**) $\alpha_3 = 180°$. (- - -) Undeformed model; (——) and (——) extreme deformed configurations during oscillation.

Note that the nodes are joined using straight lines, so that it is not possible to observe the curvature of the different links. However, it is possible to have a quite clear idea about how the linkage oscillates.

In order to check the correctness of the procedure and the quality of the approximation due to the Craig-Bampton modal reduction of the subsystems, the assembled system in the configuration with $\alpha_3 = 140°$ is analyzed using a FE commercial software. The finite element model of the entire system has 250,000 nodes, and it is obtained by assembling the finite element models of the component substructures. In Table 3, the natural frequencies of modes in the plane XZ below 200 Hz are compared with those obtained using the substructuring approach. The results highlight that the procedure is correctly implemented and that the approximation with 20 Craig–Bampton modes is acceptable. Figure 10 shows the mode shapes corresponding to modes 1, 2 and 3 of the assembled system in the configuration with $\alpha_3 = 140°$, computed using a FE commercial software. These mode shapes appear to be well correlated with those obtained using the substructuring procedure and shown in Figures 7–9.

Table 3. Natural frequencies.

Mode	FE [Hz]	Substructuring [Hz]
1	4.2	4.3
2	97.5	94.9
3	176.9	170.6

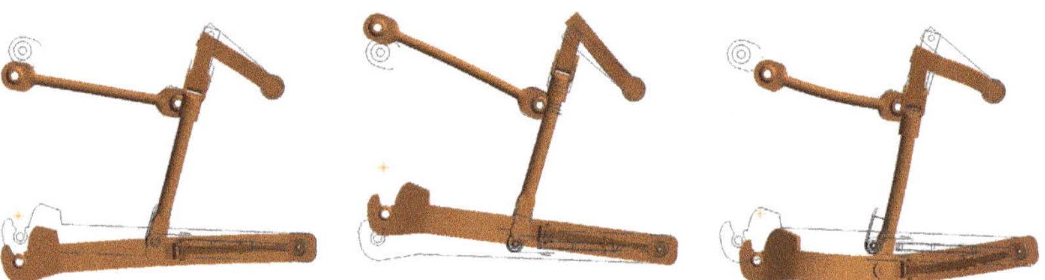

Figure 10. From left to right, mode shapes of the modes 1, 2 and 3 of the assembled system in the configuration with $\alpha_3 = 140°$, computed using a FE commercial software.

In order to quantify the computational burden of the proposed approach with respect to a more traditional approach using a commercial FE software, we calculate the computational times spent to compute a prescribed number of frequency response functions of the assembled systems in a given number of configurations and in a given frequency band with an assigned frequency step. The following parameters are used for the comparison:

- number of frequency response functions: 11;
- number of configurations: 46;
- frequency band: 0–200 Hz;
- frequency step: 0.1 Hz;

The computational time necessary to obtain the FRFs using the substructuring procedure is the sum of the following times:

1. the time spent to perform the Craig-Bampton modal reduction of all subsystem with 20 fixed interface modes using a commercial FE software (226 s);
2. the time spent to compute the full frequency response function matrix of the assembled system in all the considered configurations (33 s).

The computational time necessary to obtain the FRFs using a commercial FE software is the sum of the following times:

1. solution of the eigenvalue problem for each of the considered configuration using 20 modes (6113 s);
2. evaluation of the frequency response functions corresponding to a single excitation point (89,240 s). Note that, for every configurations, the software evaluates the frequency response functions at all the degrees of freedom.

For the considered system, the configuration-dependent substructuring approach provides frequency response functions that are 368 times faster than when using a traditional approach.

6. Conclusions

In this paper, the substructuring approach is extended to the vibrational analysis of mechanisms. The proposed approach is applied to predict the natural frequencies, the mode shapes and the frequency response functions of a three-point linkage for a set of different positions of the input link. The approach cannot be seen as an alternative to multibody system analysis but rather as a technique providing different and complementary information.

For the considered system, three frequency ranges where system resonances are possible are found; the first three mode shapes and their dependence on the configuration are shown, the frequency response functions of the system are evaluated for each configuration and the drive-point frequency response function of the end-point of the lower link is observed. This kind of result can be exploited in the preliminary design stages to highlight a possible cluster of frequencies that should be avoided as exciting frequencies or vice-versa to modify the system to move the cluster of natural frequencies of the mechanism away from the excitation frequencies.

In order to validate the effectiveness of the procedure and to quantify the computational burden, a commercial FE software is used to compute the natural frequencies, the mode shapes and the frequency response functions. The results provided by the substructuring approach are correlated with the reference results provided by the full FE model. Moreover, the computational burden required to obtain the solution using the substructuring approach is significantly lower than that using a traditional approach.

The potential applications of configuration-dependent substructuring in the dynamic analysis of a mechanism and in mechanism design are very promising.

Author Contributions: All authors (J.B., W.D. and A.F.) have equally contributed to the conceptualization, methodology, and writing. All authors have read and agreed to the published version of the manuscript.

Funding: This research was developed in the framework of the project BRIC-ID14 (2019) funded by INAIL (National Institute for Insurance against Accidents at Work and Occupational Diseases).

Data Availability Statement: The data presented in this study are included in the article except for component geometrical models that cannot be shared due to industrial ownership.

Acknowledgments: The authors acknowledge the Italian National Institute for Insurance against Accidents at Work and Occupational Diseases (INAIL) for founding this research.

Conflicts of Interest: The authors declare that they have no conflict of interest.

References

1. de Klerk, D.; Rixen, D.J.; Voormeeren, S. General Framework for Dynamic Substructuring: History, Review, and Classification of Techniques. *AIAA J.* **2008**, *46*, 1169–1181. [CrossRef]
2. Semm, T.; Rebelein, C.; Zaeh, M. Prediction of the position dependent dynamic behavior of a machine tool considering local damping effects. *CIRP J. Manuf. Sci. Technol.* **2019**, *27*, 68–77. [CrossRef]
3. Semm, T.; Nierlich, M.B.; Zaeh, M.F. Substructure Coupling of a Machine Tool in Arbitrary Axis Positions Considering Local Linear Damping Models. *J. Manuf. Sci. Eng.* **2019**, *141*, 071014. [CrossRef]
4. Brunetti, J.; D'Ambrogio, W.; Fregolent, A. Dynamic coupling of substructures with sliding friction interfaces. *Mech. Syst. Signal Process.* **2020**, *141*, 106731. [CrossRef]
5. Carassale, L.; Silvestri, P.; Lengu, R.; Mazzaron, P. Modeling Rail-Vehicle Coupled Dynamics by a Time-Varying Substructuring Scheme. In *Dynamic Substructures, Volume 4*; Linderholt, A., Allen, M.S., Mayes, R.L., Rixen, D., Eds.; Springer International Publishing: Cham, Switzerland, 2020; pp. 167–171.
6. Brunetti, J.; D'Ambrogio, W.; Fregolent, A. Friction-induced vibrations in the framework of dynamic substructuring. *Nonlinear Dyn.* **2021**, *103*, 3301–3314. [CrossRef]
7. Brunetti, J.; D'Ambrogio, W.; Fregolent, A. Evaluation of Different Contact Assumptions in the Analysis of Friction-Induced Vibrations Using Dynamic Substructuring. *Machines* **2022**, *10*, 384. [CrossRef]
8. D'Ambrogio, W.; Fregolent, A. Experimental Dynamic Substructuring: Significance and Perspectives. In *50+ Years of AIMETA*; Rega, G., Ed.; Springer International Publishing: Cham, Switzerland, 2022; pp. 305–319.
9. Wittenburg, J. *Dynamics of Multibody Systems*, 2nd ed.; Springer: Berlin/Heidelberg, Germany, 2008.
10. Shabana, A.A. *Dynamics of Multibody Systems*, 4th ed.; Cambridge University Press: Cambridge, UK, 2013.

11. Brunetti, J.; D'Ambrogio, W.; Fregolent, A. Analysis of the Vibrations of Operators' Seats in Agricultural Machinery Using Dynamic Substructuring. *Appl. Sci.* **2021**, *11*, 4749. [CrossRef]
12. Voormeeren, S.N.; Rixen, D.J. A Family of Substructure Decoupling Techniques Based on a Dual Assembly Approach. *Mech. Syst. Signal Process.* **2012**, *27*, 379–396. [CrossRef]
13. Craig, R.R., Jr.; Bampton, M. Coupling of substructures for dynamic analyses. *AIAA J.* **1968**, *6*, 1313–1319. [CrossRef]

machines

Article

Design of the Drive Mechanism of a Rotating Feeding Device

Matteo Bottin *, Riccardo Minto and Giulio Rosati

Department of Industrial Engineering, University of Padova, 35131 Padova, Italy
* Correspondence: matteo.bottin@unipd.it

Abstract: Component batching can be a source of time waste in specific industrial applications, such as kitting. Kitting operations are usually performed by hoppers, but other devices can be used to optimize the process. In a previous work, a rotary device has been proved to be more efficient than hoppers; such a device allows the kitting and releasing of the components in a single rotatory movement, while traditional hoppers require at least two movements. In this paper, an improvement of such feeding device is proposed. The movement of the rotary device is driven by a four-bar linkage mechanism which is designed through functional synthesis. Thank to the four-bar linkage mechanism, the alternate motion of the rotary distributor is derived from the constant speed of the motor.

Keywords: feeding; rotary device; functional synthesis; mechanical design

1. Introduction

The target of many companies is to either reduce production costs or increase the throughput. This is important, especially for Western countries, where the unit direct production cost of goods in the manufacturing industry is driven mainly by the cost of labor. To overcome such issue, production systems must be optimized by increasing their flexibility [1,2].

In the manufacturing industry, kitting lines play a major role. The aim of kitting lines is to create kits of different objects with predetermined quantities [3]. To achieve this, firstly, small batches are created, one for each type of object; then, the batches are grouped together to assemble the kit. Such kits may be sold to the final customer, called *sales kit*, or prepared for other assembly processes, called *production kit*. To create the batches, a common solution is given by hopper systems. Hoppers are used in series to group components, perform quality checks through weight measurements, and act as buffers. However, hoppers suffer from a non-negligible drawback, i.e., for each release operation, they need two movements, opening, and closing. To overcome such issue, hopper design has been studied in the last years, and both axial-symmetric [4–8] and eccentric [9,10] solutions have been analyzed, trying to optimize the flow of the components to reduce the drawback. However, hoppers are yet a source of inefficiency [11]. For small kit sizes, robots can be used as an alternate means of kitting [12,13]; however, robot productivity is too low with respect to hoppers when the number of parts per kit is large.

To the best of the authors' knowledge, there is a general lack of alternatives to hoppers; indeed, very few works have been published in this area in the last few years. For example, Pantyukhina et al. [14] proposed a mechanical toothed hopper-feeding device whose purpose is not component kitting, but component orientation. Gao et al. [15] studied a rotational device to be used for a continuous flow of small particles; thus, the purpose and design are different from a kitting operation, in which rather large components are kitted.

In this paper, a novel solution for component batching is proposed by means of a compartmentalized rotating device. Such a device, which follows the concepts introduced in previous works [16,17], requires only a single rotational movement for each release operation. In fact, it is made up of compartments divided by blades specifically

Citation: Bottin, M.; Minto, R.; Rosati, G. Design of the Drive Mechanism of a Rotating Feeding Device. *Machines* **2022**, *10*, 1160. https://doi.org/10.3390/machines10121160

Academic Editor: Dan Zhang

Received: 29 September 2022
Accepted: 1 December 2022
Published: 4 December 2022

Publisher's Note: MDPI stays neutral with regard to jurisdictional claims in published maps and institutional affiliations.

designed for optimized component falling. This work focuses both on the mechanical design and how it affects the blade shape to ensure the free-falling motion of the components. In particular, we propose to use a four-bar linkage mechanism to drive the movement of the blades; in this way, an alternate motion of the blades is obtained without requiring complex electronics, since it is derived from the motion of an electric motor driven at a constant speed. This design choice on the one side simplifies the control system of the rotating device, which is the main advantage provided by the solution proposed, on the other side it requires a re-design of the blades. Both aspects are addressed in the paper.

The work is structured as follows. Firstly, the mechanical design of both the mechanism and the blades is presented in Section 2. Then, the numerical results of the design principles are presented in Section 3. Finally, experimental tests are described in Section 4, and the performances are evaluated.

2. Mechanical Design

The proposed device is shown in Figure 1. The device is composed of three main parts:

- A fixed cylindrical structure, in brown in Figure 1a, which acts as a container for the components to be kitted. The cylinder axis is horizontal, and the curved surface has two openings: one at the top to receive new components, and one at the bottom to exit the components.
- The rotary distributor, in teal green in Figure 1a, coaxial with the cylinder, which divides the cylinder into two compartments via some blades. The rotation of the blades around the cylinder axis allows the components to fall within one compartment or the other (Figure 2).
- A four-bar linkage mechanism, in light grey in Figure 1a, whose rocker link is fixed with the blades (angle φ_3 of Figure 3). The crank link is controlled by means of an electric motor that rotates at constant speed, whose absolute angle is q.

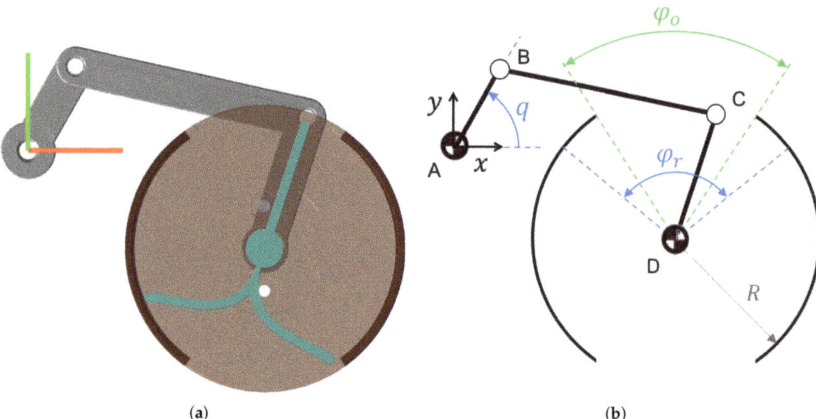

(a)   (b)

Figure 1. Design of the proposed mechanism. CAD model (**a**) with the four-bar linkage system (light grey) and the rotary distributor with the blades (teal green); mathematical model (**b**).

The four-bar linkage mechanism has been chosen for its extreme simplicity both in terms of control, manufacturing, and costs. In fact, the device can be built by means of cheap 3D rapid prototyping with great results.

Exploiting the movement of the mechanism, if the crank link rotates with a fixed speed the blades rotate periodically to the left and to the right, allowing the components to fall within one compartment or the other (an example can be seen in Figure 2). This allows the machine to be controlled by a very simple control system.

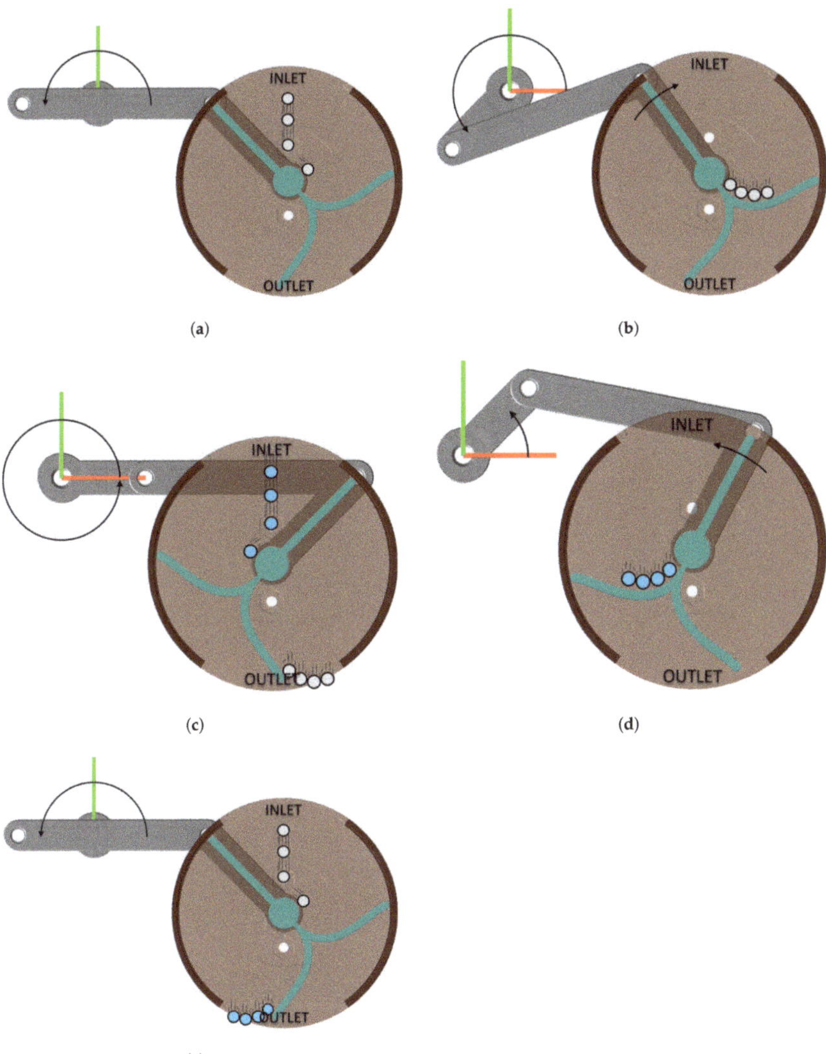

Figure 2. Functioning of the mechanism. The rotary distributor (teal green) moves to the left and to the right to batch components, while the four-bar linkage mechanism is driven by a motor fixed to the crank which rotates at constant speed. The parts fall into the compartment from the inlet (to the top) and exit from the outlet (to the bottom) under the action of gravity. In the sequence depicted above, the first set of parts (white) exits to the right (**a**–**c**), and the second set (blue) exits to the left (**c**–**e**).

2.1. Design of the Four-Bar Linkage Mechanism

The design of the four-bar linkage can be performed via precision-point synthesis [18]. Precision-point synthesis allows sizing the mechanism so that it crosses specific configurations, called precision points. Such configurations are chosen by design to exploit specific mechanism behavior.

Let us consider the vector scheme of Figure 3. The position closure equation can be written as follows:

$$\sum_{i=\{1,\dots,4\}} \pm \mathbf{z}_i = 0 \quad \rightarrow \quad \mathbf{z}_1 + \mathbf{z}_2 - \mathbf{z}_3 - \mathbf{z}_4 = 0 \tag{1}$$

which must hold for every mechanism configuration. If a configuration 0 is considered, each vector i complex form is:

$$\mathbf{z}_{i,0} = a_i e^{j\varphi_{i,0}} \tag{2}$$

where a_i is the vector length and $\varphi_{i,0}$ is the absolute angle. If a rotation $\delta_{i,k}$ is applied to vector $\mathbf{z}_{i,0}$, the resulting vector $\mathbf{z}_{i,k}$ is:

$$\mathbf{z}_{i,k} = a_i e^{j(\varphi_{i,0}+\delta_{i,k})} = \mathbf{z}_{i,0} e^{j\delta_{i,k}} \tag{3}$$

which leads to the position closure equation for configuration k:

$$\sum_{i=\{1,\dots,4\}} \pm \mathbf{z}_{i,k} = \sum_{i=\{1,\dots,4\}} \pm \mathbf{z}_{i,0} e^{j\delta_{i,k}} = 0 \tag{4}$$

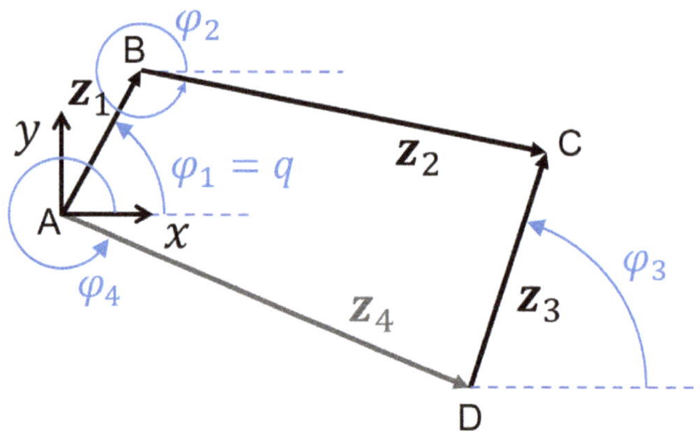

Figure 3. Schematic of the four-bar linkage mechanism. The vectors represent the position of the links of the mechanism.

Subtracting Equations (1) and (4) we get:

$$\sum_{i=\{1,\dots,4\}} \pm(\mathbf{z}_{i,k} - \mathbf{z}_{i,0}) = \sum_{i=\{1,\dots,4\}} \pm \mathbf{z}_{i,0}(e^{j\delta_{i,k}} - 1) = 0 \tag{5}$$

which represents the *position closure equation of the displacements*. For displacement k, Equation (5) becomes:

$$\mathbf{z}_{1,0}(e^{j\delta_{1,k}} - 1) + \mathbf{z}_{2,0}(e^{j\delta_{2,k}} - 1) - \mathbf{z}_{3,0}(e^{j\delta_{3,k}} - 1) = 0 \tag{6}$$

where $\mathbf{z}_{4,0}$ has been deleted since it is fixed, i.e., $\mathbf{z}_{4,k} = \mathbf{z}_{4,0}$; thus, $\delta_{4,k} = 0$ for every displacement k.

Equation (6) can be used to size the mechanism based on the design requirements. In particular, the designer is usually interested in designing the mechanism such as that a specific displacement of the crank $\delta_{1,k}$ corresponds to a specific displacement of the rocker $\delta_{3,k}$ (the so-called *functional synthesis*). As a result, the number of unknowns in Equation (6) is 7: the three vector lengths (a_1, a_2, a_3) and orientations ($\varphi_{1,0}$, $\varphi_{2,0}$, $\varphi_{3,0}$), and the displacement of the coupler link ($\delta_{2,k}$). The corresponding number of scalar equations is 2, one for the real part and one for the imaginary part of the vectorial equation. Vector z_4 length (a_4) and orientation ($\varphi_{4,0}$) are calculated by means of Equation (1) for configuration 0.

As a result, configuration 0 is not considered—since its equations are used to calculate only z_4—thus, the number of unknowns for N displacements is $6 + N$, whereas the number of equations is $2N$. Table 1 shows the number of unknown parameters for N displacements up to 6. One of the main advantages of the functional synthesis of a mechanism is that the solution to be found can be scaled and oriented according to the needs. That said, if a length of a link is fixed, all the other links can be calculated based on such length; if the length is doubled, all the other links are doubled as well. Moreover, the same mechanism can be rotated while maintaining the functionality [19].

Table 1. Number of unknowns, equations and solutions for N displacements.

N	Unknowns	Equations	Solutions
1	7 ($a_1,a_2,a_3,\varphi_{1,0},\varphi_{2,0},\varphi_{3,0},\delta_{2,1}$)	2	∞^5
2	8 ($a_1,a_2,a_3,\varphi_{1,0},\varphi_{2,0},\varphi_{3,0},\delta_{2,1},\delta_{2,2}$)	4	∞^4
3	9 ($a_1,a_2,a_3,\varphi_{1,0},\varphi_{2,0},\varphi_{3,0},\delta_{2,1},\delta_{2,2},\delta_{2,3}$)	6	∞^3
4	10 ($a_1,a_2,a_3,\varphi_{1,0},\varphi_{2,0},\varphi_{3,0},\delta_{2,1},\delta_{2,2},\delta_{2,3},\delta_{2,4}$)	8	∞^2
5	11 ($a_1,a_2,a_3,\varphi_{1,0},\varphi_{2,0},\varphi_{3,0},\delta_{2,1},\delta_{2,2},\delta_{2,3},\delta_{2,4},\delta_{2,5}$)	10	∞^1
6	12 ($a_1,a_2,a_3,\varphi_{1,0},\varphi_{2,0},\varphi_{3,0},\delta_{2,1},\delta_{2,2},\delta_{2,3},\delta_{2,4},\delta_{2,5},\delta_{2,6}$)	12	1

If $N = 3$, the corresponding equations retrieved from Equation (6) can be compacted in a matrix form as follows:

$$\mathbf{E} \cdot \mathbf{z}_0 = \begin{bmatrix} e^{j\delta_{1,1}} - 1 & e^{j\delta_{2,1}} - 1 & 1 - e^{j\delta_{3,1}} \\ e^{j\delta_{1,2}} - 1 & e^{j\delta_{2,2}} - 1 & 1 - e^{j\delta_{3,2}} \\ e^{j\delta_{1,3}} - 1 & e^{j\delta_{2,3}} - 1 & 1 - e^{j\delta_{3,3}} \end{bmatrix} \begin{Bmatrix} \mathbf{z}_{1,0} \\ \mathbf{z}_{2,0} \\ \mathbf{z}_{3,0} \end{Bmatrix} = \begin{Bmatrix} 0 \\ 0 \\ 0 \end{Bmatrix} \tag{7}$$

where, as stated previously, $z_{1,0}$ and $z_{3,0}$ displacements ($\delta_{1,k}, \delta_{3,k}$) are known by design. Systems with $N > 3$ are more difficult to solve due to the coupling of the parameters and the fact that matrix $\mathbf{E}$ is non-squared, but some methods have been proposed [20–22].

From Table 1, it can be inferred that there are ∞^3 solutions to the system of Equation (7). As a result, three parameters must be chosen. The unknowns are the three vectors and the displacements of the coupler link ($\delta_{2,1},\delta_{2,2},\delta_{2,3}$). The algebraic system admits solutions different from the trivial one if and only if the determinant of the matrix of displacements $\mathbf{E}$ is equal to zero. In other words, it is possible to choose one displacement of the coupler ($\delta_{2,k}$) to calculate the others. Then, the system can be solved by choosing the length and the orientation of one of the vectors. Finally, z_4 is calculated by means of (1) for configuration 0. Please note that by choosing one vector and one displacement, the number of unknowns reduces by 3; thus, the solution is unique.

The determinant of the matrix of displacements is:

$$\det\left(\begin{bmatrix} e^{j\delta_{1,1}} - 1 & e^{j\delta_{2,1}} - 1 & 1 - e^{j\delta_{3,1}} \\ e^{j\delta_{1,2}} - 1 & e^{j\delta_{2,2}} - 1 & 1 - e^{j\delta_{3,2}} \\ e^{j\delta_{1,3}} - 1 & e^{j\delta_{2,3}} - 1 & 1 - e^{j\delta_{3,3}} \end{bmatrix}\right) = \mathbf{E}_{21}e^{j\delta_{2,1}} + \mathbf{E}_{22}e^{j\delta_{2,2}} + \mathbf{E}_{23}e^{j\delta_{2,3}} + \mathbf{E}_0 = 0 \tag{8}$$

where

$$\mathbf{E}_{21} = e^{j\delta_{1,3}} - e^{j\delta_{1,2}} + e^{j\delta_{3,2}} - e^{j\delta_{3,3}} + e^{j\delta_{1,2}}e^{j\delta_{3,3}} - e^{j\delta_{1,3}}e^{j\delta_{3,2}} \tag{9}$$

$$\mathbf{E}_{22} = e^{j\delta_{1,1}} - e^{j\delta_{1,3}} - e^{j\delta_{3,1}} + e^{j\delta_{3,3}} - e^{j\delta_{1,1}}e^{j\delta_{3,3}} + e^{j\delta_{1,3}}e^{j\delta_{3,1}} \tag{10}$$

$$\mathbf{E}_{23} = e^{j\delta_{1,2}} - e^{j\delta_{1,1}} + e^{j\delta_{3,1}} - e^{j\delta_{3,2}} + e^{j\delta_{1,1}}e^{j\delta_{3,2}} - e^{j\delta_{1,2}}e^{j\delta_{3,1}} \tag{11}$$

$$\mathbf{E}_0 = e^{j\delta_{1,2}}e^{j\delta_{3,1}} - e^{j\delta_{1,1}}e^{j\delta_{3,2}} + e^{j\delta_{1,1}}e^{j\delta_{3,3}} - e^{j\delta_{1,3}}e^{j\delta_{3,1}} - e^{j\delta_{1,2}}e^{j\delta_{3,3}} - e^{j\delta_{1,3}}e^{j\delta_{3,2}} \tag{12}$$

where all the terms are known since are design variables.

Equation (8) is a non-linear complex equation system that can be solved numerically by choosing one of the displacements $\delta_{2,k}$. In fact, the system has two equations (one for the real part and the other for the imaginary part) but three unknowns $(\delta_{2,1}, \delta_{2,2}, \delta_{2,3})$. By choosing one of such unknowns it is possible to solve the system.

Once the displacements are found, the system of Equation (7) can be solved by removing one row (since the determinant of the 3×3 matrix is null). The resulting system is a 2×2 system with 3 unknown vectors. As a result, it is necessary to choose one of such vectors to solve the system. Without losing in generality, lets us consider that vector $z_{3,0}$ has been chosen and the third equation removed:

$$\begin{bmatrix} e^{j\delta_{1,1}} - 1 & e^{j\delta_{2,1}} - 1 \\ e^{j\delta_{1,2}} - 1 & e^{j\delta_{2,2}} - 1 \end{bmatrix} \begin{Bmatrix} \mathbf{z}_{1,0} \\ \mathbf{z}_{2,0} \end{Bmatrix} = - \begin{Bmatrix} 1 - e^{j\delta_{3,1}} \\ 1 - e^{j\delta_{3,2}} \end{Bmatrix} \mathbf{z}_{3,0} \tag{13}$$

which can be solved if the rank of the matrix of displacements is equal to 2. From Equation (13) the three vectors $\mathbf{z}_{1,0}$, $\mathbf{z}_{2,0}$, and $\mathbf{z}_{3,0}$ are fully defined, while $\mathbf{z}_{4,0}$ is calculated by means of Equation (1) for configuration 0:

$$\mathbf{z}_{4,0} = \mathbf{z}_{1,0} + \mathbf{z}_{2,0} - \mathbf{z}_{3,0} \tag{14}$$

Finally, link length a_i and orientation $\varphi_{i,0}$ are calculated from the complex vector form:

$$a_i = |\mathbf{z}_i| = \sqrt{\Re(\mathbf{z}_i)^2 + \Im(\mathbf{z}_i)^2}$$
$$\varphi_{i,0} = \mathrm{atan2}(\Im(\mathbf{z}_i), \Re(\mathbf{z}_i)) \tag{15}$$

where $\Re(\mathbf{z}_i)$ and $\Im(\mathbf{z}_i)$ are the real and imaginary part of the complex vector $\mathbf{z}_i$.

It is worth noting that the solution given from the functional synthesis allows the mechanism to fulfill the required configurations, but nothing can be said about the behavior of the mechanism in all the other configurations. In other words, it is possible that the crank is not able to perform a complete rotation around its axis. Further checks must be performed to address the validity of the solution based on specific design requirements, such as the Grashof Law. Finally, $\mathbf{z}_{3,0}$ has to be chosen in such a way that even the largest parts to be kitted can easily fall within the compartments. In this sense, the mechanism and the motor must be scaled accordingly.

Provided that the actuation torque of a four-bar linkage mechanism may be very high when a singular configuration is approached, the mechanism obtained by the synthesis must be checked to verify that the actuation torque is bounded within reasonable values. If the solution is not applicable, the design must restart and different values on the constraints (i.e., displacements $\delta_{1,k}, \delta_{2,k}, \delta_{3,k}$ and one vector) must be applied.

2.2. Design of the Blades

In a hopper system, the easiest trajectory a component should follow is the free falling motion, without any interaction with the blade [16]. Indeed, this is the trajectory that requires the smallest time to let the component exit the cylinder volume. Different component behavior could be exploited by means of more complex interaction models, such as by considering the dynamics of the components [17]. The optimal design of the blade is the one that follows the movement of the component during the free fall without interacting with it.

The steps for the design of the blades are [16]:

1. The starting position of the component falling on the blade (x_0, y_0) is calculated. This position is arbitrary and can be chosen according to specific needs (e.g., the motor shaft dimensions).

2. The free falling motion law of the component is calculated to retrieve the movement of the component in the absolute reference frame (0):

$$y(t) = y_0 - \frac{1}{2}gt^2 \tag{16}$$

3. By imposing a specific motion law of the rotary distributor it is possible to calculate the orientation of the reference frame relative to the rotary distributor. In particular, if $\psi(t)$ is the rotation angle of the rotary distributor, the transformation of coordinates between the absolute reference frame (0) and relative reference frame (r) is described by the transformation matrix:

$$\mathbf{T}_{r0}(t) = \begin{bmatrix} \cos(\psi(t)) & -\sin(\psi(t)) & 0 \\ \sin(\psi(t)) & \cos(\psi(t)) & 0 \\ 0 & 0 & 1 \end{bmatrix} \tag{17}$$

4. The position $\mathbf{P}_r(t)$ of the center of the component in the relative reference frame can be calculated by means of T_{r0}:

$$\mathbf{P}_r(t) = \left\{ \begin{matrix} x(t) \\ y(t) \\ 1 \end{matrix} \right\}_r = \mathbf{T}_{r0}(t) \left\{ \begin{matrix} x_0 \\ y(t) \\ 1 \end{matrix} \right\}_0 \tag{18}$$

5. The blade shape is the lower envelope of the circles centered in $\mathbf{P}_r(t)$ in the reference frame. In such a way, the blade will always be very close to the component during its fall.

From Equation (18) it can be inferred that the motion law of the rotary distributor $\psi(t)$ plays a major role in the design of the blade: different motion laws and rotational speeds result in very different blades. Particular attention must be paid to high-speed motion laws, since the incoming parts may rebound from the blades. However, it must be considered that the speed of the rocker (and of the blades) is very low near the dead points, right when the incoming parts enter and exit the device. In the case of a direct motor installed on the rotary distributor [16], $\psi(t)$ is described by well-known motion laws, such as third-degree polynomials or trapezoidal speed laws. Indeed, it is nearly impossible to drive the distributor at a constant speed, since it would allow a very short time for the components to fall within the compartments. Moreover, this type of control system requires control electronics which increases the costs of the device.

On the other hand, using a four-bar linkage allows the motor to be moved at a constant speed, while $\psi(t)$ is driven solely by the speed ratio between the crank and the rocker, which is given by the synthesis of the mechanism. In this case, $\psi(t) = \varphi_3(t)$, which can be derived directly by the kinematic of the four-bar linkage. In fact, the motion law can be expressed conveniently through its first derivative $\dot{\psi}(t)$:

$$\dot{\psi}(t) = \dot{\varphi}_3(t) = \frac{d\varphi_3}{dt} = \frac{d\varphi_3}{dq}\frac{dq}{dt} = w_{\varphi_3}(q)\dot{q}(t) \tag{19}$$

where $w_{\varphi_3}(q)$ is the speed ratio between the rocker and the crank and is:

$$w_{\varphi_3}(q) = \frac{a_1 \sin(\varphi_2 - q)}{a_3 \sin(\varphi_2 - \varphi_3)} \tag{20}$$

The motion law $\psi(t)$ can be derived via integration of Equation (19), where the differential variable is q, since in our case $\dot{q}$ is assumed to be constant.

3. Validation

3.1. Mechanism Synthesis

The methodology presented in the last section has been implemented to perform both the mechanism synthesis and the blade profile generation. For the mechanism synthesis, 3 precision points have been chosen. The precision points, which represent displacements from configuration 0, have been chosen to follow specific design principles:

- The angular top opening range φ_O is arbitrary but should be wide enough to allow multiple pieces to fall within the device without falling outside of the cylinder. Moreover, it is mandatory that $\varphi_O \leq \varphi_r$. Indeed, if the top opening range is wider than the blade stroke, the two compartments will always have an open top gap which could allow components to pass through. As a result, the first configuration has been chosen so that $\delta_{3,1} = \varphi_r$. Since configuration 0 must be reached once every crank full rotation, this first displacement, which represents half of the full movement of the rocker, must be performed in half rotation; thus, $\delta_{1,1} = 180°$.
- The other two displacements are chosen so that the rocker moves from the dead points by a certain amount with the same crank displacement. In other words, $\delta_{3,3} = \delta_{3,1} - \delta_{3,2}$, and $\delta_{1,3} = 180° + \delta_{1,2}$.

Vector $\mathbf{z}_3$ has been chosen so that the rotary compartment fits the cylindrical part used in a previous work [16] and $\varphi_3 = \pi/4$. All the numerical values used in the synthesis are shown in Table 2. In our case $\varphi_r = 90°$, so for every crank full rotation the rocker must move 90° to the right and 90° to the left. It is worth noting that the synthesis does not ensure that the rocker moves exactly 90° every half crank rotation, but, thanks to $\delta_{3,1} = 90°$, ensures that the rocker moves at least 90° every half crank rotations.

Table 2. Numerical values used in the synthesis of the mechanism. Design requirements to the left, fixed unknown values to the right.

Parameter	Value	Unit	Unknown	Value	Unit
$\delta_{1,1}$	180	[°]	$\delta_{2,3}$	20	[°]
$\delta_{1,2}$	45	[°]	a_3	56.6	[mm]
$\delta_{1,3}$	225	[°]	$\varphi_{3,0}$	45	[°]
$\delta_{3,1}$	90	[°]			
$\delta_{3,2}$	20	[°]			
$\delta_{3,3}$	70	[°]			

Synthesis results are shown in Table 3 and in Figure 4. The configurations shown in Figure 4 repeat every crank rotation at specific q values. The main result of the synthesis is the mechanism reliability: in fact, since $\varphi_O < \varphi_r$, there is a certain angle at which the rotary distributor can move without entering the top opening range. By imposing such angles via the precision points, it is ensured that the rocker enters the top opening range in a pre-determined timespan, which can be calculated by choosing the crank rotational speed. In our example, the mechanism is designed to rotate the rocker about 20° from the dead points ($\delta_{3,2} = \delta_{3,1} - \delta_{3,3} = 20°$) when the crank rotates about 45° about its axis ($\delta_{1,2} = \delta_{1,3} - \delta_{1,1} = 45°$). The time needed for such rotation is:

$$t_{span} = \frac{\delta_{1,2}}{\dot{q}} = \frac{\delta_{1,3} - \delta_{1,1}}{\dot{q}} \tag{21}$$

where $\dot{q}$ is the constant rotation speed of the crank.

As a result, the timing of the mechanism is not dependent on the control system, but rather on the mechanical design of the four-bar linkage. The device can be cheaper: there is no need for complex electronics or sensors; instead, the motor can be driven at a constant speed by using a single potentiometer.

Table 3. Results of the synthesis of the mechanism. Link lengths to the left, link orientation at configuration 0 to the right.

Parameter	Value	Unit	Parameter	Value	Unit
a_1	39.6	[mm]	$\varphi_{1,0}$	−2.55	[°]
a_2	104.2	[mm]	$\varphi_{2,0}$	−13.49	[°]
a_4	120.6	[mm]	$\varphi_{4,0}$	326.76	[°]

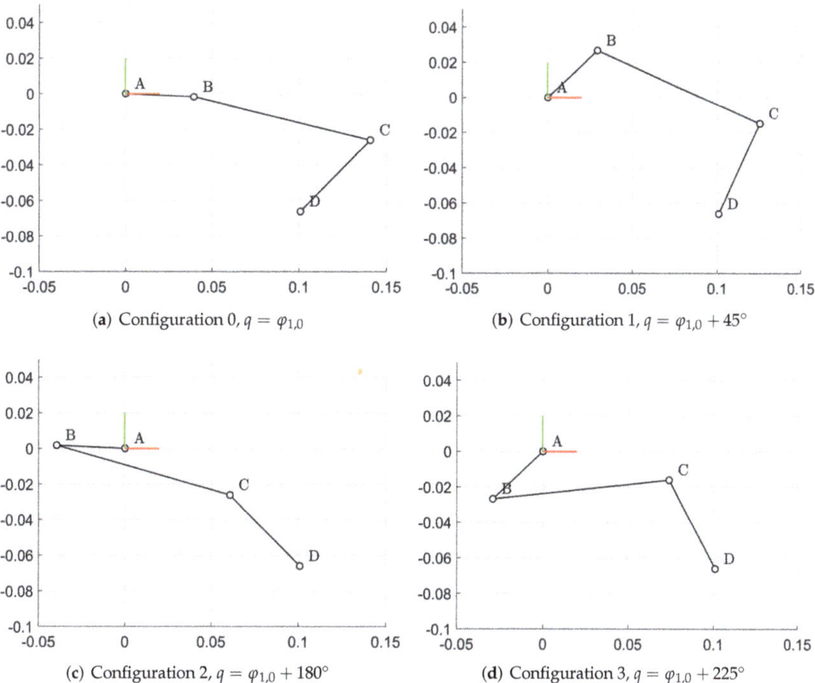

Figure 4. Configurations of the synthesis. (**a**) configuration 0; (**b**–**d**) configurations of the precision points. The same configuration numbers are reported in Figure 5.

The overall behavior of the four-bar linkage mechanism for every possible configuration is depicted in Figure 5. As expected, the movement of the rocker can not be controlled precisely between the precision points. In particular, it can be noted how the overall rocker movement range is higher than φ_r (in our example, $\max(\varphi_3) - \min(\varphi_3) = 96.32°$), but, actually, every crank half rotation the rocker is placed at exactly 90° with respect to the previous crank half rotation.

3.2. Blade Design

From the mechanism synthesis, it is possible to perform the blade design. In this case, the two blades—one for the left compartment, and the one for the right compartment—must be designed separately, since the rocker movement is uneven between the two crank half rotations.

The motion law $\psi(t)$ of Equation (17) is equal to $\varphi_3(q/\dot{q})$ since the motor is moved at constant speed $\dot{q}$, hence there is a direct proportion between the crank rotation q and the movement time t. The two motion law for the two blades are retrieved directly from Figure 5a, where:

- for the movement from left to right: $\psi(t) = \varphi_3(t) = \varphi_3(q/\dot{q})$ for $q \in [0, 180]°$;
- for the movement from right to left: $\psi(t) = \varphi_3(t) = \varphi_3(q/\dot{q})$ for $q \in [180, 360]°$.

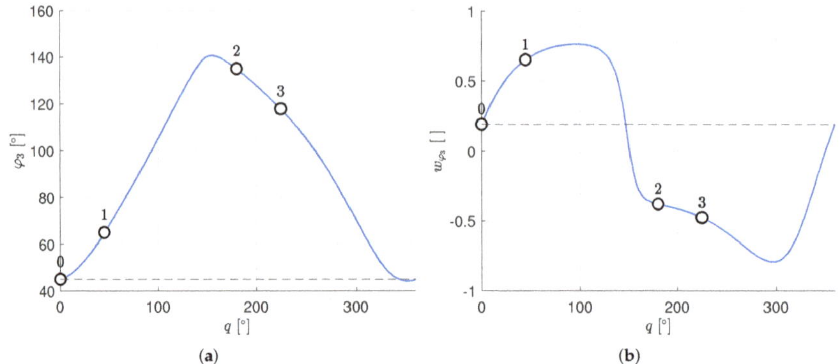

Figure 5. Rocker position (**a**) and speed ratio (**b**) for each crank configuration, where q is the displacement with respect to configuration 0. The dots represent the configuration 0 (to the left of each graph) and the three precision points (where $q = \delta_{1,k}$ with $k = 1, 2, 3$). Dashed lines show how $\varphi_3(360°) = \varphi_3(0°)$ and $w_{\varphi_3}(360°) = w_{\varphi_3}(0°)$.

All the numerical values used in the blade design are shown in Table 4, where $\dot{q}$ is calculated by choosing the time required for a crank half rotation T and r is the diameter of the component.

Table 4. Numerical values used in the design of the blades.

Parameter	Left Blade	Right Blade
x_0 [mm]	−8.7	8.7
y_0 [mm]	−5	−5
r [mm]	4	4
T [s]	0.11	0.11

The two blades are depicted in Figure 6. In light blue (Figure 6a) and orange (Figure 6b) is shown the component relative trajectory during the free fall. Since the blade must be adjacent but not interact with the piece, the top side of the blade is actually the envelope of the bottom part of the component during the fall. Such a side is depicted in the figures with a bold black line.

Please note that the blade, in the simulation, pierces the cylindrical wall. This is due to the fact that the free fall motion allows the component to exit the cylindrical structure in a time t_{out} lower than T. As a result, the physical blade should be cut to fit the cylindrical dimensions. This cutting does not influence the component motion: in fact, it is the expression of the fact that the component has yet left the cylinder while the movement of the rotary distributor from one side to the other is still ongoing; thus, T could be further reduced. The value of t_{out} is the lower limit of T, and can be calculated by solving Equation (16) for $y(t) = -R$, where R is the radius of the cylinder (Figure 1a):

$$t_{out} = \sqrt{\frac{2(R + y_0)}{g}} \tag{22}$$

which, for our example, with $|R| = a_3$, $t_{out} = 0.1026$ s.

Varying T has an important influence on the final blade shape. In Figure 7 are shown different blades for different T values where, for simplicity, the component trajectories are stopped at $t = t_{out}$. The more the time T increases, the more the shape of the blade resembles a single-curvature arc, and its initial position (the one depicted in Figure 7) tends to align to the vertical direction. In this way, components that fall from the inlet are not blocked; thus, they simply fall through the device without grouping on the blade.

Since the motion law of the blade $\psi(t)$ differs between the left-right and right-left movements (Figure 5a), the blades of the two compartments are different (Figure 5a), although they look very similar. A more intuitive comparison is depicted in Figure 8, where the right blade has been mirrored with respect to the vertical axis. The left blade, which corresponds to the range $q \in [0, 180]°$ presents an evident sharp bend right before the cylinder edges, which corresponds to the last part of the rotary movement, during which the speed rapidly changes (range $q \in [120, 160]°$ of Figure 5b). Such change is less evident in the other movement ($q \in [180, 360]°$), where can be seen a small spike in the rocker speed before returning to configuration 0. Moreover, the first movement presents a generally higher rocker speed, which increases the distance of the component from the horizontal axis, resulting in a blade that bends towards the top of the absolute reference system.

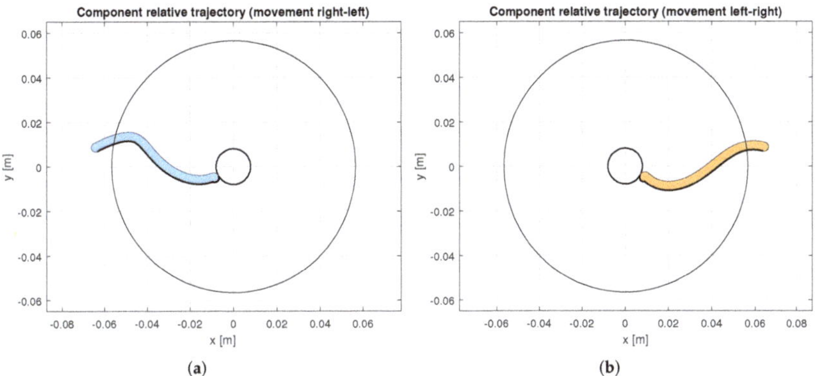

Figure 6. Shape of the two blades -left (**a**) and right (**b**)- for the two compartments. Please note that the shape of the blade pierces the cylinder walls since in the movement time T the component is able to fall outside of the cylinder via free fall.

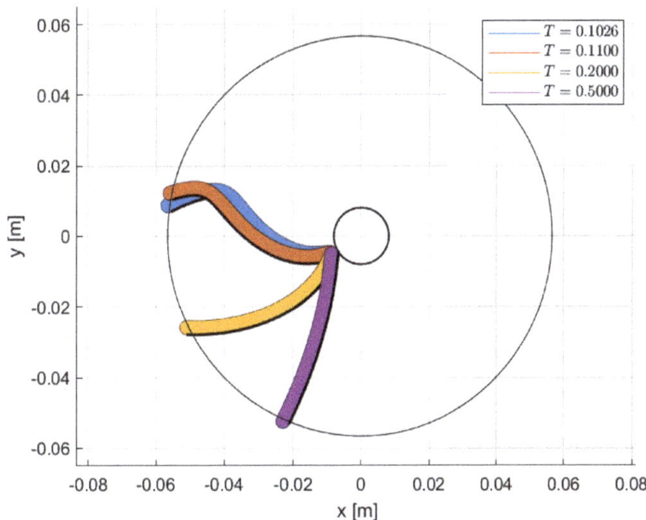

Figure 7. Blade shapes for different values of T. The trajectories of the components have been stopped evenly at $t = t_{out}$.

From a manufacturing point of view, it is more convenient to produce a single blade rather than two distinct blades. In this sense, it is mandatory, for our example, to consider only the right blade. In fact, if the left blade is considered, during the left-right movement (which would require the right blade) the trajectory of the component in the relative reference frame would still be the orange of Figure 8. As a result, since the trajectory pierces the left blade, the component would interact with the blade, producing unexpected behavior. On the contrary, if the right blade is considered, in the right-left movement the component would fall producing a certain distance to the right blade. Nonetheless, the free fall motion would still be guaranteed.

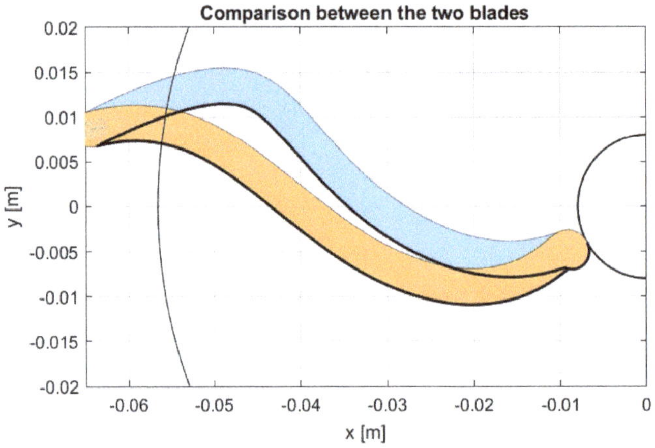

Figure 8. Comparison of the two blades.

Finally, it has to be noted that the blade has been designed to be as close to the rotary distributor shaft to be as flexible as possible. In fact, in terms of speed, the initial part of the free falling motion is the slowest, since speed increases linearly with time; thus, the trajectory performed by a component adjacent to the shaft is the worst-case scenario in terms of possible interactions with the blade. Once the blade is designed with the component adjacent to the shaft, any other starting point on the blade ensures no interaction with the blade. An example is provided by Figure 9, in which a different starting point on the blade is considered. In here it can be seen that the component on the left, after a rotation of the blade, has a certain distance from the blade itself; the component whose falling starts in (x_0, y_0) (component to the right), instead, perfectly follows the blade profile. As a result, any component that is placed on any other point on the blade will never interact with the blade and will fall by means of gravity.

3.3. Motor Torques

The four-bar linkage comes with a drawback which is related to the variability of the torques that the motor must apply to the crank to move the entire mechanism, especially when it is close to the singular configurations. To this regard, the proposed design has been compared to the direct driven solution shown in [16]. In the previous work, the motion of the blades is directly driven by an electric motor.

To simplify the comparison, only inertial forces are considered, while friction is neglected. Figure 10 represents the motor torques for three different scenarios, the proposed design (blue) and two cases for the direct drive solution, with the motor following a trapezoidal speed law (orange) and a trapezoidal acceleration law (yellow). The speed coefficient c_v is set equal to 1.333; it is defined as

$$c_v = \frac{T}{T - T_a} \tag{23}$$

where T is the motion time for one kitting operation (i.e., 180° for the proposed system and 120° for direct drive solution), and T_a is the acceleration time, which is considered equal to deceleration time. T is set to 0.11 s. As to the trapezoidal acceleration law, the ratio between the time at constant acceleration and the total acceleration time T_a, is set equal to 1/3.

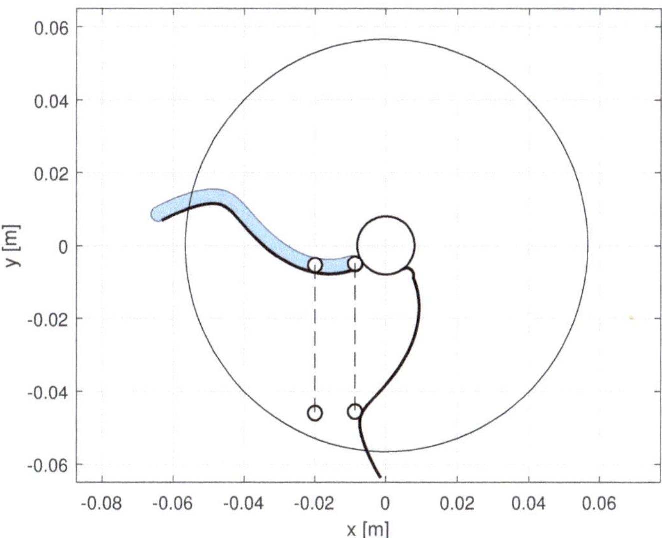

Figure 9. Comparison of trajectories of components with different starting points, at the beginning of the blade movement and after a blade rotation of about 140°. A component which starts the fall away from the rotary compartment shaft will not interact with the blade.

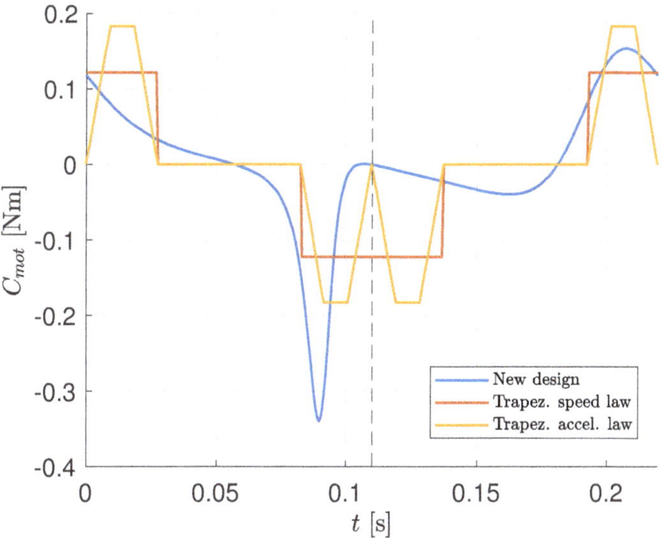

Figure 10. Motor torques for different design solutions. Proposed design (blue) and direct motor installed on the rotary distributor: trapezoidal speed law (orange) and trapezoidal acceleration law (yellow).

The proposed system presents a higher value of maximum torque (0.339 Nm) due to the proximity to a singular configuration. However, the root mean square of the motor torque (0.0874 Nm) is comparable to that of the other two solutions (0.0859 Nm for the trapezoidal velocity law and 0.0961 Nm for the trapezoidal acceleration law, respectively), so that they can be regarded as substantially equivalent in terms of torque requirements.

4. Experimental Results

To test the effectiveness of the method, an experimental setup has been developed at the University of Padova, as seen in Figure 11.

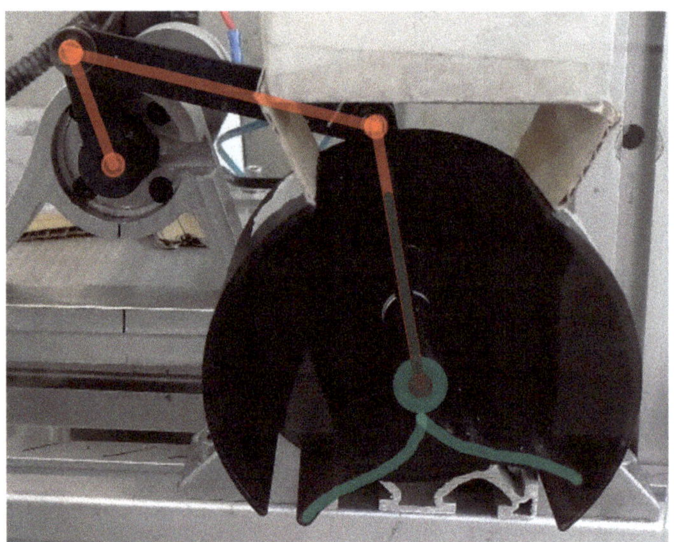

Figure 11. Experimental setup. The four-bar linkage mechanism is highlighted in red, while the blades are highlighted in green.

The links, the cylinder, and the blades have been rapidly prototyped by using the plastic material PLA. For the latter, only the right blade has been considered for simplicity. The crank is driven by an MAE M543-0900 motor.

The device has proved its effectiveness with one component and has been tested with multiple components all at once, to simulate the kitting operation, as in Figure 12. Results of the effectiveness of the device with multiple components and different T values are shown in Table 5. The same movement has been performed multiple times to have an empirical information about any possible problems. The components are fed from a grouping station at the top. In Table 5 a check sign ($\checkmark$) indicates that the device, during the tests, have not shown any issues; an approx sign ($\approx$) indicates that only in a minor set of tests ($<5\%$) some issues occurred; finally, a cross sign ($\times$) indicates that with a specific T and number of components the blade is not reliable for industrial use. It has to be noted that, in terms of reliability, only the cases with the check mark can be considered suitable for industrial use.

As expected, the design of the blade is more effective with few components. Indeed, increasing the number of components to be kitted increases the chance of interference between the components; thus, they may not follow the predicted trajectory. With many components the tests have shown pieces stuck between the blades and the cylinder walls, either during the loading or the unloading phase. In particular, with $T = 0.11$ s and 6 components, some pieces have fallen within the wrong compartment.

Please note that most of the issues at low T values occur at inlet side. Optimization of the cylinder walls and/or of the blade dividing the two compartments is likely to reduce failures, but needs further investigation and testing. Also the influence on failures of the direction and timing of incoming parts is worth to be investigated, focusing on the dynamic interaction between the dividing blade and the falling parts.

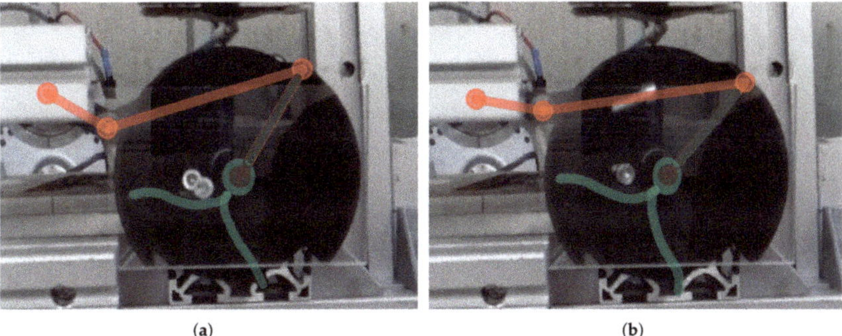

(a)   (b)

Figure 12. Example of interaction with multiple components. (**a**) the two components are perfectly aligned on the blade before the free fall; (**b**) one component is falling on the other which is starting the free falling motion. The four-bar linkage system is highlighted in red, while the blades are highlighted in green.

Table 5. Experimental results of multiple rotations with different T and a different number of components. The blades are proven to be very effective by design with few components, but for small T and a high number of components some problems may occur.

Components	$T = 0.11$ s	$T = 0.2$ s	$T = 0.5$ s
1	✓	✓	✓
2	≈	✓	✓
6	×	≈	✓

5. Conclusions

In this paper, the design of a novel feeding device has been proposed. The device is composed of a rotary distributor, divided into two compartments by a linear blade, and a four-bar linkage mechanism used to drive the distributor. Each compartment is used alternatively to group components and let them fall into a following kitting station, while the alternate left-right movements are driven by the kinematics of the four-bar linkage mechanism. The components are accompanied by blades specifically designed to support the components' free-falling motion.

Both the four-bar linkage mechanism and the blade design have been presented in this paper. The mechanism design is performed by a specific functional synthesis, which requires three parameters to be performed algebraically. The blade shape, on the other hand, is designed exploiting the four-bar linkage kinematic results.

The main advantage of the proposed mechanism is its simplicity in the control system. In fact, the movement of the blade is driven solely by the kinematics of the four-bar linkage, where the crank is fixed to the shaft of an electric motor controlled at a fixed speed, thus requiring a very simple control system.

The device has proved to be reliable for very few components, whereas multiple pieces can interact with each other, resulting in unexpected behavior, especially at high speed. Nonetheless, the fact that loading and unloading of the compartments can be performed in a single movement—since while one compartment is unloading the other can be loaded—shows very promising expectations for future developments.

Author Contributions: Conceptualization, M.B. and G.R.; methodology, G.R.; software, M.B. and R.M.; validation, M.B. and R.M.; formal analysis, M.B. and G.R.; investigation, R.M.; resources, G.R.; writing—original draft preparation, M.B. and R.M.; writing—review and editing, R.M. and G.R.; visualization, M.B. and R.M.; supervision, G.R. All authors have read and agreed to the published version of the manuscript.

Funding: This research received no external funding.

Institutional Review Board Statement: Not applicable.

Informed Consent Statement: Not applicable.

Data Availability Statement: Not applicable.

Conflicts of Interest: The authors declare no conflict of interest.

References

1. Heilala, J.; Montonen, J.; Väätäinen, O. Life cycle and unit-cost analysis for modular reconfigurable flexible light assembly systems. *Proc. Inst. Mech. Eng. Part B J. Eng. Manuf.* **2008**, *222*, 1289–1299. [CrossRef]
2. Faccio, M.; Bottin, M.; Rosati, G. Collaborative and traditional robotic assembly: A comparison model. *Int. J. Adv. Manuf. Technol.* **2019**, *102*, 1355–1372. [CrossRef]
3. Limère, V.; Landeghem, H.V.; Goetschalckx, M.; Aghezzaf, E.; McGinnis, L.F. Optimising part feeding in the automotive assembly industry: Deciding between kitting and line stocking. *Int. J. Prod. Res.* **2012**, *50*, 4046–4060. [CrossRef]
4. Höhner, D.; Wirtz, S.; Scherer, V. A numerical study on the influence of particle shape on hopper discharge within the polyhedral and multi-sphere discrete element method. *Powder Technol.* **2012**, *226*, 16–28. [CrossRef]
5. Huang, X.; Zheng, Q.; Yu, A.; Yan, W. Optimised curved hoppers with maximum mass discharge rate—An experimental study. *Powder Technol.* **2021**, *377*, 350–360. [CrossRef]
6. Cox, G.; McCue, S.; Thamwattana, N.; Hill, J. Perturbation solutions for flow through symmetrical hoppers with inserts and asymmetrical wedge hoppers. *J. Eng. Math.* **2005**, *52*, 63–91. [CrossRef]
7. Liu, H.; Jia, F.; Xiao, Y.; Han, Y.; Li, G.; Li, A.; Bai, S. Numerical analysis of the effect of the contraction rate of the curved hopper on flow characteristics of the silo discharge. *Powder Technol.* **2019**, *356*, 858–870. [CrossRef]
8. Huang, X.; Zheng, Q.; Liu, D.; Yu, A.; Yan, W. A design method of hopper shape optimization with improved mass flow pattern and reduced particle segregation. *Chem. Eng. Sci.* **2022**, *253*, 117579. [CrossRef]
9. Ketterhagen, W.R.; Hancock, B.C. Optimizing the design of eccentric feed hoppers for tablet presses using DEM. *Comput. Chem. Eng.* **2010**, *34*, 1072–1081. [CrossRef]
10. Timoleanov, K.; Savenkov, D.; Gorgadze, L. Features of designing feeding hoppers of loose materials of low productivity in agriculture. In Proceedings of the International Conference on Modern Trends in Manufacturing Technologies and Equipment (ICMTMTE 2018), Sevastopol, Russia, 10–14 September 2018; Volume 224. [CrossRef]
11. Comand, N.; Minto, R.; Boschetti, G.; Faccio, M.; Rosati, G. Optimization of a Kitting Line: A Case Study. *Robotics* **2019**, *8*, 70. [CrossRef]
12. Roshanbin, A.; Tirmizi, A.; Raeymaekers, S.; Verhees, D.; Afzal, M.R. Towards Easing Automation of Robotic Kitting Operations. In Proceedings of the 2022 IEEE 17th International Conference on Control & Automation (ICCA), Naples, Italy, 27–30 June 2022; Volume 2022, pp. 424–431. [CrossRef]
13. Fager, P.; Hanson, R.; Fasth-Berglund, A. Dual Robot Kit preparation in batch preparation of component kits for mixed model assembly. *IFAC PapersOnLine* **2020**, *53*, 10627–10632. [CrossRef]
14. Pantyukhina, E.; Preis, V.; Khachaturian, A. Feed rate evaluation of mechanical toothed hopper-feeding device with ring orientator for parts, asymmetric at the ends. *IFAC PapersOnLine* **2019**, *53*, 1260. [CrossRef]
15. Gao, X.; Xie, G.; Xu, Y.; Yu, Y.; Lai, Q. Application of a staggered symmetrical spiral groove wheel on a quantitative feeding device and investigation of particle motion characteristics based on DEM. *Powder Technol.* **2022**, *407*, 117650. [CrossRef]
16. Comand, N.; Bottin, M.; Rosati, G. Improving Components Feeding: A Rotatory Device. *Mech. Mach. Sci.* **2021**, *91*, 468–475. [CrossRef]
17. Comand, N.; Bottin, M.; Rosati, G.; Agrawal, S. A dynamic model for the optimization of rotary feeding devices. *Mech. Mach. Theory* **2021**, *166*, 104479. [CrossRef]
18. Rosati, G.; Cenci, S.; Boschetti, G.; Zanotto, D.; Masiero, S. Design of a single-dof active hand orthosis for neurorehabilitation. In Proceedings of the 2009 IEEE International Conference on Rehabilitation Robotics, Kyoto, Japan, 23–26 June 2009; pp. 161–166. [CrossRef]
19. Jianwei, S.; Jinkui, C.; Baoyu, S. A unified model of harmonic characteristic parameter method for dimensional synthesis of linkage mechanism. *Appl. Math. Model.* **2012**, *36*, 6001–6010. [CrossRef]
20. Cabrera, J.; Simon, A.; Prado, M. Optimal synthesis of mechanisms with genetic algorithms. *Mech. Mach. Theory* **2002**, *37*, 1165–1177. [CrossRef]

21. Acharyya, S.; Mandal, M. Performance of EAs for four-bar linkage synthesis. *Mech. Mach. Theory* **2009**, *44*, 1784–1794. [CrossRef]
22. Huang, Q.; Yu, Y.; Zhang, K.; Li, S.; Lu, H.; Li, J.; Zhang, A.; Mei, T. Optimal synthesis of mechanisms using repellency evolutionary algorithm. *Knowl.-Based Syst.* **2022**, *239*, 107928. [CrossRef]

Review

Recent Advancements in the Tribological Modelling of Rough Interfaces

Nicola Menga [1,2], Carmine Putignano [1,2,*] and Giuseppe Carbone [1,2]

1   Department of Mechanics, Mathematics and Management, Politecnico of Bari, V.le Japigia, 182, 70126 Bari, Italy
2   Department of Mechanical Engineering, Imperial College London, Exhibition Road, London SW7 2AZ, UK
*   Correspondence: carmine.putignano@poliba.it; Tel.: +39-0805963512

Abstract: This paper analyses some effective strategies proposed in the last few years to tackle contact mechanics problems involving rough interfaces. In particular, we present Boundary Element Methods capable of solving the contact with great accuracy and, at the same time, with a marked computational efficiency. Particular attention is paid to non-linearly elastic constitutive relations and, specifically, to a linearly viscoelastic rheology. Possible implications deal with all the tribological mechanical systems, where contact interactions are present, including, e.g., seals, bearings and dampers.

Keywords: applied mechanics; tribology; Boundary Elements Methods

1. Introduction

In the last few decades, a variety of analytical, numerical and experimental approaches have been implemented to understand what happens when two solids, with real rough interfaces, come into contact. The reasons for such a marked attention in the rough contact mechanics are different and both theoretical and applicative. In fact, there is certainly a genuine academic interest in the problem due to its specific intricacy: the presence of the roughness, whose spectrum covers several orders of magnitude and goes down to the atomistic scales, introduces a huge number of space and time scales. On the other hand, rough contact mechanics has a practical importance in the optimized design of engineering systems and components: classical industrial applications include seals and dampers ([1,2]), but a constantly increasing interest is rising in the frontiers' fields, such as bio-adhesive ([3–6]), cellular scaffolds ([7,8]) and even touch-screen devices ([9]).

The first theoretical answer to the rough contact problem was given by Greenwood and Williamson in 1966 in a pioneering paper where the first of the so-called multiasperity theories has beeen introduced. Following such a contribution, a variety of models have been proposed (see, e.g., [10–13]): basically, these models reduce the surface roughness to a discrete distribution of asperities behaving as independent Hertzian punches, thus neglecting the reciprocal interaction between the contact clusters. Due to this assumption, multiasperity theories cannot match experimental results in terms of applied load and contact area ([14]). In the last twenty years, Persson has developed a totally different approach, where the contact pressure probability distribution is demonstrated to be determined by a diffusive process, being dependent on the magnification at which the contact interface is observed. This theory can be considered exact in full contact conditions, but provides still qualitatively accurate information for partial contacts. On the other hand, to provide quantitativelly reliable predictions as needed in applications, a number of numerical methods and, in particular, several Boundary Element approaches, implemented either in the real ([15–20]) and in the Fourier space ([21–24]), have been developed. Nowadays, these techniques are extremely accurate, but at the same time, in most cases were affected by a significant limitation: they were developed for linear elastic contact mechanics.

Citation: Menga, N.; Putignano, C.; Carbone, G. Recent Advancements in the Tribological Modelling of Rough Interfaces. *Machines* **2022**, *10*, 1205. https://doi.org/10.3390/machines10121205

Academic Editor: Francisco J. G. Silva

Received: 23 September 2022
Accepted: 1 December 2022
Published: 12 December 2022

Publisher's Note: MDPI stays neutral with regard to jurisdictional claims in published maps and institutional affiliations.

However, it should be pointed out that, in many cases, and, crucially, when soft materials are involved in the problem, a marked non-elastic time-dependent mechanical behaviour is evident. Specifically, in a number of systems of applicative interest, including civil engineering ([2,25]) and biomechanics ([7,8,26]), soft matter can be described, with good approximation, by a linearly viscoelastic rheology. Recently, due to the theoretical and practical prominence of viscoelastic contact mechanics, a large amount of research activities was dedicated to the topic ([27–35]): different cases, including constant sliding velocity, reciprocating motion and lubricated contacts, have been investigated by some of the authors of this paper and other research groups in the world ([27–41]). There are still multiple issues to point out, but these studies have already properly demonstrated how viscoelasticity alters the contact solution in terms of contact area, load, separation, stiffness and, ultimately, friction compared to the elastic case. Results not only differ from the orders of magnitude from purely elastic conditions, but also have a qualitative impact on the global solution. To this extent, in this paper, we will review, *inter alia*, a point that, given its theoretical and practical importance, has to be properly accounted for: this is the anisotropy induced by the material viscoelastic rheology on the contact solution. As shown in Ref. [33,42], when a rigid isotropic rough punch slides over a viscoelastic layer, the contact solution, in terms of both contact clusters and displacement, is strongly anisotropic. In fact, in each contact cluster, we have a different behaviour between the leading edge and the trailing edge, where the material is not yet relaxed. Clearly when the contacting surfaces are already anisotropic, as it often happens in real interfaces due to the manufacturing treatment, the problem further complicates. All this has significant practical consequences: for example, this is crucial for sealing systems as it intervenes on the percolation phenomenon. In fact, most of the theories in the field ([43–46]) assume that the contact patches distribution is perfectly isotropic, but clearly this assumption fails when dealing with viscoelastic interfaces, thus leading to a potential underestimation of the percolating flow in rubber-based viscoelastic seals.

Another issue to properly account for when dealing with rough contacts is the case of coated bodies. Soft coatings offer the chance to tailor the resulting interface behavior in terms of adhesive toughness, local contact stiffness, frictional behavior, etc. Similarly, biological systems have also often evolved in order to exploit specific features of multi-layer tissues such as, for instance, human skin. For these reasons, besides the aforementioned classical investigations focusing on both adhesive [47–53] and adhesiveless [17,28,54–61] contacts involving half-spaces, detailed studies have been led specifically focusing on contacts of thin layers [35,62–66].

As a matter of fact, material dissimilarity between the contacting bodies (i.e., *material coupling*) is the only source of interaction between the displacement fields in the directions normal and tangential to the surface for half-space contacts [67,68]. This has been clearly pointed out in a series of studies dealing with both homogeneous [69–71] and graded [72,73] elastic materials. However, very little has been conducted with respect to the case of thin bodies. In Refs. [74–76], the aforementioned coupling description has been proven to hold true only for the case of semi-infinite contacts, whereas contacts involving deformable layers of finite thickness behave differently. Indeed, in this case, a thickness-related source of normal-tangential coupling exists, which has been defined as *geometric* coupling [77,78] (notably, the latter term is negligible for very thick bodies). In Refs [77,78], it has been demonstrated that the contact area is strongly affected by the geometric coupling in sliding frictional contacts, with predicted values up to 10% larger than the expected values for the uncoupled case. Electrical conductivity prediction [79], and wear estimation [80] are side problems which could result to be worsened by neglecting geometric coupling effects. Ref. [78] also shows that geometric coupling affects the overall frictional beahvior of the interface by inducing an asymmetric normal pressure distribution also in purely eleastic contacts. Similarly, the finite thickness of the deformable substrate is also expected to impact the adhesive performance of the contact interface. It is the case, for instance, of thin coatings in orthopedic implants [81,82], medical adhesive bands [83], and pressure sensitive adhesives [84–86], where the layer is comparable in thickness to the surface features size. It has been demonstrated that, in this case, the specific boundary conditions applied to the thin deformable layer may play a key role in determining the adhesive strength and toughness at the interface, both in the case of adhesive contact mechanics against wavy counterparts [64], and peeling detachment from flat substrates [87–90].

In this paper, we review the main approaches developed to tackle all the aforementioned issues and, in particular, to understand what happens in terms of pressure distribution, displacement and friction when rough interfaces are in contact. The paper is structured as follows. Section 2 include the methodology developed for elastic interfaces, while the following one is dedicated to the viscoelastic one. Results and final remarks complete the manuscript.

2. BEM Formulation for Elastic Sliding Contacts

The system under investigation is shown in Figure 1, where a randomly rough rigid surface is in steady-state sliding contact with a deformable solid backed onto a rigid substrate. In the same figure, $r(\mathbf{x})$ represents the surface roughness (with $\mathbf{x}$ being the in-plane position vector) with periodicity λ, and h is, in general, the deformable thickness. Notably, for $h \to \infty$, the behavior of the contacting solids asymptotically approaches the half-space. Morevoer, as shown in Figure 1, we assume the rigid rough indenter to penetrate the solid surface by a quantity δ_z, whereas with $\bar{u}_z$ and Δ we indicate the mean normal displacement of the solid surface and the mean penetration, respectively, so that:

$$\delta_z = \Delta + \bar{u}_z. \tag{1}$$

In addition, Λ and λ indicate the roughness peak and fundamental wave-length, respectively.

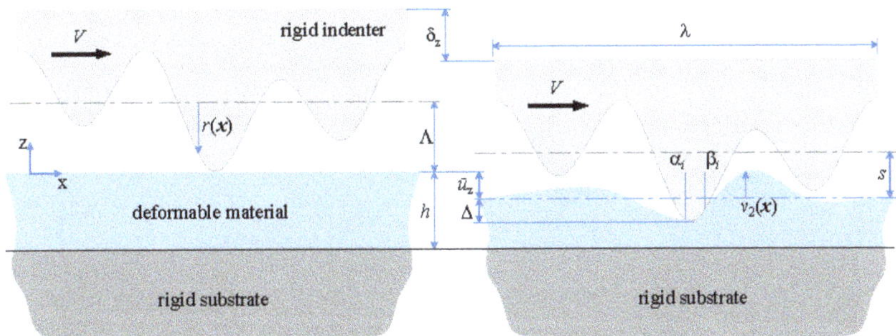

Figure 1. The sliding contact between a rigid rough profile and a deformable layer of thickness h backed onto a rigid substrate.

In order to generalize our contact model, we assume the presence of Amonton/Coulomb friction at the contact interface, with friction coefficient μ_c, all over the contact domain Ω. Indeed, the stress distribution acting along the x direction is given by

$$\tau_x(\mathbf{x}) = \mu_c p(\mathbf{x}); \quad \mathbf{x} \in \Omega, \tag{2}$$

where $p(\mathbf{x})$ is the normal pressure distribution.

We assume that the relative sliding speed only occurs in the x direction. And that, μ_c does not depend on the relative sliding speed. Notably, the frictionless contact mechanics behavior (i.e., purely normal indentation) is easily reseambled for $\mu_c = 0$.

Following the contact formulation given in Refs. [17,28,63,64], we formulate the contact problem in terms of interfacial surface displacement vector $\mathbf{v} = (v_x, v_y, v_z)$ and stress vector $\sigma = (\mu_c p, 0, -p)$. In the reference system joint to the moving indenter, we have

$$\mathbf{v}(x) = \mathbf{u}(x) - \bar{\mathbf{u}} = \int_\Omega d^2 s \Theta(x - s, h) \sigma(s); \quad x \in \Omega, \tag{3}$$

where we applied the the coordinate transformation $x - Vt \to x$. In Equation (3), $\mathbf{u}$ and $\bar{\mathbf{u}}$ represent the total and mean surface displacement vectors, respectively, which can be calculated following the procedure indicated in Refs. [17,28,63,64]. Notably,

$$\bar{\mathbf{u}} = \frac{1}{\lambda^2} \int_\Omega \mathbf{u}(x) d^2 x \tag{4}$$

with λ being the size of the periodic calculation domain. Moreover, the term $\Theta(x, h)$ in Equation (3) represents the Green's tensor, which depends on the specific thickness h of the elastic body. Moreover, we define the mean contact pressure as

$$p_m = \frac{1}{\lambda^2} \int_\Omega p(x) d^2 x \tag{5}$$

Notably, by means of Equation (3), the contact problem can be reduced to a Fredholm equation of the first kind. Presently, let us focus our attention on the specific case of an half-space and a frictionless contact. To numerically solve the contact problem, the penetration depth Δ is controlled, while the computational domain D is discretized with small squares of non-uniform size. The unknown stress in each single square is assumed to be uniformly distributed on it. Thus, we can discretize Equation (3) as the following linear system:

$$v_i = L_{ij} \sigma_j \tag{6}$$

where σ_i is the normal stress uniformly acting on the square, v_i is the normal displacement at the centre of each square, and L_{ij} is the elastic response matrix, that is, the matrix provided by the discretized version of Equation (3). In the case of , L_{ij} can be computed by employing the Love solution (see [91]), which furnishes the elastic displacement due to a uniform pressure on a rectangular area, and adding up the contribution of each elementary cell D to account for the periodicity of the problem. Thus $L_{ij} = l_{ij} - l_m$, where:

$$
\begin{aligned}
l_{ij} = \frac{1 - \nu^2}{\pi E} \sum_{k=-\infty}^{+\infty} \sum_{h=-\infty}^{+\infty} \Bigg\{ & (\xi_{ij} + d_j) \ln \left(\frac{(\eta_{ij} + d_j) + \left[(\xi_{ij} + d_j)^2 + (\eta_{ij} + d_j)^2 \right]^{1/2}}{(\eta_{ij} - d_j) + \left[(\xi_{ij} + d_j)^2 + (\eta_{ij} - d_j)^2 \right]^{1/2}} \right) \\
+ & (\eta_{ij} + d_j) \ln \left(\frac{(\xi_{ij} + d_j) + \left[(\eta_{ij} + d_j)^2 + (\xi_{ij} + d_j)^2 \right]^{1/2}}{(\xi_{ij} - d_j) + \left[(\xi_{ij} - d_j)^2 + (\eta_{ij} + d_j)^2 \right]^{1/2}} \right) \\
+ & (\xi_{ij} - d_j) \ln \left(\frac{(\eta_{ij} - d_j) + \left[(\xi_{ij} - d_j)^2 + (\eta_{ij} - d_j)^2 \right]^{1/2}}{(\eta_{ij} + d_j) + \left[(\xi_{ij} - d_j)^2 + (\eta_{ij} + d_j)^2 \right]^{1/2}} \right) \\
+ & (\eta_{ij} - d_j) \ln \left(\frac{(\xi_{ij} - d_j) + \left[(\xi_{ij} - d_j)^2 + (\eta_{ij} - d_j)^2 \right]^{1/2}}{(\xi_{ij} + d_j) + \left[(\xi_{ij} - d_j)^2 + (\eta_{ij} - d_j)^2 \right]^{1/2}} \right) \Bigg\}
\end{aligned} \tag{7}
$$

and

$$l_m = \frac{1-\nu^2}{\pi E}(\frac{d_j}{\lambda})^2 \sum_{k=-\infty}^{+\infty} \sum_{h=-\infty}^{+\infty} \left\{ \lambda(h+1) \ln\left(\frac{k+1+[(h+1)^2+(k+1)^2]^{1/2}}{k-1+[(h+1)^2+(k-1)^2]^{1/2}} \right) \right.$$

$$+\lambda(k+1) \ln\left(\frac{h+1+[(k+1)^2+(h+1)^2]^{1/2}}{h-1+[(k+1)^2+(h-1)^2]^{1/2}} \right)$$

$$+\lambda(h-1) \ln\left(\frac{k-1+[(h-1)^2+(k-1)^2]^{1/2}}{k+1+[(h-1)^2+(k+1)^2]^{1/2}} \right)$$

$$\left. +\lambda(k-1) \ln\left(\frac{h-1+[(k-1)^2+(h-1)^2]^{1/2}}{h+1+[(k-1)^2+(h+1)^2]^{1/2}} \right) \right\} \tag{8}$$

where d_j is the size of the elementary boundary element and $\xi_{ij} = |x_j - x_i| + \lambda h$ and $\eta_{ij} = |y_j - y_i| + \lambda k$.

Clearly, Equation (6) will be exploited to compute the interfacial stresses once the displacements v_i are known. As the problem under investigation belongs to the class of mixed boundary problems, we need to determine the real contact area. This can be performed iteratively by implementing the following procedure: (1) fix the displacement Δ_i, (2) evaluate the so-called bearing area as the intersection between the deformed elastic layer, calculated with respect to the elastic solution determined previously for the penetration $\Delta_{i-1} < \Delta_i$, and the rigid rough punch, (3) compute the displacements in the contact clusters as $v_i = h_i - h_{max} + \Delta$, where $h_i = h(x_i)$, h_{max} the maximum height of the rough profile, (4) solve Equation (6) to assess the stress distribution σ_j in the contact areas, (5) determine the displacements $v_i = L_{ij}\sigma_j$ out of the contact areas, (6) update the contact area at each iterative step by deleting the elements with negative pressure and adding those where there exists compenetration. It should be noted that we invert the matrix L_{ij} only for those points belonging to the contact area: this leads to a strong reduction of the computational efforts. The numerical inversion of the Equation (6) is made by means of an iterative method based on a Gauss-Seidel scheme.

It should be noted that the scheme previously described can be employed only in the adhesiveless case as negative values for the pressure distribution are discarded during the iterative case. In the adhesive case, based on the energy balance defined in Refs. [63], under isothermal and frictionless (i.e., $\mu_c = 0$) conditions, for any given value of the contact penetration Δ, the contact domain Ω can be calculated by requiring that, at equilibrium,

$$\left(\frac{\partial \mathcal{F}}{\partial \Omega}\right)_\Delta = 0 \tag{9}$$

where $\mathcal{F} = \mathcal{E} + \mathcal{A}$ is the total free energy, with

$$\mathcal{E} = \frac{1}{2} \int_\Omega p(\mathbf{x}) v_z(\mathbf{x}) d^2 x \tag{10}$$

being the interfacial elastic energy stored into the deformable body, and

$$\mathcal{A} = -\Delta\gamma A \tag{11}$$

being the adhesion energy with $\Delta\gamma$ being the work of adhesion, also referred to as the Duprè energy of adhesion.

Notably, Equation (9) can be numerically calculated across infinitesimal variations of the discretized contact domain Ω in the direction normal to the local boundary $\partial\Omega$.

We also observe that the present formulation can be extended to the case of frictional contacts by following the procedure defined in Refs. [49], and the fundamental solution derived in Ref. [92]. To this extent, let us focus on the case of a rigid 1D rough profile in contact with a thin elastic layer of thickness h.

Morevoer, the term $\Theta(x)$ in Equation (3) takes the form

$$\Theta(x) = \frac{1}{E}\begin{pmatrix} G_{xx} & G_{xz} \\ G_{xz} & G_{zz} \end{pmatrix},$$ (12)

where E is the Young's modulus fo the elastic material, and according to Ref. [77] we have

$$G_{xx}(x) = -\frac{2(1-\nu^2)}{\pi}\left[\log\left|2\sin\left(\frac{q_0 x}{2}\right)\right| + \sum_{m=1}^{\infty} B(mq_0 h)\frac{\cos(mq_0 x)}{m}\right],$$ (13)

$$G_{xz}(x) = -G_{zx}(x) = \frac{1+\nu}{\pi}\left[\frac{1-2\nu}{2}[\operatorname{sgn}(x)\pi - q_0 x] - \sum_{m=1}^{\infty} C(mq_0 h)\frac{\sin(mq_0 x)}{m}\right],$$ (14)

$$G_{zz}(x) = -\frac{2(1-\nu^2)}{\pi}\left[\log\left|2\sin\left(\frac{q_0 x}{2}\right)\right| + \sum_{m=1}^{\infty} A(mq_0 h)\frac{\cos(mq_0 x)}{m}\right],$$ (15)

with $q_0 = 2\pi/\lambda$, and

$$A(mq_0 h) = 1 + \frac{2mq_0 h - (3-4\nu)\sinh(2mq_0 h)}{5 + 2(mq_0 h)^2 - 4\nu(3-2\nu) + (3-4\nu)\cosh(2mq_0 h)},$$ (16)

$$B(mq_0 h) = 1 - \frac{2mq_0 h + (3-4\nu)\sinh(2mq_0 h)}{5 + 2(mq_0 h)^2 - 4\nu(3-2\nu) + (3-4\nu)\cosh(2mq_0 h)},$$ (17)

$$C(mq_0 h) = \frac{4(1-\nu)\left[2 + (mq_0 h)^2 - 6\nu + 4\nu^2\right]}{5 + 2(mq_0 h)^2 - 4\nu(3-2\nu) + (3-4\nu)\cosh(2mq_0 h)}.$$ (18)

3. BEM Formulation for Viscoelastic Sliding Contacts

In multiple applications, it is necessary to account for soft contacts and, specifically, for a linear viscoelastic rheology for the solids into contact. To this end, let us briefly recall the main features of linear viscoelasticity [93,94] throughout the following integral equation, which correlates two time-dependent quantities, that is, the strain $\varepsilon(t)$ and the stress $\sigma(t)$:

$$\varepsilon(t) = \int_{-\infty}^{t} d\tau \mathcal{J}(t-\tau)\dot{\sigma}(\tau),$$ (19)

where $\mathcal{J}(t)$ is the creep function and the symbol '·' refers to the time derivative. Now, if we define the real quantities E_0 and E_∞ respectively as the rubbery and glassy elastic moduli of the viscoelastic material, $C(\tau)$ as a the positive defined as creep spectrum [93], and τ is the relaxation time distribution, the creep function $\mathcal{J}(t)$ can be written as:

$$\mathcal{J}(t) = \mathcal{H}(t)\left[\frac{1}{E_0} - \int_0^{+\infty} d\tau C(\tau)\exp(-t/\tau)\right] = \mathcal{H}(t)\left[\frac{1}{E_\infty} + \int_0^{+\infty} d\tau C(\tau)(1-\exp(-t/\tau))\right]$$ (20)

where $\mathcal{H}(t)$ is the Heaviside step function introduced so that $\mathcal{J}(t)$ can satisfy the principle of causality and, thus, $\mathcal{J}(t<0) = 0$.

Now, Equation (20), let us understand what happens when viscoelastic bodies are into contact. Due to the linearity of the system, as the geometrical domain is non-finite and, thus, translational invariant [91], let us focus on the following integral equation formulated to correlate the normal surface displacement $u(x,t)$ and the normal interfacial stress derivative $\dot{\sigma}(x',\tau)$:

$$u(x,t) = \int_{-\infty}^{t} d\tau \int d^2x' G_{tot}(\mathbf{x}-\mathbf{x}',t-\tau)\dot{\sigma}(\mathbf{x}',\tau),$$ (21)

where $\mathbf{x}$ is the position vector, t is the time, G_{tot} is a global Green's function. If we assume that we are dealing with a perfectly homogenous solid, we can factorize the integral equation kernel in Equation (21) in two terms, thus writing:

$$u(x,t) = \int_{-\infty}^{t} d\tau \int d^2x' \, \mathcal{J}(t-\tau) G(\mathbf{x}-\mathbf{x}') \dot{\sigma}(\mathbf{x}',\tau), \tag{22}$$

where $G(\mathbf{x})$ is a spatial Green's function, which is defined in detail in Ref. [95]. It is crucial to observe that, in Ref. [28], under steady-state assumptions, that means the punch is sliding at constant v over the viscoelastic substrate, Equation (22) has been further developed. In fact, we can introduce a speed parametrically dependent Green's function $G(x,v)$, thus strongly simplifying Equation (22) as:

$$u(x) = \int d^2x' \mathcal{G}(x-x',v)\sigma(x') \tag{23}$$

Similarly, in Ref. [36], reciprocating conditions are exploited: the rough rigid punch is assumed in sinusoidal motion over the viscoelastic substrate. Indeed, in this case, we can introduce a Green's function, being parametrically dependent on the time t, and we can write the following Equation:

$$u(x,t) = \int d^2x' \mathcal{G}(x-x',t)\sigma(x',t) \tag{24}$$

When we reduce Equation (22) to Equation (23) or to Equation (24), respectively for the steady-state or the reciprocating conditions, a dramatic reduction in the computational complexity is obtained. As a result, multi-scale problems, where the roughness spectrum covers several orders of magnitude, can be investigated. On the other side, as these simplifications affect the generality of the kinematic conditions which can be investigated, in Ref. [95] Equation (22) has been directly tackled. Although the computational complexity has allowed the solution just for the 1D case, it has been possible to explore different conditions, including normal indentation and transient contacts [95].

Another aspect to consider when approaching viscoelastic contact problems deals with the case of thin viscoelastic layers. Let us focus here on the case of a 1D rough profile in sliding contact with a thin viscoelastic layer under the assumptions of frictional interactions at the interface (i.e., $\mu_c > 0$). In this case, again other steady conditions, we can define a speed parametrically dependent tensor Θ^V as conducted previously in Equation (3) for the purely elastic case. In particular, $\Theta^V(x)$, with x being the position coordinate, can be defined as follows:

$$\Theta^V(x) = J(0^+) \begin{bmatrix} G_{xx}(x) & G_{xz}(x) \\ G_{xz}(x) & G_{zz}(x) \end{bmatrix} + \int_{0^+}^{+\infty} \begin{bmatrix} G_{xx}(x+Vt) & G_{xz}(x+Vt) \\ G_{xz}(x+Vt) & G_{zz}(x+Vt) \end{bmatrix} \dot{J}(t)dt, \tag{25}$$

which, as expected, parametrically depends on the sliding velocity V. Specifically, in Equation (25), the viscoelastic creep function $J(t)$ is given by (20), and the G terms are given by Equations (13)–(15).

Since the interface is adhesiveless, the solution strategy adopted to calculate the unknow contact area domain is the same as previously introduced.

Notably, a very recent study has also demonstrated that adhesive viscoelastic contacts in sliding conditions can be addressed by relying on an energy balance approach.

4. Results and Discussion

4.1. Two-Dimentional Viscoelastic Sliding Rough Contacts: The Role of Anisotropy

An aspect with really important implications from both a theoretical and an applicative point of view is the anisotropy related to the contact area and on the deformation field when a rigid surface is in sliding contact with a viscoelastic half-space [42]. A field, where this can become crucial, is the fluid leakage: percolation is really influenced by the contact anisotropic solution. To quantify all these effects, we can employ self-affine fractal rigid surfaces generated with spectral components in the interval $q_r < q < q_c$, where $q_r = 2\pi/\lambda$, with λ being the side of the square punch equal to $\lambda = 0.01$ m, $q_r = Nq_0$ and N number of scales (or wavelengths) [17]. Computations are carried out with $N = 64$. Moreover, regarding the material properties, in the following developments, we employ a linear viscoelastic material with a single relaxation time $\tau = 0.1$ s, with a high frequency modulus E_∞ equal to $E_\infty = 10^8$ Pa, the ratio $E_\infty/E_0 = 11$ and the Poisson ratio equal to $\nu = 0.5$.

Let us start showing in Figure 2 this anisotropic effect underlined for the first time in Ref. [36]. Basically, when we focus our attention on the sliding contact mechanics of a perfectly isotropic rough surface over a linearly viscoelastic solid, the contact area solution results are anisotropic: in detail, each contact cluster, as zoomed in the inset, tends to shrink at the trailing edge, where the material is still relaxing.Consequently, each contact area is stretched perpendicularly to the speed.

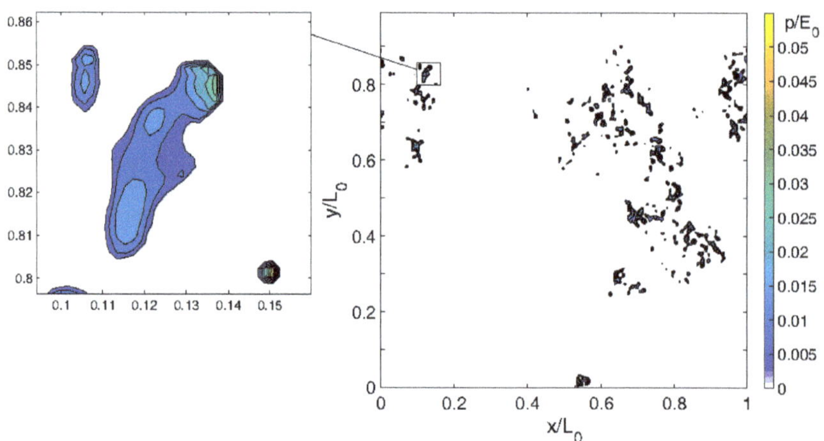

Figure 2. Contour plot of a pressure distribution for a perfectly isotropic surface sliding over a linearly viscoelastic layer. In the zoomed inset, a single patch of the contact region: this results in being stretched perpendicularly to the speed.

We can quantify the anisotropy degree by introducing, in a certain range of wave vectors $\zeta_1 q_0 < |\mathbf{q}| < \zeta_2 q_0$, the symmetric anisotropy tensor for the deformed surface $\mathcal{M}(\zeta_1, \zeta_2)$:

$$\mathcal{M}(\zeta_1, \zeta_2) = \int_{\zeta_1 q_0 < |\mathbf{q}| < \zeta_2 q_0} d^2 q \, \mathbf{q} \otimes \mathbf{q} C_d(\mathbf{q}, \zeta_1, \zeta_2) \tag{26}$$

where $C_d(\mathbf{q}; \zeta_1, \zeta_2) = (2\pi)^{-2} \int d^2 x \langle u(0; \zeta_1, \zeta_2) u(\mathbf{x}; \zeta_1, \zeta_2) \rangle \exp(-i\mathbf{q} \cdot \mathbf{x})$ is the power spectral density of the deformed surface $u(\mathbf{x}; \zeta_1, \zeta_2)$ that is filtered. Crucially, the band-pass filter in the interval $[\zeta_1, \zeta_2]$ is defined to highlight the frequencies where the viscoelastic effects are higher.

Now, let us notice that the quantity $\mathcal{M}_{ij} = \int_{\zeta_1 q_0 < |\mathbf{q}| < \zeta_2 q_0} d^2 q \, q_i q_j C_d(q)$, with i and $j = 1, 2$, is the second order moments of the filtered surface power spectral density of the , i.e., $\mathcal{M}_{11} = \mu_{20} = \langle u_x^2 \rangle$, $\mathcal{M}_{22} = \mu_{02} = \langle u_y^2 \rangle$, $\mathcal{M}_{12} = \mu_{11} = \langle u_x u_y \rangle$, where $u_x = \partial u / \partial x$, $u_y = \partial u / \partial y$ (see also Ref. [96]). Thus, once we have defined the symmetric tensor $\mathcal{M}$, the quadratic form $\mathcal{Q}(\mathbf{x}) = \mathcal{M}_{ij} x_i x_j$ can be introduced. In a polar reference system with $x = r \cos\theta$, and $y = r \sin\theta$, one may obtain:

$$\mathcal{Q}(\mathbf{x}) = r^2 |\nabla u \cdot \mathbf{e}(\theta)|^2 = r^2 \mu_2(\theta)$$

where $\mathbf{e}(\theta)$ is the unit vector $(\cos\theta, \sin\theta)$ and

$$\mu_2(\theta) = \mu_{20} \cos^2(\theta) + 2\mu_{11} \sin(\theta)\cos(\theta) + \mu_{02} \sin^2(\theta) \tag{27}$$

is, thus, the average square slope of the profile defined by carrying out a cut of the deformed surface $u(\mathbf{x}; \zeta_1, \zeta_2)$ along the direction θ [33]. Thus, when plotting the quantity $\mu_2(\theta)$ in a polar diagram, if the system is isotropic, $\mu_2(\theta)$ has to be circular; otherwise, there exists a different elliptical shape. Furthermore, we can quantify the degree of anisotropy by looking at the ratio $\gamma_d = \mu_{2\,\min}/\mu_{2\,\max}$ between the minimum $\mu_{2\,\min}$ and the maximum $\mu_{2\,\max}$ eigenvalues of the tensor $\mathcal{M}$; furthermore, the principal direction of anisotropy can be introduced by detecting the value of the angle θ_d maximizing $\mu_2(\theta)$, i.e., $\mu_2(\theta_d) = \mu_{2\,\max}$.

Similarly, we can introduce the roughness anisotropy tensor $\mathbf{M}$ for the rigid surface as $\mathbf{M}(\zeta_1, \zeta_2) = \int_{\zeta_1 q_0 < |\mathbf{q}| < \zeta_2 q_0} d^2 q \mathbf{q} \otimes \mathbf{q} C(\mathbf{q}, \zeta_1, \zeta_2)$: its components are ,then, $M_{11} = m_{20} = \langle h_x^2 \rangle$, $M_{22} = m_{20} = \langle h_y^2 \rangle$, $M_{12} = m_{11} = \langle h_x h_y \rangle$ with $h_x = \partial h/\partial x$ and $h_y = \partial h/\partial y$. As conducted before, it is possible to associate to the tensor $\mathbf{M}$ the quadratic form Q and the parameters γ_s and θ_s for the quantity $m_2(\theta)$. The latter is the average square slope for the profile cut on the rough surface $h(\mathbf{x}; \zeta_1, \zeta_2)$ along the direction θ [33].

If we focus on the isotropic surface sliding over a viscoelastic half-space as pointed out in Figure 2, it is possible to plot $m_2(\theta)$ and $\mu_2(\theta)$. In Figure 3, we observe that $m_2(\theta)$ is perfectly circular (with $\gamma_s = 1$), while crucially $\mu_2(\theta)$ is elliptical with $\gamma_d = 0.37$ and θ_d is approximately $\pi/2$. We quantify in this way what was clear in Figure 2: as the contact clusters are perpendicular to the velocity direction, the maximum anisotropy angle must be close to $\pi/2$. In Ref. [28], it was demonstrated that the contact area shrinkage and the anisotropic shape for the spectral moment $\mu_2(\theta)$ are correlated: the shrinkage of each contact cluster at the trailing edge applies a high-pass filter to the frequencies which corresponds to the scales along the velocity direction: thus, as observed in Figure 3, the spectral moment of the deformed profile reduces in such a direction. More details on the generalization of the anisotropy induced by viscoelasticity can be found in Ref. [42].

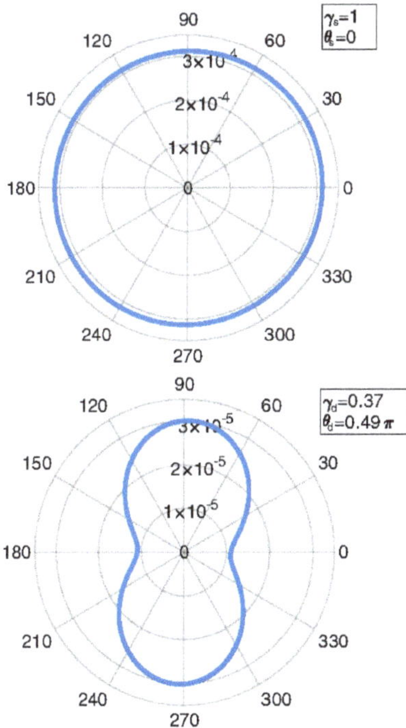

Figure 3. Polar plots of $m_2(\theta)$ for the rigid surface (**on the top**) and $\mu_2(\theta)$ for the deformed half-space (**on the bottom**). Calculations are carried out for a constant normal pressure p equal to $p = 32$ kPa and a dimensionless speed equal to $v\tau/L_0 = 0.13$.

4.2. Adhesion in Elastic Contacts of Thin Layers

Another paradigmatic condition to consider when dealing with rough contact is the presence of thin layers. In this case, in order to simplify the calculations, without any loss of generality of the method employed, adhesive elastic conditions are investigated in the case of a 1D rigid wavy profile (with single wavelength λ and amplitude Λ) in adhesive contact against a linear elastic layer of thickness h. Moreover, since the elastic layer presents a finite thickness, specific boundary conditions can be imposed to the layer surface opposed to the contact interface. In this regard, we consider two different boundary conditions as reported in the insets of Figures 4: the *confined* case, where a rigid constraint is applied to the layer boundary (see Figure 4a); and the *remote pressure* case, which consider a free layer boundary with uniform pressure applied (see Figure 4b).

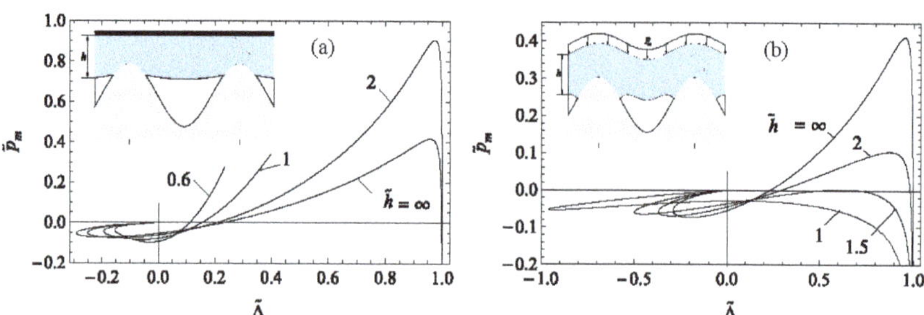

Figure 4. The dimensionless mean pressure $\tilde{p}_m = 2(1 - \nu^2)p/(Eq_0\Lambda)$ vs. the dimensionless mean penetration $\tilde{\Delta} = \Delta/\Lambda$ for elastic thin layers under (**a**) confined and (**b**) remote pressure boundary conditions. The dimensionless thickness is $\tilde{h} = q_0 h$. Results refer to $(1 - \nu^2)q_0\Delta\gamma/(\pi E) = 0.05$, and $q_0\Lambda = 2$.

Figure 4 shows the dimensionless mean pressure $\tilde{p}_m$ as a function of the $\tilde{\Delta}$.

Referring to the *confined* configuration, Figure 4a shows that the contact behaves stiffer as the dimensionless thickness $\tilde{h}$ is reduced because of the presence of the upper rigid constraint which hampers the elastic deformation of the layer. A different scenario holds in the case of *remote pressure* configuration, which, at a relatively small value of $\tilde{h}$, behaves like an Euler-Bernoulli beam. Indeed, as shown in Figure 4b, the contact stiffness significantly reduces with $\tilde{h}$ reducing. This leads to a peculiar behavior as, in presence of external loads, a threshold thickness h_{th} exists below which partial contact cannot occur (i.e., jump into full-contact occurs). Notably, in the *confined* configuration, complete contact can occur only for $h > \Lambda$.

Figure 4 also allows to appreciate the different behavior of the two configurations at pull-off, under load controlled conditions. Indeed, in the case of *remote pressure* configuration, reducing the layer dimensionless thickness $\tilde{h}$ leads to lower pull-off pressure and smaller (more negative) pull-off penetration. As shown in Ref. [64], this entails larger adhesive toughness, thus suggesting safety applications, where large amount of energy needs to be absorbed. On the contrary, the *confined* case shows increasing pull-off pressure with $\tilde{h}$ decreasing, which is peculiarly suited for structural applications (e.g., adhesives).

4.3. Frictional Elastic/Viscoelastic Sliding Rough Contacts of Thin Layers

In the case of sliding contacts with frictional interactions at the sliding interface, the elastic field in the deformable body depends on the distribution of the normal and tangential tractions at the interface. In the usual assumption of half-space contacts (i.e., $h \gg \lambda$) with rigid against incompressible (i.e., $\nu = 0.5$) materials, the presence of in-plane stress (e.g., frictional) does not play any role, and the frictionless normal contact solutions still holds true (uncoupled case). However, Equation (3) shows that in the case of non-vanishing out-of-diagonal terms in the Green's tensor Θ, the elastic fields caused by normal and in-plane stress interact with each other, and the linear superposition of the contact solutions is no longer possible (coupled case). In this case, the contact behavior depends on the specific distribution of both normal and tangential stress. This is the case, for instance, of frictional sliding contacts involving thin layers and/or compressible (i.e., $v < 0.5$) materials.

The frictional contact of thin elastic layers can be investigated by exploiting the formalism given by Equations (3) and (12)–(18). The same approach can be adopted for the viscoelastic case, providing that the Green's tensor component are calculated by means of Equation (25). Specifically, in what follows we focus on the case of a 1D rigid rough profile in contact with a thin layer of thickness h backed onto a rigid substrate (i.e., *confined* configuration). For most of the calculation, we assume incompressible material for the layer, so that the only source of normal-tangential coupling arises from the finite thickness of the deformable layer. In agreement with Refs. [77,78], we qualitatively refer to this as to *geometric* coupling. The rigid profile presents a self-affine roughness spanning over 100 scales with a periodicity wavelength λ and Hurst exponent $H = 0.8$. We also assume $r_{rms} = 10$ μm, and $\bar{g} = 0.13$, as the profile root mean square height and slope, respectively.

Due to normal-tangential coupling, different interfacial displacements are expected for frictional and frictionless contact conditions, given the same contact configuration. In turn, this may give rise to a different value of the contact area a in the presence of interfacial friction compared to the value a_0 of the frictionless case. Indeed, Figure 5a shows the contact area ratio a/a_0 as a function of the dimensionless mean contact pressure $\tilde{p}_m$, in the presence of *geometric* coupling. We observe that, at low values of $\tilde{p}_m$, the effect of coupling on the contact area ratio is poor; whereas, for $\tilde{p}_m > 2$, the contact area ratio significantly increases by increasing the value of $\tilde{p}_m$. Finally, at very high contact pressure, a saturation of the value of a occurs, as the full contact conditions is approached both for frictional and frictionless contacts. As expected, increasing the interfacial friction coefficient, leads to enhanced coupling effects. Indeed, in highly frictional contacts (e.g., rubber contacts with $\mu_c \approx 1$) the predicted contact area increase may raise up to 10% compared to the expected value in frictionless case. Moreover, a closer look at Equations (14) and (17) shows that the geometric coupling term is fast decaying with $\tilde{h} = q_0 h$ increasing. This is confirmed by the data shown in Figure 5b, where we observe that the actual contact area increase predicted for $\tilde{h} = 1.5$ is of only 2 %, compared to the uncoupled conditions. A further increase in the layer thickness leads to vanising coupling effects, as the confinement offered by the rigid substrate is very poor.

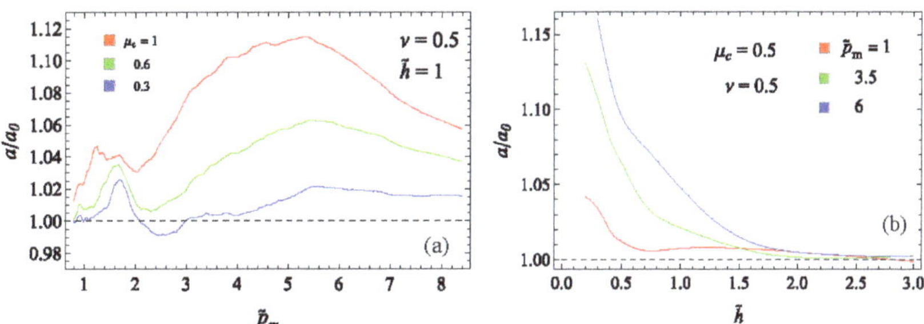

Figure 5. The contact area ratio (a_0 is the contact area for uncoupled case, i.e., with $\mu_c = 0$) as a fuction of (**a**) the dimensionless contact mean pressure $\tilde{p}_m = 2(1 - v^2)p/(Eq_0\Lambda)$, and (**b**) the dimensionless layer thickness $\tilde{h} = q_0 h$.

The frictional behavior of the interface is also affected by coupling. Indeed, from Equation (14) we observe that, in the presence of geometric coupling (i.e., for thin layers), the normal displacements on the layer surface under a normal point force are asymmetric even in the case of a purely elastic material. Hence, in rough contacts' conditions, an asymmetric pressure distribution is expected on each contacting asperity, which eventually entails an additional friction force opposing the relative sliding between the indenter and the deformable body. In the case of viscoelastic materials, as clearly shown in Refs. [62,66], due to the delayed material relaxation, an asymmetric contact pressure distribution is expected even in the case of vanishing coupling (i.e., for $h \gg \lambda$). Therefore, in contacts involving thin viscoelastic layers, the pressure asymmetry can be ascribed to the combined actions of both coupling and viscoelasticity [78]. The friction term resulting from the degree of contact pressure asymmetry is usually referred to as interlocking friction. The overall friction coefficient μ experienced by the contacting bodies can be calculated as

$$\mu = \mu_c + \mu_a$$

where μ_c is the Coulomb friction coefficient at the sliding interface, and

$$\mu_a = \frac{1}{\lambda p_m} \int_\Omega p(x) u_z'(x) dx$$

is the interlocking friction coefficient due to either coupling and/or viscoelasticity, with u_z' being the spatial first derivative of u_z.

Figure 6a shows the normalized friction coefficient $\mu_a / \mu_c \bar{g}$ induced by the asymmetry of the contact pressure distribution in purely elastic contacts as a function of the dimensionless mean contact presssure $\tilde{p}_m$, for different values of ν. For $\nu = 0.5$, due to geometric coupling, the degree of asymmetry of the contact pressure distribution is the highest possbile; moreover, since the pressure eccentricity is shifted in the direction of sliding, the resulting normalized friction coefficient $\mu_a / \mu_c \bar{g} > 0$. Consequently, under these conditions, regardless of the specific value of $\tilde{p}_m$, the overall contact friction is higher than for uncoupled contacts (i.e., for incompressible half-space). For $\nu < 0.5$, also *material coupling* occurs between normal and tangential displacements fields. The contact pressure eccentricity depends on the value of the contact mean pressure, therefore the frictional behavior of the contact may result in being increased or decreased with respect to the uncoupled corresponding case.

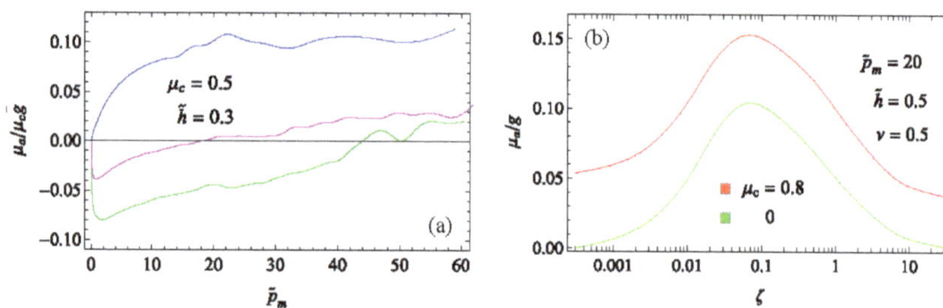

Figure 6. (a) the normalized elastic friction coefficient $\mu_a / \mu_c \bar{g}$ as a function of the dimensionless contact mean pressure $\tilde{p}_m = 2(1 - \nu^2) p / (E q_0 \Lambda)$, for different values of the layer Poisson's ratio ν. (b) the normalized viscoelastic friction coefficient $\mu_a / \bar{g}$ as a function of the dimensionless sliding velocity $\zeta = V t_c q_0$, in the frictional and frictionless case.

Figure 6b shows the normalized friction coefficient $\mu_a / \bar{g}$ as a function of the dimensionless sliding velocity $\zeta = V t_c q_0$, for the frictional ($\mu_c = 0.8$) and frictionless case ($\mu_c = 0$). In this case, the layer is assumed lineraly viscoelastic (with single relaxation time t_c). Firstly, we observe that at very high and very low values of the dimensionless sliding velocity, the normalized friction coefficient $\mu_a / \bar{g}$ presents its minima. Indeed, under these conditions, the viscoelastic hysteresis vanishes and the material response is barely elastic. We also observe that the presence of geometric coupling leads to higher values of $\mu_a / \bar{g}$ compared to uncoupled conditions. Moreover, since the coupling terms in Equations (14) and (17) do not explicitly depend on ζ, non-vanishing values of $\mu_a / \bar{g}$ are reported even for $\zeta \to 0$ and $\zeta \to \infty$ when coupling occurs.

5. Conclusions

In this paper, we review a variety of methodologies dealing with rough contact mechanics. In particular, we focus on Boundary Element methods: these have really impacted the tribology community in the last fifteen years as they have provided a solution for both purely elastic and viscoelastic normal and sliding contacts. In particular, for viscoelastic solids, we find important practical implications in all the systems, where there occurs relative motion between the contacting viscoelastic bodies: these include countless possible applications, such as, for example, vibration isolators, dynamic seals, pick and place devices. From a numerical point of view, given the translation invariance and the linearity of these systems, these problems have been tackled by means of a convolution integral, possibly accounting for the time and the space domains or relying on particular kinetic conditions, such as steady-state or reciprocating motions. This has required one to develop the *ad hoc* defined Green's functions. In a wider sense, and specifically, when thin layers are considered, an entire Green's tensor has to be considered to account for the coupling between normal and tangential actions: this can lead to an increase in the friction force.

This paper demonstrates the necessity of developing proper numerical strategies, such as the BEM introduced in this paper, to have accurate interfacial information, while preserving computational efficiency.

Author Contributions: Conceptualization, N.M.; C.P. and G.C.; methodology, N.M.; C.P. and G.C.; validation, N.M.; C.P. and G.C.; formal analysis, N.M.; C.P. and G.C.; investigation, N.M.; C.P. and G.C.; resources, N.M.; C.P. and G.C.; data curation, N.M.; C.P. and G.C.; writing—original draft preparation, N.M.; C.P. and G.C.; writing—review and editing, N.M.; C.P. and G.C.; visualization, N.M.; C.P. and G.C.; supervision, N.M.; C.P. and G.C. All authors have read and agreed to the published version of the manuscript.

Funding: This research was funded by Italian Ministry of Education, University and Research (Program Department of Excellence: Legge 232/2016 Grant No. CUP-D94I18000260001).

Conflicts of Interest: The authors declare no conflict of interest.

References

1. Thatte A.; Salant, R.F. Effects of multi-scale viscoelasticity of polymers on high-pressure, high-frequency sealing dynamics. *Tribol. Int.* **2012**, *52*, 75–86. [CrossRef]
2. Shukla, A.; Datta, T. Optimal Use of Viscoelastic Dampers in Building Frames for Seismic Force. *J. Struct.* **1999**, *125*, 401–409. [CrossRef]
3. Geim, A.K.; Dubonos, S.V.; Gricorieva, I.V.; Novoselov, K.S.; Zhukov, A.A.; Shapoval, S.Y. Microfabricated adhesive mimicking gecko foot-hair. *Nat. Mater.* **2003**, *2*, 461–463. [CrossRef] [PubMed]
4. Dening, K.; Heepe, L.; Afferrante, L.; Carbone, G.; Gorb, S. Adhesion control by inflation: Implications from biology to artificial attachment device. *Appl. Phys. A Mater. Sci. Process.* **2014**, *116*, 567–573. [CrossRef]
5. Arzt, E.; Gorb, S.; Spolenak R. From micro to nano contacts in biological attachment devices. *Proc. Natl. Acad. Sci. USA* **2003**, *100*, 10603–10606. [CrossRef] [PubMed]
6. Lin, S.; Ryu, S.; Tokareva, O.; Gronau, G.; Jakobsen, M.M.; Huang, W.; Rizzo, D.J.; Li, D.; Stai, C.; Pugno, N.; et al. Predictive modelling-based design and experiments for synthesis and spinning of bioinspired silk fibres. *Nat. Commun.* **2015**, *6*, 6892. [CrossRef]
7. Angelini, T.E.; Dunn, A.C.; Uruena, J.M.; Dickrell, D.J.; Burris, D.L.; Sawyer, W.G. Cell Friction. *Faraday Discuss.* **2012**, *156*, 31–39. [CrossRef]
8. Dunn, A.C.; Cobb, J.A.; Kantzios, A.N.; Lee, S.J.; Sarntinoranont M.; Tran-Son-Tay, R.; Sawyer, W.G. Friction Coefficient Measurement of Hydrogel Materials on Living Epithelial Cells. *Tribol. Lett.* **2008**, *30*, 13–19. [CrossRef]
9. Chortos, A.; Liu, J.; Bao, Z. Pursuing prosthetic electronic skin. *Nature* **2016**, *15*, 937. [CrossRef]
10. Greenwood, J.A.; Williamson, J.B.P. Contact of Nominally Flat Surfaces. *Proc. R. Soc. Lond. A* **1966**, *295*, 300–319.
11. Greenwood, J.A.; Putignano, C.; Ciavarella, M. A Greenwood & Williamson theory for line contact. *Wear* **2011**, *270*, 332–334.
12. Bush, A.W.; Gibson, R.D.; Thomas, T.R. The elastic contact of a rough surface. *Wear* **1975**, *35*, 87–111. [CrossRef]
13. Carbone, G.; Bottiglione, F. Asperity contact theories: Do they predict linearity between contact area and load? *J. Mech. Phys. Solids* **2008**, *56*, 2555–2572. [CrossRef]
14. Sahlia, R.; Pallares, G.; Ducottet, C.; Ben Alid, I.E.; Al Akhrassd, S.; Guiberta, M.; Scheibert, J. Evolution of real contact area under shear and the value of static friction of soft materials. *Proc. Natl. Acad. Sci. USA* **2017**, *115*, 471–476. [CrossRef] [PubMed]

15. Carbone, G.; Scaraggi, M.; Tartaglino, U. Adhesive contact of rough surfaces: Comparison between numerical calculations and analytical theories. *Eur. Phys. J. E—Soft Matter* **2009**, *30* , 65–74. [CrossRef]
16. Putignano, C.; Afferrante, L.; Carbone, G.; Demelio, G. The influence of the statistical properties of self-affine surfaces in elastic contact: A numerical investigation. *J. Mech. Phys. Solids* **2012**, *60*, 973–982. [CrossRef]
17. Putignano, C.; Afferrante, L.; Carbone, G.; Demelio, G. A new efficient numerical method for contact mechanics of rough surfaces. *Int. J. Solids Struct.* **2012**, *49*, 338–343. [CrossRef]
18. Paggi, M.; Ciaverella, M. The coefficient of proportionality k between real contact area and load, with new asperity models. *Wear* **2010**, *268*, 1020–1029. [CrossRef]
19. Yastrebov, V.A.; Anciaux, G.; Molinari, J.F. Contact between representative rough surfaces. *Phys. Rev. E* **2012**, *86*, 035601(R). [CrossRef]
20. Yastrebov, V.A.; Anciaux, G.; Molinari, J.F. The role of the roughness spectral breadth in elastic contact of rough surfaces. *J. Mech. Phys. Solids* **2017**, *107*, 469–493. [CrossRef]
21. Ju, Y.; Farris, T.N. Spectral Analysis of Two-Dimensional Contact Problems. *J. Tribol.* **1996**, *118*, 320–328. [CrossRef]
22. Stanley, H.M.; Kato, T. An FFT-Based Method for Rough Surface Contact. *J. Tribol.* **1997**, *119*, 481–485. [CrossRef]
23. Campana, C.; Muser, M.H. Contact Mechanics of Real vs Randomly Rough Surfaces: A Green's Function Molecular Dynamics Study. *Europhys. Lett.* **2007**, *77* , 38005. [CrossRef]
24. Dapp, W.B.; Lücke, A.; Persson, B.N.J.; Müser, M.H. Self-affine elastic contacts: Percolation and leakage. *Phys. Rev. Lett.* **2012**, *108*, 244301. [CrossRef] [PubMed]
25. Takewaki, I. *Building Control with Passive Dampers: Optimal Performance-Based Design for Earthquakes*; John Wiley & Sons: Hoboken, NJ, USA, 2011.
26. Bao, G.; Suresh, S. Cell and molecular mechanics of biological materials. *Nat. Mater.* **2003**, *2*, 715–725. [CrossRef]
27. Hunter S.C. The rolling contact of a rigid cylinder with a viscoelastic half space. *Trans. ASME Ser. E J. Appl. Mech.* **1961**, *28*, 611–617. [CrossRef]
28. Carbone, G.; Putignano, C. A novel methodology to predict sliding/rolling friction in viscoelastic materials: Theory and experiments. *J. Mech. Phys. Solids* **2013**, *61*, 1822–1834. [CrossRef]
29. Grosch, K.A. The Relation between the Friction and Visco-Elastic Properties of Rubber. *Proc. R. Soc. Lond. Ser. A Math. Phys.* **1963**, *274*, 21–39.
30. Putignano, C.; Reddyhoff, T.; Carbone, G.; Dini, D. Experimental investigation of viscoelastic rolling contacts: A comparison with theory. *Tribol. Lett.* **2013**, *51*, 105–113. [CrossRef]
31. Putignano, C.; Reddyhoff, T.; Dini, D. The influence of temperature on viscoelastic friction properties. *Tribol. Int.* **2016**, *100*, 338–343. [CrossRef]
32. Kusche, S. Frictional force between a rotationally symmetric indenter and a viscoelastic half-space. *J. Appl. Math. Mech.* **2017**, *97*, 226–239. [CrossRef]
33. Carbone, G.; Putignano, C. Rough viscoelastic sliding contact: Theory and experiments. *Phys. Rev. E* **2014**, *89*, 032408. [CrossRef] [PubMed]
34. Koumi, K.E.; Chaise, T.; Nelias, D. Rolling contact of a rigid sphere/sliding of a spherical indenter upon a viscoelastic half-space containing an ellipsoidal inhomogeneity. *J. Mech. Phys. Solids* **2015**, *80*, 1–25. [CrossRef]
35. Putignano, C.; Carbone, G.; Dini, D. Mechanics of Rough Contacts in Elastic and Viscoelastic Thin Layers. *Int. Solids Struct.* **2015**, *69–70*, 507–517. [CrossRef]
36. Putignano, C.; Carbone, G.; Dini, D. Theory of reciprocating contact for viscoelastic solids. *Phys. Rev. E* **2016**, *93*, 043003. [CrossRef]
37. Putignano, C.; Carbone, G. Viscoelastic Damping in alternate reciprocating contacts. *Sci. Rep.* **2017**, *7*, 8333. [CrossRef]
38. Putignano, C.; Carbone, G. Viscoelastic reciprocating contacts in presence of finite rough interfaces: A numerical investigation. *J. Mech. Phys. Solids* **2018**, *114*, 185–193. [CrossRef]
39. Putignano, C.; Dini, D. Soft matter lubrication: Does solid viscoelasticity matter? *Acs Appl. Mater. Interfaces* **2017**, *9*, 42287–42295. [CrossRef]
40. Putignano, C. Soft lubrication: A generalized numerical methodology. *J. Mech. Phys. Solids* **2020**, *134*, 103748. [CrossRef]
41. Putignano, C.; Burris, D.L.; Moore, A.; Dini, D. Cartilage rehydration: The sliding-induced hydrodynamic triggering mechanism. *Acta Biomater.* **2021**, *125*, 90–99. [CrossRef]
42. Putignano, C.; Menga, N.; Afferrante, L.; Carbone, G. Viscoelasticity induces anisotropy in contacts of rough solids *J. Mech. Phys. Solids* **2019**, *129*, 147–159. [CrossRef]
43. Bottiglione, F.; Carbone, G.; Mangialardi, L.; Mantriota, G. Leakage mechanism in flat seals. *J. Appl. Phys.* **2009**, *106*, 104902. [CrossRef]
44. Lorenz, B.; Persson, B.N.J. Leak rate of seals: Effective-medium theory and comparison with experiment. *Eur. Phys. J. E* **2010**, *31*, 159–167. [CrossRef] [PubMed]
45. Dapp, W.B.; Müser, M.H. Fluid leakage near the percolation threshold. *Sci. Rep.* **2016**, *6*, 19513. [CrossRef] [PubMed]
46. Vladescu, S.; Putignano, C.; Marx, N.; Keppens, T.; Reddyhoff, T.; Dini, D. The percolation of liquid through a compliant seal—An experimental and theoretical study. *J. Fluids Eng.* **2019**, *141*, 031101. [CrossRef]
47. Persson, B.N.J. Adhesion between an elastic body and a randomly rough hard surface. *Eur. Phys. J. E* **2002**, *8*, 385–401. [CrossRef]

48. Pastewka, L.; Robbins, M.O. Contact between rough surfaces and a criterion for macroscopic adhesion. *Proc. Natl. Acad. Sci. USA* **2014**, *111*, 3298–3303. [CrossRef]

49. Menga, N.; Carbone, G.; Dini, D. Do uniform tangential interfacial stresses enhance adhesion? *J. Mech. Phys. Solids* **2018**, *112*, 145–156; Corrigendum to: *J. Mech. Phys. Solids* **2019**, *133*, 103744. [CrossRef]

50. Violano G.; Afferrante, L. Size effects in adhesive contacts of viscoelastic media. *Eur. J. Mech.-A/Solids* **2022**, *96*, 104665. [CrossRef]

51. Violano, G.; Chateauminois, A.; Afferrante, L. Rate-dependent adhesion of viscoelastic contacts, Part I: Contact area and contact line velocity within model randomly rough surfaces. *Mech. Mater.* **2021**, *160*, 103926. [CrossRef]

52. Violano, G.; Chateauminois, A.; Afferrante, L. Rate-dependent adhesion of viscoelastic contacts. Part II: Numerical model and hysteresis dissipation. *Mech. Mater.* **2021**, *158*, 103884. [CrossRef]

53. Medina, S.; Dini, D.; A numerical model for the deterministic analysis of adhesive rough contacts down to the nano-scale. *Int. J. Solids Struct.* **2014**, *51*, 2620–2632. [CrossRef]

54. Persson, B.N.J. Theory of rubber friction and contact mechanics. *J. Chem. Phys.* **2001**, *115*, 3840–3861. [CrossRef]

55. Yang, C.; Persson, B.N.J. Molecular Dynamics Study of Contact Mechanics: Contact Area and Interfacial Separation from Small to Full Contact. *Phys. Rev. Lett.* **2008**, *100*, 024303 . [CrossRef]

56. Hyun, S.; Pei, L.; Molinari, J.-F.; Robbins, M.O. Finite-element analysis of contact between elastic self-affine surfaces. *Phys. Rev. E* **2004**, *70*, 026117 . [CrossRef] [PubMed]

57. Campana, C.; Mueser, M.H.; Robbins, M.O. Elastic contact between self-affine surfaces: Comparison of numerical stress and contact correlation functions with analytic predictions. *J. Phys. Condens. Matter* **2008**, *20*, 354013. [CrossRef]

58. Menga, N.; Putignano, C.; Carbone, G.; Demelio, G.P. The sliding contact of a rigid wavy surface with a viscoelastic half-space. *Proc. R. Soc. A* **2014**, *470*, 20140392. [CrossRef]

59. Müser, M.H.; Dapp, W.B.; Bugnicourt, R.; Sainsot, P.; Lesaffre, N.; Lubrecht, T.A.; Persson, B.N.J.; Harris, K.; Bennett, A.; Schulze, K.; et al. Meeting the contact-mechanics challenge. *Tribol. Lett.* **2017**, *65*, 118. [CrossRef]

60. Papangelo, A.; Putignano, C.; Hoffmann, N. Self-excited vibrations due to viscoelastic interactions. *Mech. Syst. Signal Process.* **2020**, *144*, 106894. [CrossRef]

61. Papangelo, A.; Putignano, C.; Hoffmann, N. Critical thresholds for mode-coupling instability in viscoelastic contacts. *Nonlinear Dyn.* **2021**, *104*, 2995–3011. [CrossRef]

62. Menga, N.; Afferrante, L.; Carbone, G. Effect of thickness and boundary conditions on the behavior of viscoelastic layers in sliding contact with wavy profiles. *J. Mech. Phys. Solids* **2016**, *95*, 517–529. [CrossRef]

63. Carbone, G.; Mangialardi, L. Analysis of the adhesive contact of confined layers by using a Green's function approach. *J. Mech. Phys. Solids* **2008**, *56*, 684–706. [CrossRef]

64. Menga, N.; Afferrante, L.; Carbone, G. Adhesive and adhesiveless contact mechanics of elastic layers on slightly wavy rigid substrates. *Int. J. Solids Struct.* **2016**, *88*, 101–109. [CrossRef]

65. Menga, N.; Foti, D.; Carbone, G. Viscoelastic frictional properties of rubber-layer roller bearings (RLRB) seismic isolators. *Meccanica* **2017**, *52*, 2807–2817. [CrossRef]

66. Menga, N.; Afferrante, L.; Demelio, G.P.; Carbone, G. Rough contact of sliding viscoelastic layers: Numerical calculations and theoretical predictions. *Tribol. Int.* **2018**, *122*, 67–75. [CrossRef]

67. Sackfield, A.; Hills, D.A.; Nowell, D. *Mechanics of Elastic Contacts*; Elsevier: Amsterdam, The Netherlands, 2013.

68. Barber, J.R. *Contact Mechanics*; Springer: Berlin/Heidelberg, Germany, 2018; Volume 250.

69. Nowell, D.; Hills, D.A.; Sackfield, A. Contact of dissimilar elastic cylinders under normal and tangential loading. *J. Mech. Phys. Solids* **1988**, *36*, 59–75. [CrossRef]

70. Chen, W.W.; Wang, Q.J. A numerical model for the point contact of dissimilar materials considering tangential tractions. *Mech. Mater.* **2008**, *40*, 936–948. [CrossRef]

71. Chen, W.W.; Wang, Q.J. A numerical static friction model for spherical contacts of rough surfaces, influence of load, material, and roughness. *J. Tribol.* **2009**, *131*, 021402. [CrossRef]

72. Wang, Z.J.; Wang, W.Z.; Wang, H.; Zhu, D.; Hu, Y.Z. Partial slip contact analysis on three-dimensional elastic layered half space. *J. Tribol.* **2010**, *132*, 021403. [CrossRef]

73. Elloumi, R.; Kallel-Kamoun, I.; El-Borgi, S. A fully coupled partial slip contact problem in a graded half-plane. *Mech. Mater.* **2010**, *42*, 417–428. [CrossRef]

74. Bentall, R.H.; Johnson, K.L. An elastic strip in plane rolling contact. *Int. J. Mech.* **1968**, *10*, 637–663. [CrossRef]

75. Nowell, D.; Hills, D.A. Contact problems incorporating elastic layers. *Int. J. Solids Struct.* **1988**, *24*, 105–115. [CrossRef]

76. Nowell, D.; Hills, D.A. Tractive rolling of tyred cylinders. *Int. J. Mech. Sci.* **1988**, *30*, 945-957. [CrossRef]

77. Menga, N. Rough frictional contact of elastic thin layers: The effect of geometric coupling. *Int. J. Solids Struct.* **2019**, *164*, 212–220. [CrossRef]

78. Menga, N.; Carbone, G.; Dini, D. Exploring the effect of geometric coupling on friction and energy dissipation in rough contacts of elastic and viscoelastic coatings. *J. Mech. Phys. Solids* **2021**, *148*, 104273. [CrossRef]

79. Kogut, L.; Komvopoulos, K. Electrical contact resistance theory for conductive rough surfaces. *J. Appl. Phys.* **2003**, *94*, 3153–3162. [CrossRef]

80. Menga, N.; Ciavarella, M. A Winkler solution for the axisymmetric Hertzian contact problem with wear and finite element method comparison. *J. Strain Anal. Eng.* **2015**, *50*, 156–162. [CrossRef]

81. Piveteau, L.-D.; Girona, M.I.; Schlapbach, L.; Barboux, P.; Boilot, J.-P.; Gasser, B. Thin films of calcium phosphate and titanium dioxide by a sol-gel route: A new method for coating medical implants. *J. Mater. Sci. Mater. Med.* **1999**, *10*, 161–167. [CrossRef]

82. Allen, M.; Myer, B.; Rushton, N. In Vitro and In Vivo Investigations into the Biocompatibility of Diamond-Like Carbon (DLC) Coatings for Orthopedic Applications. *J. Biomed. Mater.* **2001**, *58*, 319–328. [CrossRef]

83. Kwak, M.K.; Jeong, H.-E.; Suh, K.Y. Rational design and enhanced biocompatibility of a dry adhesive medical skin patch. *Adv. Mater.* **2011**, *23*, 3949–3953. [CrossRef]

84. Bacon, K.D.; Cummins C.F. Pressure-Sensitive Adhesive. U.S. Patent US2285570 A, 9 June 1942.

85. Peterson, D.R.; Stupp, S.I. Poly(amino Acid) Adhesive Tissue Grafts. U.S. Patent US5733868 A, 31 March 1998.

86. Al-Harthi, M.A.; Bakather, O.Y.; De, S.C. Pressure Sensitive Adhesive. U.S. Patent US8697821 B1, 15 April 2014.

87. Menga, N.; Afferrante, L.; Pugno, N.M.; Carbone, G. The multiple V-shaped double peeling of elastic thin films from elastic soft substrates. *J. Mech. Phys. Solids* **2018**, *113*, 56–64. [CrossRef]

88. Menga, N.; Dini, D.; Carbone, G. Tuning the periodic V-peeling behavior of elastic tapes applied to thin compliant substrates. *Int. J. Mech. Sci.* **2020**, *170*, 105331. [CrossRef]

89. Putignano, C.; Afferrante, L.; Mangialardi, L.; Carbone, G. Equilibrium states and stability of pretensioned adhesive tapes. *Beilstein J. Nanotechnol.* **2014**, *5*, 1725–1731. [CrossRef] [PubMed]

90. Ceglie, M.; Menga, N.; Carbone, G. The role of interfacial friction on the peeling of thin viscoelastic tapes. *J. Mech. Phys. Solids* **2022**, *159*, 104706. [CrossRef]

91. Johnson, K.L.J. *Contact Mechanics*; Cambridge University Press: Cambridge, UK, 1985.

92. Menga, N.; Carbone, G. The surface displacements of an elastic half-space subjected to uniform tangential tractions applied on a circular area. *Eur. J. Mech.-A/Solids* **2019**, *73*, 137–143. [CrossRef]

93. Christensen, R.M. *Theory of Viscoelasticity*; Academic Press: New York, NY, USA, 1982.

94. Williams M.L.; Landel R.F.; Ferry J.D. The Temperature Dependence of Relaxation Mechanisms in Amorphous Polymers and Other Glass-forming Liquids. *J. Am. Chem. Soc.* **1955**, *77*, 3701. [CrossRef]

95. Putignano, C. Oscillating viscoelastic periodic contacts: A numerical approach. *Int. J. Mech. Sci.* **2021**, *208*, 106663. [CrossRef]

96. Longuet-Higgins, M.S. The statistical analysis of a random, moving surface. *Philos. Trans. R. Soc. Lond. A* **1957**, *249*, 321.

MDPI
St. Alban-Anlage 66
4052 Basel
Switzerland
Tel. +41 61 683 77 34
Fax +41 61 302 89 18
www.mdpi.com

Machines Editorial Office
E-mail: machines@mdpi.com
www.mdpi.com/journal/machines